NEW DIRECTIONS IN
TECHNICAL SERVICES

TRENDS AND SOURCES
(1993-1995)

Peggy Johnson, Editor

In cooperation with the
Association for Library Collections &
Technical Services

American Library Association
Chicago and London
1997

While extensive effort has gone into ensuring the reliability of information appearing in this book, the publisher makes no warranty, express or implied, on the accuracy or reliability of the information, and does not assume and hereby disclaims any liability to any person for any loss or damage caused by errors or omissions in this publication.

Project editor: Louise D. Howe

Cover design: Image House

Text design: Dianne M. Rooney

Indexer: Janet Russell

Compositor: Clarinda in Bodoni and Palatino on Penta DeskTopPro

Printed on 50-pound Victor Offset, a pH-neutral stock, and bound in 12-point coated Bristol cover stock by Victor Graphics, Inc.

The paper used in this publication meets the minimum requirements of American National Standard for Information Sciences—Permanence of Paper for Printed Library materials, ANSI Z39. 48-1992.∞

ISSN# 1091-9066

Copyright © 1997 by the American Library Association. All rights reserved except those which may be granted by Sections 107 and 108 of the Copyright Revision Act of 1976.

Printed in the United States of America.

01 00 99 98 97   5 4 3 2 1

CONTENTS

ACRONYMS xiii

Introduction | 1

A
Technical Services: An Overview | 4
Sheila S. Intner

- **AA** General Works 6
- **AB** Textbooks, Guides, and Manuals 10
- **AC** Directories 11
- **AD** Bibliographies 11
- **AE** Handbooks 12
- **AF** Periodicals 12
- **AG** Sources of Expertise 13
  - **AGA** Professional Associations and Organizations 14
  - **AGB** Conferences 14
- **AH** Administration 15
  - **AHA** General Works 15
  - **AHB** Outsourcing 19
  - **AHC** Reorganization and Restructuring 21

AI	Database Management 23	
	AIA Bibliographic Networks	23
	AIB Expert Systems 24	
	AIC Hardware and Software	25
	AID Local Systems 26	

B
Acquisitions | 28
Karen A. Schmidt

BA	General Works 32
	BAA Textbooks, Guides, and Manuals 33
	BAB Bibliographies 34
	BAC Periodicals 34
	BAD Sources of Expertise 35
	Professional Associations 35
	Conferences 35
	World Wide Web Sites 35
	Acquisitions Departments Home Pages 36
BB	Approval Plans 36
BC	Automation of Acquisitions 38
BD	Out-of-Print Material 39
BE	Gifts and Exchanges 40
BF	Vendor Selection and Evaluation 40
BG	Administration 41
BH	Publishing 43
BI	Ethics 43

C
Descriptive Cataloging | 46
Janet Swan Hill

CA	General Works 48
	CAA Textbooks, Guides, and Manuals 48
	CAB Bibliographies 49
	CAC Basic Issues and Reconsiderations 49

	CAD	Periodicals, Including Electronic Journals 51
	CAE	Sources of Expertise 52
		Technical Services Home Pages 52
		Other World Wide Web Sites and Home Pages 55

- **CB** Management Issues 56
  - **CBA** Catalogers 56
  - **CBB** Training 59
  - **CBC** Workstations 60
  - **CBD** Organization of Cataloging 62
  - **CBE** Levels and Types of Cataloging 62
  - **CBF** Authority Work 65
  - **CBG** Outsourcing 68
  - **CBH** Backlogs 70
  - **CBI** Cooperative Projects and Databases, Networks 71
- **CC** Standards and Their Application 74
  - **CCA** Standards for Descriptive Content 74
  - **CCB** Descriptive Content Standards: Special Topics 75
  - **CCC** Independent Cumulations of Rule Interpretations 77
  - **CCD** Application of the Standards to Special Types of Materials 78
    - Rare Books 78
    - Manuscripts, Archives 79
    - Music and Sound Recordings 79
    - Motion Pictures and Videorecordings 79
    - Electronic and Online Resources 80
    - Serials, Including Newspapers 82
    - Nonbook Materials 82
    - Audiovisual Materials 83
    - Realia 83
    - Special Categories Not Tied to Physical Format 84
  - **CCE** Standards for Content Designation 84

D

Subject Analysis Systems | 86

Nancy J. Williamson

- **DA** General Works 88
  - **DAA** Conference Proceedings 92

Contents

- **DAB** Textbooks, Guides, and Manuals   93
- **DAC** Periodicals   94
- **DAD** Bibliographies   95
- **DAE** Sources of Expertise   95

DB Classification Systems   96
- **DBA** General Works   96
  - Textbooks, Guides, and Manuals   97
  - Classification of Special Types of Literature   98
  - Classification in Online Catalogs   98
  - Standards   99
- **DBB** Dewey Decimal Classification   99
  - General Works   99
  - Schedules   100
  - Textbooks, Guides, and Manuals   101
- **DBC** Library of Congress Classification   101
  - General Works   101
  - Schedules   102
  - Special Schedules   102
- **DBD** Other Classification Systems   103
  - Book Numbers   103
  - Universal Decimal Classification   104
  - Faceted Classification Systems   105
  - Special Classification Systems   105

DC Subject Headings and Related Systems   105
- **DCA** General Works   105
  - Standards   107
- **DCB** Library of Congress Subject Headings   108
  - General Works   108
  - Textbooks, Guides, and Manuals   108
  - Lists and Files   109
  - Problems in LCSH Special Subject Areas   109
- **DCC** Other Subject Heading Lists   110
- **DCD** Subject Authority Control   112
- **DCE** Thesauri   113
  - General Works   113
  - Standards   114
  - Lists and Files   114
  - Textbooks, Guides, and Manuals   115
  - Thesaurus Design and Construction   115

DD Internet Subject Access   117

E
Filing and Indexing | 119
Susan Morris

- **EA** General Works 121
  - **EAA** General Works: Filing 121
  - **EAB** General Works: Indexing 121
  - **EAC** Directories 126
  - **EAD** Sources of Expertise 127
    - Electronic Discussion Groups 127
    - Electronic Periodicals 128
    - World Wide Web Sites and Home Pages 128
- **EB** Standards 128
- **EC** Indexing and the Library Catalog 130
- **ED** Indexing Online Documents and Other Special Format Materials 132
- **EE** Thesauri and Controlled Vocabularies 133
- **EF** Automated Indexing 136

F
Serials Management | 139
Deborah E. Burke

- **FA** General Works 142
  - **FAA** Textbooks, Guides, and Manuals 142
  - **FAB** Directories 142
  - **FAC** Bibliographies 143
  - **FAD** Periodicals 144
  - **FAE** Sources of Expertise 145
    - Electronic Discussion Groups 145
    - Professional Associations 146
  - **FAF** Conference Proceedings 146
- **FB** Management of Serials Units 147
  - **FBA** Organization 147
  - **FBB** Relationships with External Organizations 149
- **FC** Serials Publishing 150
  - **FCA** Printed Serials 150

	FCB	Electronic Serials 150
FD		Serials Processing 152
	FDA	World Wide Web Sites and Home Pages 152
	FDB	Management of Serials Processing 152
	FDC	Subscription Agent Selection and Evaluation 153
	FDD	Subscription Agents' Databases 153
	FDE	Claiming and Other Acquisition Processes 155
	FDF	Automation of Serials Processing 156
FE		Serials Pricing 158
FF		Resource Sharing, Union Listing, Serials Holdings 159

G

Collection Management | 161

Genevieve S. Owens

GA	General Works 164
	GAA Textbooks, Guides, and Manuals 164
	GAB Bibliographies 165
	GAC Standards 166
	GAD Sources of Expertise 166
	GAE Overview 167
	GAF Automation and the Practice of Collection Management 168
GB	Organization and Staffing 169
	Responsibilities of Bibliographers and Collection Development 169
	Education and Training 170
GC	Policies 171
GD	Collection Building 173
	Theories of Selection 173
	Mass Buying 173
	Diverse Literatures and Communities 174
	Grey Literature 175
	Censorship and Intellectual Freedom 175
	Nonprint Media, Including Computer Files 177
	Serials 180
	Using Selection Tools 182
	Academic and Research Libraries 182
GE	Collection Maintenance 184
	Academic and Research Libraries 184

Applied and Interdisciplinary Fields   185
Area Studies and Foreign Materials   185
Government Publications   186
Humanities, Social Sciences, and Sciences   186
Public Libraries   187
Reference Collections   188
School Library Media Centers   188
Special Collections, Archives, and Special Libraries   189
Women's Studies   189
Weeding and Storage   190
Preservation   190
Serials Cancellation   191
GF Budget and Finance   192
Allocation   192
Budgeting   193
Fund-Raising   193
GG Assessment and Evaluation   194
Qualitative Methods   194
Conspectus   195
OCLC/AMIGOS Collection Analysis CD   195
GH Cooperative Collection Management and Resource Sharing   196

H
Preservation | 198
Wesley L. Boomgaarden

HA General Works   201
  HAA Periodicals   203
  HAB Sources of Expertise   204
    Electronic Discussion Groups   204
    World Wide Web Sites   207
  HAC Bibliographies   208
HB Commercial Library Binding   208
HC Conservation of Books and Paper; Paper Deacidification and Quality   209
HD Cooperative Preservation Efforts   211
    Statewide Preservation Planning   213

Contents

HE Copyright   214
HF Disaster Control, Recovery, Insurance, and Security   215
HG Environmental Control and Pest Management   216
HH Imaging and Preservation Reproduction   218
HI Management and Organization of Preservation Programs   224
HJ Media (Nonbook and/or Nonpaper) Preservation and Reproduction   227
HK Microform Publications   230
HL Photocopiers and Photocopying   231
HM Staff Training and User Awareness   232
HN Standards, Specifications, and Guidelines   233
      Library Binding   234
      Paper   234
      Micrographics   234
      Other Media   236
      Equipment   236

I

Access Services | 237

Julie Wessling

IA General Works   239
    IAA Copyright Resources   240
    IAB Periodicals   242
    IAC Sources of Expertise   242
        Professional Organizations   242
        Conferences   244
        Electronic Discussion Groups   244
    IAD Bibliographies   245
IB Collection Maintenance   246
IC Circulation   247
        Guidelines and Standards   250
ID Interlibrary Loan   251
    IDA General Works   252
    IDB Textbooks, Guides, and Manuals   256
    IDC Directories   257
    IDD Guidelines and Standards   258

IDE	Application Software Resources for ILL Management and Communication 259	
IE	Document Delivery 261	
	Directories 261	
	General Works 261	
IF	Library Security 266	

ABOUT THE EDITORS 269

INDEXES

    Author/Title 273

    Subject 339

ACRONYMS

AACR	*Anglo-American Cataloging Rules*
AACR2	*Anglo-American Cataloguing Rules, Second Edition*
AACR2R	*Anglo-American Cataloguing Rules, Second Edition, 1988 Revision*
AALL	American Association of Law Librarians
AAT	*Art & Architecture Thesaurus*
ACM	Association of Computer Machinery
ACRL	Association of College and Research Libraries
ADA	Americans with Disabilities Act
AIC	American Institute for Conservation
AIIM	Association for Information and Image Management
ALA	American Library Association
ALCTS	Association for Library Collections and Technical Services
ALISE	Association for Library and Information Science Education
AMC	Archival and manuscripts control
AMIA	Association of Moving Image Archivists
ANSI	American National Standards Institute
APIS	Advanced Papyrological Informatin System
ARL	Association of Research Libraries
ASI	American Society of Indexers

Acronyms

ASIS	American Society for Information Scientists
ASTED	Association pour l'advancement des sciences et des techniques de la documentation
ATG	*Against the Grain*
BC2	*Bliss Bibliographic Classification*, 2nd ed.
BISAC	Book Industry Systems Advisory Committee
BL	Bliss Classification
BSO	Broad System of Ordering
CACD	*OCLC/AMIGOS Collection Analysis CD-ROM*
CAL	Conservation Analytical Laboratory (at the Smithsonian Institution)
CARL	Colorado Alliance of Research Libraries
CC	Colon Classification
CCC	Copyright Clearance Center; also, Cooperative Cataloging Council
CCQ	*Cataloging & Classification Quarterly*
CCS	Cataloging and Classification Section (of ALCTS)
CD-ROM	compact-disc read-only memory
CENL	Conference of European National Libraries
CIC	Committee on Institutional Cooperation
CIP	Cataloging-in-publication
CJK	Chinese, Japanese, Korean
CLR	Council on Library Research
CMDS	Collection Management and Development Section (of ALCTS)
CNI	Coalition for Networked Information
COM	computer output microfilm
CONSER	Cooperative Online Serials project
COTS	Commercial-off-the-Shelf
CPA	Commission on Preservation and Access
CRL	Center for Research Libraries
CSISAC	Canadian Serials Industry Systems Advisory Committee

DDC	Dewey Decimal Classification
DPC	Digital Preservation Consortium
ECPA	European Commission on Preservation and Access
EDD	electronic document delivery
EDI	electronic data interchange
EPIC	European Preservation Information Center
ERIC	Education Resource Information Center
FAQ	frequently asked questions
FI	format integration
FID	International Federation for Information and Documentation
FID/CR	FID Committee on Classification Research
FTP	file transfer protocol
GODORT	Government Document Round Table (of ALA)
GPO	Government Printing Office
GTSR	*Guide to Technical Services Resources*
IAC	Information Access Company
IAML	International Association of Music Libraries
IASC	Indexing and Abstracting Society of Canada
IESCA	Interactive Electronic Serials Cataloging Aid
IFLA	International Federation of Library Associations and Institutions
IFM	ILL Fee Management
IIC	International Institute for Conservation
ILL	interlibrary loan
IPI	Image Permanence Institute
ISI	Institute for Scientific Information
ISKO	International Society for Knowledge Organization
ISO	International Standards Organization
ISSN	International Standard Serial Number
JACKPHY	Japanese, Arabic, Chinese, Korean, Persian, Hebrew, Yiddish

Acronyms

JAL		*The Journal of Academic Librarianship*
KWIC		key word in context
LAN		local area network
LAMA		Library Administration and Management Association
LAPT		*Library Acquisitions: Practice & Theory*
LC		U.S. Library of Congress
LCC		Library of Congress Classification
LCCDG		Library Collections Conservation Discussion Group
LCCN		Library of Congress card number
LCRI		Library of Congress Rule Interpretation
LCSH		Library of Congress subject headings
LRTS		*Library Resources & Technical Services*
MAGERT		Map and Geography Round Table (of ALA)
MAI		machine-aided indexing
MARBI		Committee on Machine-Readable Bibliographic Information
MARC		Machine-Readable Cataloging
MeSH		Medical Subject Headings
MIME		Multipurpose Internet Mail Extension
MIS		management information system
MLA		Modern Language Association; also, Music Library Association
NACO		Name Authority Cooperative
NAILLD		North American Interlibrary Loan and Document Delivery project
NAL		U.S. National Agricultural Library
NASIG		North American Serials Interest Group
NCCP		National Coordinated Cataloging Program
NCSACC		National Committee to Save America's Cultural Collections
NFAIS		National Federation of Abstracting and Information Services

NFPA	U.S. National Film Preservation Board
NII	National Information Infrastructure
NISO	National Information Standards Organization
NLC	National Library of Canada
NLM	National Library of Medicine
NML	National Media Laboratory
NOTIS	originally Northwestern On-Line Totally Integrated System; now NOTIS Systems, Inc.
O.P.	out-of-print
OCLC	Online Computer Library Center, Inc.
OCR	optical character recognition
OLUC	OCLC's Online Union Catalog
OMS	Office of Management Services (at ARL)
OPAC	online public access catalog
OSI	Open Systems Interconnections
PARS	Preservation and Reformatting Section (of ALCTS)
PCC	Program for Cooperative Cataloging
PI	preservation index
PLMS	Preservation of Library Materials Section (of ALCTS)
PMG	Photographic Materials Group (of AIC)
PVLRC	Publisher/Vendor/Library Relations Committee (of ALCTS)
RAMEAU	Repertoir d'authorite matiere encyclopedique et alpaba unifie
RAMP	Records and Archives Management Programme (of UNESCO)
RASD	Reference and Adult Services Division (of ALA)
RBMS	Rare Books and Manuscripts Section (of ACRL)
RLG	Research Libraries Group
RLIN	Research Libraries Information Network
RLMS	Reproduction of Library Materials Section (of ALCTS)
RUSA	Reference and User Services Division (of ALA; formerly RASD)

Acronyms

RVM	*Répertoire de vedettes-matière*
SAA	Society of American Archivists
SCAD	Société canadienne pour l'analyse de documents
SGML	Standard Generalized Markup Language
SICI	SISAC Item Contribution Identifier
SISAC	Serials Industry Systems Advisory Committee
SLA	Special Libraries Association
SPEC	Systems and Procedures Exchange Center (of ARL)
STM	scientific, technical, and medical
SUNY	State University of New York
TCP/IP	Transmission Control Protocol/Internet Protocol
TEI	Text Encoding Initiative
TGM	*Thesaurus for Graphic Materials*
TQM	Total Quality Management
TSWs	technical services workstations
TWPI	time-weighted preservation index
UCRI	Usage/cost relational index
UDC	Universal Decimal Classification
URL	universal resource locator
USAIN	United States Agricultural Information Network
WAA	Western Association of Art
WAIS	wide area information system
WAN	wide area network
WISPPR	Wisconsin Preservation Program
WLN	Western Library Network
WWW	World Wide Web

Introduction

New Directions in Technical Services has two aims. It continues, in one volume, the annual bibliographic essays previously published in *Library Resources & Technical Services* and supplements the *Guide to Technical Services Resources (GTSR)* published in 1994. We have sought to build on the strengths of the two previous publications and to remedy gaps and other problems brought to our attention by practitioners and reviewers. Through annotated and evaluative references, this work lists and describes the best and most useful tools appearing between 1993 and 1995 for technical service practitioners, students, and educators.

Scope

The most recent "Year's Work in Technical Services" essays, published in 1993, covered resources appearing in 1992. The *Guide to Technical Services Resources* included resources published through 1993. Building on these two guides, we have attempted to cover resources for technical services practice that became available from 1993 through 1995, though coverage will, of necessity, have some latitude at either end. Resources published before 1993 listed here are included because they are particularly significant, they were missed in the earlier *GTSR*, they have been revised or modified, or

Introduction

additional forms of access (usually electronic) are now available. This work is not merely a literature review. Selections have been included because they serve as a basic tool or resource, contribute in a substantial way to theory and practice in the field, or identify and address a unique problem. Where appropriate, new or revised standards and resources aiding in their application are included.

New Directions in Technical Services has over 1,200 entries and includes several types of information resources. While journal articles predominate, monographs and journal titles are also represented. We suggest avenues for professional support and expertise, such as professional associations, conferences, vendor support groups, and electronic discussion groups. Readers familiar with the "Year's Work . . ." articles from *Library Resources & Technical Services* and *GTSR* will notice a significant increase in electronic resources, particularly World Wide Web sites. Many print resources are also available via the Internet; we have sought to include access information in the form of URLs (Universal Resource Locators) for these resources. Continuing accuracy for electronic resources is a problem because of their mutability, both in content and access. Nevertheless, online resources are increasingly important for practitioners and provide the most up-to-date information in many areas. Readers are advised to check the currency of the electronic version being consulted and to research (using a Web browser) any resource for which the access information provided here is no longer valid.

This compilation is intended primarily as a tool for technical services as practiced in North American libraries and contains a preponderance of English-language, North American materials. However, selected resources outside this general scope are included when compilers found them particularly relevant.

Style and Structure

New Directions in Technical Services is divided into nine chapters. Database management is addressed in the first (Overview) chapter and automation issues also are included, as appropriate, in subsequent chapters. Resources on authority work are cited primarily in Chapter C, Descriptive Cataloging, though some resources on this important topic are found in other chapters. A single chapter (H) combines preservation and reproduction of library materials, reflecting the convergence of these two responsibilities. The chapter subjects loosely parallel the organization in many library operations, though such divisions seem increasingly artificial as functional separations become less distinct in many libraries. Several topics (copyright and outsourcing are two examples) are of compelling interest throughout technical service operations. Consequently, many resources could be listed in several chapters. Cross-references are provided to resources described in detail in another chapter. A subject index directs the reader to topical entries throughout this work.

We have used alphanumeric codes for entries, a system familiar to readers of the *Guide to Reference Books* and *Guide to Technical Services Resources*. This has proved an effective way to organize resources within chapters and to link annotations between chapters. The specific codes, however, stand alone in this book and do not reflect the application of codes in earlier publications.

Each chapter begins with an introduction to the topic it covers, focusing on trends and developments from 1993 to 1995, as evidenced in the resources cited. Many subsections within chapters provide introductions to the topics that follow. Chapters vary considerably in length. This reflects the nature of the topic covered, the volume of resources published and available, and the combination into one chapter of topics treated separately in *GTSR*. Some chapters include a substantial number of guides, manuals, and standards; these tools may be nonexistent for other areas or unchanged since *GTSR* was published. Readers are advised to consult the *Guide to Technical Services Resources* for a complete listing of the major resources for and about library technical services.

Acknowledgments

Several contributors to this volume were chapter compilers for *GTSR*—Sheila S. Intner, Karen A. Schmidt, Janet Swan Hill, Nancy J. Williamson, and Wesley L. Boomgaarden; other contributors—Deborah E. Burke, Susan Morris, Genevieve S. Owens, and Julie Wessling—are new with this publication. I am indebted to their considerable expertise, familiarity with the literature, and diligence in bringing this project to completion. I also wish to thank a group of advisors from the ALCTS Publications Committee—Carleen Ruschoff, Bill Robnett, and Suzanne Freeman (committee chair)—who assisted with planning. I wish to express my deepest thanks to my husband, Lee, for his help and support, and my children, Carson and Amelia, whose enthusiasm for everything is a constant source of encouragement.

Peggy Johnson
Editor

A

Technical Services: An Overview

Sheila S. Intner

The shape of the literature of technical services in the years from 1993 through 1995, the time frame encompassed by this book, is surprising in that it silhouettes only a few key issues of concern. It is quite unlike the *Guide to Technical Services Resources*, which revealed a broad spectrum of issues with numbers of entries distributed fairly evenly among the literary genres covered by this bibliography. If the content were mapped out on a bar graph with topics on the x-axis and numbers of entries on the y-axis, there would be a few spikes indicating large numbers of entries and many gaps where very few appear. In this compilation, the topics of concern include the following:

(1) the fact of change—paradigm shifts being caused by progress in the development of electronic access;

(2) downsizing of technical services budgets, staffs, etc.;

(3) outsourcing—in some cases, describing how it was or should be done, or offering options for contracting with vendors to perform technical services tasks, and in others, asking if technical services departments will survive into the twenty-first century as the drive to outsource functions builds;

(4) reorganizing to accommodate changes in library service, to utilize networked services more effectively, etc.;

(5) retraining technical services staff members to effect the reorganization; and, finally,

(6) electronic access, which is the driving force behind most of the other issues.

Of the six issues, outsourcing and reorganizing the technical services department have been written about the most. Outsourcing, while it is just an old idea in new garb, is affecting technical services librarians profoundly, because the differences in cost between local library products and outsourced vendor products sometimes are extraordinarily dramatic. No rational library director can ignore them, even if outsourcing means losing some of the expertise technical services librarians can supply. In the literature presented here, reorganizing technical services also is termed *restructuring*, perhaps to imply difference in kind as well as in magnitude. Clearly, halfhearted changes such as splitting a job between two staff members in order to avoid letting one go, which has been done to keep promises not to fire anyone as jobs dwindle, are not adequate answers to the *Big Question:* How should libraries reorganize and restructure? It may be ironic, but it is no comfort that public services staffs are going through a similar syndrome.

The six issues named above are not mutually exclusive. All are closely related to the continuing maturation of computer technologies, especially networking technologies, which has made it possible to link databases of all kinds with ease. This phenomenon is operating even when observers fail to recognize its influence. For example, downsizing usually is seen as a result of declining economics, but this appraisal might be too simplistic and shortsighted. This is not to say declining economics is not a factor. In academic institutions, the bad bottom line can be attributed to declining enrollments and research grants; in cities, towns, and smaller communities, the problem is lost jobs and declining tax bases; and in the corporate world, it is declining profits, although people talk about other worrisome factors such as loss of market share or overblown expectations of profit margins. All these things are, indeed, happening today, and downsizing is a logical reaction to them, but downsizing can be a response to positive factors as well as negative ones. Experience with automation shows it results in increased productivity (sometimes exponentially increased!) and changes in the mix of products and services wanted. Thus, downsizing reflects the ability of fewer people to handle stable or slow-growing numbers of transactions, once they have computer systems in place with which to work. It reflects the changing mix of information products and services librarians provide—more self-guided electronic databases, but fewer books and mediated services.

The latter years of the 1990s eventually may be seen as the moment in which great new strides were taken in information services, much as we now view the 1890s as the origin of Library of Congress catalog cards and the 1970s as the rise of bibliographic utilities. At this writing, however, they only can be deemed a prime *unmoment:* the information world is unstable and the future toward which it heads is unclear. Libraries and librarians are uneasy and unhappy as they acknowledge that the need for change in technical services is intensifying and they search for answers to questions many wish they did not have to articulate.

Chapter Content and Organization

Database Management, which appeared as a separate chapter in *GTSR,* has been included in this chapter as a subtopic, designated AI. The topic has been defined narrowly to include only writings for general audiences that discuss hardware and software, or network systems for technical services. "Database management" items referring to only one technical services area were excluded as too specific for the Overview.

Every effort was made to obtain all the documents cited here, and all the annotations were written after the compiler read the material. Any errors of omission or commission are entirely the fault of the compiler. To the reader goes a sincere apology for all of them.

AA
General Works

AA1. Beehler, Sandra A., and Patricia G. Court. "Speaking in Tongues: Communications between Technical Services and Public Services in an Online Environment." Sound cassette recorded at the AAL Eighty-seventh Annual Meeting, Seattle, July 9–13, 1994. Valencia, Calif.: American Association of Law Libraries, 1994.

AA2. Bierbaum, Esther Green. "Searching for the Human Good: Some Suggestions for a Code of Ethics for Technical Services." *Technical Services Quarterly* 11, no. 3 (1994): 1–18.

After examining the American Library Association's general code of ethics for librarians, the author states it does not serve the technical services situation fully. She investigates the meaning and implications of ethical considerations in many contexts, then describes the context peculiar to technical services. She suggests nine elements for code of ethics for technical services: ethics as the guide to difficult decisions; professional purpose and mission; prevalence of change; influence of external factors; right of all to unbiased decisions; danger of self-serving acts; necessity of fiscal accountability; absence of favoritism; and legal responsibilities.

AA3. Biossonnas, Christian M. "Darwinism in Technical Services: Natural Selection in an Evolving Information Delivery Environment." *Library Acquisitions: Practice & Theory* 19, no. 1 (1995): 21–32.

Discusses the evolution of technical services in the manner of the biological theory of evolution, namely, natural selection—the struggle to survive and pass on one's genes to future generations. Examines the forces shaping libraries currently and suggests how technical services departments and technical services librarians might adapt and evolve to meet an uncertain future, using the example of Cornell University to illustrate the possibilities.

AA4. Dillon, Martin. *Measuring the Impact of Technology on Libraries: A Discussion Paper.* Dublin, Ohio: OCLC, Inc., 1993.

A paper which, according to the author's preamble, deals with the possibility of tracking the movement now under way in the library world, focusing on what statistics should be counted to tell librarians where we are and where we are going.

AA5. Fons, Theodore A., and Wendy Sistrunk. "The Future of Technical Services: A Report on the NETSL Spring Conference." *ALCTS Newsletter* 6, no. 4 (1995): 52–54.

Four invited papers presented to the New England Technical Services Librarians explore the tools and skills technical services librarians need to contribute to the evolving library. Sheila Intner (Simmons College Graduate School of Library and Information Science) addressed the state of library education; Marshal Keys (NELINET) described the three most significant forces at work in the world at large affecting libraries and technical services; Barbara Tillet (LC) described new and changed tools from LC; and Cynthia Watters (Middlebury College) discussed the importance of communication and cooperation within the cataloging community. After these presentations, Michael Gorman gave the closing address on his provocative views on general and specific issues relating to the future of technical services and technical services librarians.

AA6. Howarth, Lynne C. "Modelling Technical Services in Libraries: A Microanalysis Employing Domain Analysis and Ishikawa ('Fishbone') Diagrams." *Technical Services Quarterly* 12, no. 3 (1995): 1–16.

Reports the preliminary results of a three-year study begun in 1993, funded by the Social Sciences and Humanities Research Council of Canada, which attempts to construct a theoretical basis for defining "technical services" in libraries using domain analysis and Ishikawa diagrams. Findings indicate that the domain (as reflected in articles in the two primary journals, *Library Resources & Technical Services* and *Technical Services Quarterly*) has changed in each decade from the 1950s to the 1990s. The 1980s appeared to include the largest number of functional areas (cataloging, serials, collection development, preservation, acquisitions, and circulation), while the 1950s included the fewest (cataloging, acquisitions, and serials).

AA7. Intner, Sheila S. *Interfaces: Relationships between Library Technical and Public Services.* Englewood, Colo.: Libraries Unlimited, 1993.

This book, a compilation of thirty-nine columns written by the author for *Technicalities* **(AF4)**, is divided into four parts: an overview that looks at general issues; the catalog and cataloging problems; overlaps in interest, responsibility, and function of the two administrative areas; and policies for the future. While these are, basically, opinion pieces, a good many attempt to analyze current problems and suggest solutions, often with attention to the findings of research. Some, however, are meant to provoke thought and discussion, while others merely express the author's views on highly debatable issues.

AA8. ———. "Interfaces." [Column in] *Technicalities* 13, no. 2 (Feb. 1993)–15, no. 10 (Oct. 1995).

Continues and completes Intner's bimonthly columns following those included in the book of the same title. The column is continued by Jean Weihs with the January 1996 issue.

AA9. McCombs, Gillian M. "The Internet and Technical Services: A Point Break Approach." *Library Resources & Technical Services* 38, no. 2 (April 1994): 169–77.

Examines development of the Internet and its library applications in terms of how these are used by technical services and how they may affect its future operations as development progresses. Suggests five strategies for dealing successfully with the Internet: explore it; promote change; spread the word; collaborate; and lobby for standards. This thoughtful article is far from being a "Note on Operations," as it was classified by the *LRTS* editors, and should be read as the scholarly analysis it is.

AA10. McCoy, Patricia Sayre. "Technical Services and the Internet." *Wilson Library Bulletin* 69 (March 1995): 37–40.

Excerpt of a paper presented at the American Association of Law Libraries Annual Meeting, July 1994, giving an overview of ways in which technical services librarians can utilize Internet resources in their daily work. Sections cover acquisitions, listservs, serials, networks, and staff training, and a sidebar includes a list of twenty listservs and two gopher resources with their online addresses and subscription and searching information. The full version of the paper presented at the meeting, titled "The Internet as a Library-Wide Resource: Tools for Acquisitions, Serials, Cataloging, ILL, and Reference," can be accessed on the University of Chicago D'Angelo Law Library WWW page; URL: http://www-law.lib.uchicago.edu/lib/seattle/

AA11. Mumm, James A., and Ann Sitkin. "The Internet in Technical Services: The Impact for Acquiring Resources and Providing Bibliographic Access on Technical Services." Sound cassette recorded at the AALL eighty-seventh Annual Meeting, Seattle, Wash., July 9–13, 1994. Valencia, Calif.: American Association of Law Libraries, 1994.

Mumm and Sitkin were the planners of this program; presentations were by Taylor Fitchett (on the future of technical services from perspective of library administration), Mary Lu Linnane (on uses of the Internet in acquisitions), and Diane Hillman (on bibliographic control of electronic resources and MARBI's activities in this area).

AA12. Philips, Phoebe F. "Computers and Technical Services." [Opinion piece] *Computers in Libraries* 14, no. 3 (March 1994): 11.

A very brief nonscholarly review of current issues in technical services that is, nevertheless, useful because it is addressed to a different audience.

AA13. Reich, Vicky. "A Future for Technical Services." *Library Acquisitions: Practice & Theory* 18, no. 4 (1994): 359–61.

An editorial that speculates on the changing roles and functions of technical services in making information available, organizing it, and preserving it in the research library of the future. Reich concludes that "back-room" services will be needed more than ever, but the skills required to do what will need to be done will change dramatically.

AA14. Taylor, Arlene G. "The Information Universe: Will We Have Chaos or Control?" *American Libraries* 25 (July/August 1994): 629–32.

Addresses the notion that, in the electronic libraries of the future, there will be no need for technical services librarians. Argues that the four processes of identifying information-bearing entries, acquiring them, providing access to them, and locating copies for those who need them will always be necessary.

AA15. Weihs, Jean. "Interfaces." [Column in] *Technicalities* 16, no. 1 (Jan. 1996–).

Continues the bimonthly column written by Sheila S. Intner between 1987 and 1995, focusing on issues of interest to both librarians who prepare bibliographic data and those who search and retrieve it.

AA16. Younger, Jennifer A. "Virtual Support: Evolving Technical Services." In *The Virtual Library: Visions and Realities,* edited by Laverna M. Saunders, 71–86. Westport, Conn.: Meckler, 1993.

Characterizes the role of technical services as creating the infrastructure that supports library services. In the case of virtual library services, technical services' task is to create the appropriate infrastructure to support them. The author gives a detailed, thoughtful assessment of the activities that will be required to accomplish this task.

AB
Textbooks, Guides, and Manuals

AB1. Evans, G. Edward, and Sandra M. Heft. *Introduction to Technical Services.* 6th ed. Libraries Science Text Series. Englewood, Colo.: Libraries Unlimited, 1994.

Now addressing professional librarians as well as those without the master's degree credential who work in library technical service departments, authors Evans and Heft provide a thorough and carefully organized explanation of three components that typically comprise technical service departments: acquisitions, cataloging, and serials. The book begins with a general section giving an overview of technical service activities and organizational issues, computers, and bibliographic networks. This is followed by three more sections devoted to detailed descriptions of acquisitions, cataloging, and serials work. Acquisitions work receives the most attention, with ten chapters; cataloging is next, with eight; and serials is given just one chapter. Although computerized systems are described and analyzed, the breakdown of chapters within each section follows traditional lines, e.g., the cataloging specifics are descriptive cataloging using the rules of the *Anglo-American Cataloguing Rules* for books and "other formats," then choice of access points, subject cataloging, and classification. Serials cataloging is treated separately in the final chapter. Similarly, acquisitions is divided into verification procedures and source, order procedures, fiscal issues, receiving, bindery, and gifts and exchange. Each chapter closes with a list of review questions. A lengthy bibliography is divided by topic and concludes with a list of selected periodicals. The bibliography has been updated, with most imprints from the late 1980s and early 1990s, but still includes a selection of older works (e.g., the 1982 second edition of Donald Foster's *Managing the Cataloging Department* and the 1974 ALA guidelines for handling orders for various types of library materials). An appendix discussing automating small libraries, a glossary of terms, and an index complete the book.

AB2. Nevin, Susanne. "Minnesota Opportunities for Technical Services Excellence (MOTSE): An Innovative CE Program for Technical Services Staff." *Library Resources & Technical Services* 38, no. 2 (April 1994): 195–98.

Briefly describes an innovative continuing education program established in the state of Minnesota to improve technical services. MOTSE involves regular workshop sessions taught by teams of trainers who are supplied with complete teaching materials to support their efforts. Covers the background, goals, and initial implementation of the program, and may serve as a guide to others wishing to initiate similar efforts in other states.

AC
Directories

AC1. Reich, Vicky, Connie Brooks, Willy Cromwell, and Scott Wicks. "Electronic Discussion Lists and Journals: A Guide for Technical Services Staff." *Library Resources & Technical Services* 39 (July 1995): 303–19.

An introduction to electronic resources aimed at the technical services librarian that can be accessed via the Internet. Gives a brief overview explaining the nature of various types of electronic "publication" in general, followed by lists of the resources. For each title, the subscription address, message to subscribe, editor, subscriber base, scope, traffic, and archiving policy are provided, as well as informative and evaluative comments by the compilers. General interest titles are followed by specific areas of technical services, including acquisitions, serials, and collection development (one grouping); cataloging; and preservation. Information is accurate as of March 15, 1995.

AD
Bibliographies

AD1. Drabenstott, Karen M. *Analytical Review of the Library of the Future.* Washington, D.C.: Council on Library Resources, 1994.

A report funded by the Council on Library Resources intended to assist librarians in keeping abreast of new developments and, therefore, empower them to influence the future of information services. The project not only identifies the literature published between 1984 and 1994, creates a digital database of document surrogates, and generates a bibliographic resource but also synthesizes its main themes.

AD2. Johnson, Peggy, ed. *Guide to Technical Services Resources.* Chicago: American Library Association, 1994.

A bibliographic survey of technical services. Intended to be a comprehensive and practical guide to the principal information resources for practitioners, educators, and students, the book covers all media from about 1985 to 1992. A few earlier landmark works are included as well as a few later works available at the time of publication. Although most entries are for printed monographs, serial titles, and articles, there are sections in each chapter for "Sources of Expertise" as well as numerous entries for electronic discussion groups, listservs, etc. Contains an overview chapter by Sheila S. Intner, with the assistance of Samson C. Soong and Judy Jeng, followed by chapters on Acquisitions by Karen A. Schmidt, Descriptive Cataloging by Janet Swan Hill, Subject Analysis Systems by Nancy J. Williamson, Authority Work by

Stephen S. Hearn, Filing and Indexing by Sarah E. Thomas, Serials by Marcia L. Tuttle, Collection Management by Peggy Johnson, Preservation by Wesley L. Boomgaarden, Reproduction of Library Materials by Erich J. Kesse, Database Management by Christina Perkins Meyer, and Access Services by Farideh Tehrani.

AD3. Karp, Rashelle S. "Technical Services." Chapter 5 in *The Academic Library of the 90s*, edited by Rashelle Karp, 199–222. Bibliographies and Indexes in Library and Information Science, no. 9. Westport, Conn.: Greenwood Press, 1994.

More than 160 entries for articles, papers, and some monographs on topics in technical services. Furnishes some coverage of state and regional library association conference proceedings. Each entry includes a citation and very brief descriptive annotation and gives research findings.

AD4. Smiraglia, Richard P., and Gregory H. Leazer. "Reflecting the Maturation of a Profession: Thirty-Five Years of *Library Resources & Technical Services*." *Library Resources & Technical Services* 38, no. 1 (1994): 27–46.

Reviews and evaluates the content of the articles published in *LRTS* over its thirty-five-year history. The authors, editor, and assistant editor of the journal, respectively, conclude that *LRTS* reflects the growth of a maturing scholarly discipline within the scope of its areas of interest.

AE
Handbooks

AE1. Association for Library Collections & Technical Services. "Guidelines for ALCTS Members to Supplement the ALA Code of Ethics." *ALCTS Newsletter* 5, no. 2 (1994): 24.

An addition of nine brief guidelines intended to assist ALCTS members in the interpretation and application of the ALA Code of Ethics for librarians to their own work areas. The ALCTS Board of Directors voted on February 7, 1994, to officially adopt these guidelines.

AF
Periodicals

AF1. *Library Acquisitions: Practice & Theory.* New York: Pergamon Press, v. 1– , 1977– . Quarterly. ISSN 0364-6408.

Currently edited by Carol Pitts Diedrichs, *LAPT* is principally devoted to papers focusing on issues in acquisitions and collection development.

However, articles with broader perspectives appear in its pages frequently enough to warrant inclusion in this chapter, as many of this chapter's other entries clearly show.

AF2. *Library Mosaics.* Culver City, Calif.: Yenor, v. 1– , 1989– . Bimonthly. ISSN 1054-9676.

Known by its subtitle as "The Magazine for Support Staff," this popular periodical is aimed at paraprofessional librarians and contains articles, features, conference reports, and news of interest to this large and growing audience, whose self-awareness also is growing as more and more library tasks once assigned solely to professional librarians are being reassigned to them.

AF3. *OCLC Systems & Services.* Westport, Conn.: Meckler, v. 9– , 1993– . Quarterly. ISSN 1065-075X. Continues *OCLC Memo.*

Practical periodical aimed at OCLC users with information and news about the bibliographic utility, its staff members, products, and services, as well as first-person experiences in problem solving, new applications of OCLC utility programs, etc.

AF4. *Technicalities.* Kansas City, Mo.: Media Periodicals, a division of Westport Publishers, v. 1– , 1981– . Monthly. ISSN 0272-0884.

Edited by Brian Alley until March 1995 and now edited by Sheila S. Intner, current issues include three new columns: a monthly column on finance ("Money Matters," by Murray S. Martin), a bimonthly column on serials ("Making Sense of Serials," by Tony Stankus), and a quarterly column on preservation ("Practical Preservation Practice," by Barbra B. Higginbotham). The newsletter has a new emphasis on products and publications with several signed, evaluative reviews of about 500–1,000 words in length in most issues ("Worth Noting").

AG
Sources of Expertise

AG1. Baker, Barry B., ed. "Technical Services Report." *Technical Services Quarterly* [begins with] 12, no. 1 (1994)– .

A quarterly report consisting of sometimes brief, sometimes lengthy descriptions by contributing authors of developments in the technical services world: news of the bibliographic utilities and networks; happenings at conferences and meetings; innovations in relevant publications; and research projects, etc.

AGA
Professional Associations and Organizations

AGA1. Association of Research Libraries (ARL), Office of Management Services. ARL Executive Director: Duane E. Webster, 21 Dupont Circle, Suite 800, Washington, DC 20036. Phone: 202-296-2296; fax: 202-872-0884; URL: http://arl.cni.org/

Publishes a series of "SPEC" Kits (i.e., Systems and Procedures Exchange Center Kits) several times a year on topics of immediate interest to the membership. Each kit includes the results of a survey of members and documents relating to the topic submitted by respondents. Also sponsors training institutes on effective management practices for large research libraries.

AGB
Conferences

These conferences, though their titles may imply a narrow focus, frequently address issues of interest to all technical service areas.

AGB1. Allerton Park Institute. Allerton House, Robert Allerton Park, Monticello, Ill. Once a year; October.

Sponsored by the Graduate School of Library and Information Science, University of Illinois at Urbana-Champaign. Proceedings are published.

AGB2. The College of Charleston Conference: Issues in Book and Serial Acquisitions. Charleston, S.C. Once a year; November.

Papers from the conference may appear in *Library Acquisitions: Practice & Theory* or other journals.

AGB3. Clinic on Library Applications of Data Processing. Urbana-Champaign, Ill. Once a year; April.

Sponsored by the Graduate School of Library and Information Science, University of Illinois at Urbana-Champaign. Proceedings are published.

AGB4. The Feather River Institute on Acquisitions and Collection Development. Feather River Conference Center, Calif. Once a year; May.

Sponsored by the University of the Pacific Libraries, Stockton, Calif. Although intended as a forum for acquisitions librarians, vendors, and publishers to discuss issues of mutual concern, the Feather River Institute discusses subjects that extend to other areas of technical services; e.g., the program for the 1996 Institute included papers titled "Realities of the Team Model in Library Technical Services" and "Applying Serials & Acquisitions Skills on the 'Other Side.'"

AGB5. New York State Education and Research Network [NYSERNet] Conference. Once a year; usually September or October.

NYSERNet is the network that connects libraries and educational institutions in New York State to the Internet. Provides educational opportunities for librarians. NYSERNet's Web home page is found at URL http://www.nysernet.org

AGB6. Snowbird Leadership Institute. Once a year; August.

Founded by the late Dennis Day, Library Director, Salt Lake City Public Library. Sponsored in 1996 by Ameritech Library Services.

AGB7. Special Libraries Association Annual Conferences. Twice a year; January and June.

SLA Executive Director: David R. Bender. Headquarters: 1700 18th Street, NW, Washington, DC 20009-2508. Phone: 202-234-4700; fax: 202-265-9317.

AH
Administration

AHA
General Works

AHA1. Allen, Nancy H., and James F. Williams II. "The Future of Technical Services: An Administrative Perspective." *Advances in Librarianship* 19 (1995): 159–89.

Starting with a brief but thoughtful analysis of current societal and academic trends, the authors examine library technical services to determine what changes have occurred and why. Then, extrapolating from the discussion, they suggest what will happen in the future. Six change agents are identified: the search for savings; pressure on library space; greater electronic publication; user demands for access; pressures on the organization; and advances in software. Each of these is seen as prompting moves to alter technical service activities and operations, including outsourcing some, combining some with public and administrative services, and incorporating some into user-controlled functions. Contains an extensive bibliography.

AHA2. Coffey, James R. "Competency Modelling for Hiring in Technical Services: Developing a Methodology." *Library Administration & Management* 6 (Fall 1992): 162–72.

Puts together an ideal profile of attitudes and experiences common to successful technical services staff members by interviewing two groups of people who enjoy these jobs and perform well in them. Respondents in both groups felt that behavior always should be purposeful, that harmonious personal relationships in the workplace were a necessity, and that order and

15

productivity were related. More specific attitudes also are described, along with the author's application of his ideal attitudinal profile in job searches for two new technical services staff members.

AHA3. Davis, Susan, and Deanna Iltis [presenters], and Judy Chandler Irvin [recorder]. "Integrating Documents Processing into Traditional Technical Services." *The Serials Librarian* 25, nos. 3/4 (1995): 289–94.

In a paper presented at the NASIG conference, the authors discuss two case studies in shifting the responsibility for processing government documents from separate document departments to general technical services departments. Covers the reorganization, staffing, and work flow changes needed to accomplish the change.

AHA4. Dougherty, Richard M., and Ann P. Dougherty. "The Academic Library: A Time of Crisis, Change, and Opportunity." *Journal of Academic Librarianship* 18 (Jan. 1993): 342–46.

Reports the results of a poll of *JAL* Board of Editors members and expert referees about their principal concerns for academic libraries. Commonly held themes that emerged included concern for the future of library education, copyright, budgets and funding, access to information, electronic publishing, and library infrastructure. Of these, access to information and library infrastructure fall directly into the area of technical services. The brief discussion of the opinions of this group of fifteen leading librarians highlights the trends driving change in technical services today.

AHA5. Flood, Susan. "ALCTS Commercial Technical Services Committee, June 28, 1994." [Conference report in] *Library Acquisitions: Practice & Theory* 19 (Spring 1995): 123–24.

Reports on a program held at the 1994 ALA annual conference in Miami, Florida, in which four invited speakers addressed the subject: "User Groups: Effective Communication with Vendors?" Jerry Brock, past president of the Dynix Users Group, addressed elements needed for effective communication, including the autonomy of the user group. Anita Cook, past president of Innovative Interfaces User Group, discussed the enhancement process of Innovative. Jennifer Meldrum, director of marketing for MARCIVE, Inc., spoke about forms of communication that can be used by any group, emphasizing the Internet and listservs. Bob Walton, vice-president of Innovative Interfaces, Inc., spoke about the schizophrenic relationship of user groups and their companies, discussing both the advantages and disadvantages of domination by the company versus autonomy of the user group.

AHA6. Frost, Carolyn O. "Quality in Technical Services: A User-Centered Definition for Future Information Environments." In "The Visible College" [column], edited by Ling Hwe Jeng. *Journal of Education for Library and Information Science* 35 (Summer 1994): 229–32.

The author attempts to define quality in terms of user needs for intellectual access to information. She suggests that the concept of bibliographic services and the domain of bibliographic control must be broadened to include affording users the ability to navigate the domain and empowering them to manage their own information resources as well as providing such services for libraries and library materials.

AHA7. Hirshon, Arnold, ed. *After the Electronic Revolution, Will You Be the First to Go?* Chicago: American Library Association, 1993.

Proceedings of the 1992 ALCTS President's Program held at the ALA Annual Conference in San Francisco. Hirshon introduces the idea of a paradigm shift in information access based on the premise that it is changing from a two-stage process of finding bibliographic information first followed by obtaining the documents they represent, to a single-step process in which bibliographic information and documents are integrated into one unit delivered electronically. Authors of papers include Theodor Holm Nelson, who addresses changes in electronic communication; Peter S. Graham, who discusses intellectual preservation; Thomas Duncan, who writes about electronic information systems and their impact on scholarly research; and Susan K. Martin, who explains the "Strategic Visions Statement" funded by the Council for Library Resources.

AHA8. Hunt, Caroline C. "Technical Services and the Faculty Client in the Digital Age." *Library Acquisitions: Practice & Theory* 19, no. 2 (1995): 185–89.

Paper delivered at the 1994 Charleston Conference exploring the impact on scholars of the changes occurring in library services and delivery. In particular, the author, a professor of English at the College of Charleston, discusses questions of access, issues raised by the changing nature of the information hierarchy, new definitions of "publishing," and problems of communication between library staff and faculty.

AHA9. Pierce, Darlene M., and Eileen Theodore-Shusta. "Automation: The Bridge between Technical Services and Government Documents." *Cataloging & Classification Quarterly* 18, nos. 3/4 (1994): 75–84.

Describes how procedures previously performed in the Olson Library documents unit were transferred to the technical services unit after bibliographic records for approximately 130,000 government documents were entered into the library's online catalog.

AHA10. Rogers, Shelley L. "Technical Services Librarians in the 21st Century." *Technicalities* 12 (Feb. 1992): 14–15.

Visualizes technical services activities being performed electronically by a librarian of the future, then briefly examines the trends at work in the 1990s that underlie the scenario.

AHA11. Silberman, Richard M. "A Mandate for Change in the Library Environment." *Library Administration & Management* 7, no. 3 (1993): 145–52.

Argues that the poor air quality in modern sealed library environments causes symptoms that resemble the classic cold. This building-related syndrome is the result of the circulation by air-handling systems of pollutants that have their source in books and paper, photocopies, computers, and other hardware (equipment that typically is found in technical services departments in large numbers). Library directors are urged to pay attention to the problem and ensure that staff and patrons breathe healthy air.

AHA12. Smith, Terry. "Training Technical Services OCLC Users." *OCLC Systems & Services* 10 (Summer/Fall 1994): 49–53.

Reviews the reasons why a library might wish to train uninitiated staff or end users in using OCLC and the resources available to a trainer attempting to do so. Suggests many different types of resources: OCLC bibliographic formats, standards, and technical bulletins; computer-based training modules; workshops, guides, and reference cards prepared by the regional networks; online help files; workbooks for offline users; locally prepared guides and other training aids; professional literature, including OCLC and regional network newsletters, etc.; OCLC Users Group meetings; informal contact with other librarians; programs at professional meetings; listservs on the Internet; and library school programs.

AHA13. Thawley, John, and Philip G. Kent, eds. *Amalgamations & the Centralisation of Technical Services: Profit or Loss: Papers Presented at a Seminar Held in Melbourne, November 13, 1992.* [Melbourne?]: Australian Library and Information Association, [1993?].

Describes the effects on library technical services on the consolidation of Australia's seventy universities into thirty, implemented in order to improve performance and accountability. Includes case studies from six of the consolidated institutions that pull no punches about the stresses experienced by the institutions while finding the results worthwhile.

AHA14. Wilson, Karen A. "Redesigning Technical Services Work Areas for the 21st Century: A Report of the ALCTS Creative Ideas in Technical Services Discussion Group, American Library Association, Midwinter Meeting, Los Angeles, February 1994." *Technical Services Quarterly* 12, no. 2 (1994): 55–60.

Reports on four roundtable discussions facilitated by Tatiana Barr of Stanford University, Louise Leonard of the University of Florida, Aline Soules of the University of Michigan, and Linda Thompson of the University of Houston, involving approximately fifty-participants. Current trends prompting reorganization, the nature of reorganization in various institutions, and the effects of new administrative structures were explored and illustrated in the discussions. Some specific topics covered were integrating copy cataloging with acquisitions, outsourcing copy cataloging, using collection development staff in cataloging, and integrating serials and monographic acquisitions units. Includes brief bibliography of related sources.

AHB
Outsourcing

AHB1. Bush, Carmel C., Margo Sassé, and Patricia Smith. "Toward a New World Order: A Survey of Outsourcing Capabilities of Vendors for Acquisitions, Cataloging and Collection Development Services." *Library Acquisitions: Practice & Theory* 18, no. 4 (1994): 397–416.

Bush et al. show, with their survey of selected library suppliers, that there are new opportunities for outsourcing various technical service functions, including pre-order searching and verification, claiming, copy cataloging, and original cataloging. The authors find that collection development seems to be the least affected by new vendor services. They suggest librarians work together with suppliers in the development of new contract services.

AHB2. Dwyer, Jim. "Does Outsourcing Mean 'You're Out'?" *Technicalities* 14 (June 1994): 1, 6.

Humorous opinion piece from the "Reader's Soapbox" column challenges the points made by Sheila S. Intner in her March 1994 "Interfaces" column on outsourcing **(AHB5)**. The author identifies problems he has observed as a result of downsizing technical services and substituting computing for human judgments.

AHB3. *The Future Is Now: The Changing Face of Technical Services: Proceedings of the OCLC Symposium, ALA Midwinter Conference, February 4, 1994.* Dublin, Ohio: OCLC, Inc., 1994.

Six invited presentations examine various aspects of outsourcing, focusing mainly on OCLC-related services such as shared cataloging. Issues relating to college/university libraries, public libraries, and large research libraries are covered. Following the papers, a question-and-answer session and several OCLC policy documents are reproduced. The proceedings also are available online at OCLC's Web site at URL ftp://ftp.rsch.oclc.org/pub/documentation/ala_symposium/cataloging1994/

AHB4. Gibbs, Nancy J. "ALCTS/Role of the Professional in Academic Research Technical Services Departments Discussion Group." *Library Acquisitions: Practice & Theory* 18 (Fall 1994): 321–22.

Summary of three presentations about technical services outsourcing given at the Discussion Group's meeting in Los Angeles. Speakers included Gary Shirk, Chief Information Officer at Yankee Book Peddler, Keith Schmiedl, President of Coutts Library Services, and Seno Laskowski, Head of Cataloging, University of Alberta in Calgary. Describes what outsourcing is and that it is not a new idea, how it is being used currently to downsize technical services departments, and how new vendor-library partnerships might be used to benefit libraries and librarians.

AHB5. Intner, Sheila S. "Outsourcing: What Does It Mean for Technical Services?" *Technicalities* 14 (March 1994); 3–5.

Examines outsourcing and the motives behind its renewed popularity in libraries. In addition to cost-cutting, outsourcing provides a means of getting rid of tasks that library directors believe are unimportant, routine, and even unnecessary. Such tasks include cataloging, serials control, and the acquisition of monographs. Suggests the potential benefits and pitfalls that can result from outsourcing.

AHB6. Johnson, Marda L. "Technical Services Productivity Alternatives." *Library Acquisitions: Practice & Theory* 19, no. 2 (1995): 215-17.

Written by the manager of OCLC's Product Implementation Department, this brief paper delivered at the 1994 Charleston Conference describes three recently implemented services offered to libraries by OCLC intended to supply products or services previously performed in-house by library technical services departments: TechPro cataloging service; PromptCat copy cataloging of vendor-supplied materials; and PromptSelect materials ordering system.

AHB7. Ogburn, Joyce L., ed. "Special Section on Outsourcing." *Library Acquisitions: Practice & Theory* 18, no. 4 (1994): 363–95.

Four papers introduce and discuss outsourcing as a business strategy for various aspects of library practice. Ogburn provides "An Introduction to Outsourcing," followed by papers titled "Catalog Outsourcing at Wright State University: Implications for Acquisitions Manager," by Barbara A. Winters; "Outsourcing from the A/V Vendor's Viewpoint: The Dynamics of a New Relationship," by Linda F. Crismond; and "Outsourced Library Technical Services: The Bookseller's Perspective," by Gary M. Shirk. All contributions are thoughtful and balanced, with issues and concerns reviewed in some detail. Important reading for those seeking basic, yet practical, information about outsourcing.

AHB8. Propas, Sharon W. "Ongoing Changes in Stanford University Libraries Technical Services." *Library Acquisitions: Practice & Theory* 19, no. 4 (1995): 431–33.

Increased outsourcing is driving major changes in technical services work flow and organization at Stanford University. The University's goals in making the changes and the process employed are discussed briefly.

AHB9. Varner, Carroll H. "Outsourcing Library Production: The Leader's Role." In *Continuity & Transformation: The Promise of Confluence: Proceedings of the ACRL Seventh National Conference, March 29–April 1, 1995, Pittsburgh, Pennsylvania,* edited by Richard AmRhein, 445–48. Chicago: Association of College & Research Libraries, 1996.

Suggests that outsourcing of technical services activities provides an opportunity for leadership in creating new roles for librarians along with the advantages to the library in cost and efficiency.

AHC
Reorganization and Restructuring

AHC1. Bazirjian, R. "Automation and Technical Services Organization." *Library Acquisitions: Practice & Theory* 17 (1993): 73–77.

Describes the experience of Syracuse University in reorganizing technical services. Four issues are highlighted: (1) streamlining functions and eliminating duplication of effort, (2) ensuring cost-effectiveness, (3) providing immediate access to all technical services work, and (4) using the integrated database to bind all functions together.

AHC2. Bevis, Mary D., and Sonja L. McAbee. "NOTIS as an Impetus for Change in Technical Services Departmental Staffing." *Technical Services Quarterly* 12, no. 2 (1994): 29–43.

Report of a survey of NOTIS users on the technical services staffing and compensation patterns before automation and afterwards. The authors found that most respondents described shifts in responsibilities with the major impact on nonprofessional staff members, involving increased emphasis on technical skills and authority maintenance in all areas. The authors believe that survey results argue for upgrading nonprofessional job descriptions after automation. Provides a useful perspective on staffing issues for any library using support staff in serials management.

AHC3. Cook, Eleanor I., and Pat Farthing. "A Technical Services Perspective of Implementing an Organizational Review while Simultaneously Installing an Integrated Library System." *Library Acquisition: Practice & Theory* 19, no. 4 (1995): 445–61.

Describes the process by which reorganization was planned and implemented at Appalachian State University at the same time a new integrated online system (INNOPAC) was being implemented. Provides new and old organizational charts, the timetable of activities, statement of goals and objectives of the reorganization, and lists of concerns that emerged during the process, as well as a list of "hot topics" that technical services personnel believed were most important.

AHC4. Coulter, Cynthia, and Lola Halpin [workshop leaders], with Rita Broadway, [recorder]. "Who Needs to Know What? Essential Communication for Automation Implementation and Effective Reorganization." In *A Kaleidoscope of Choices: Reshaping Roles and Opportunities for Serialists*, edited by Beth Holley and Mary Ann Sheble, 311–18. Binghamton, N.Y.: Haworth Press, 1995. Also published as *The Serials Librarian* 25, nos. 3/4 (1995).

Report of a workshop held at a NASIG conference investigating the need for and procedures designed to facilitate effective decision making, communication, and the involvement of all staff in a reorganization process for the technical services. Workshop leaders illustrate their points with incidents from personal experiences at Emory University (Lola Halpin) and the University of Northern Iowa (Cynthia M. Coulter).

AHC5. Davis, Trisha L. "Blurring the Lines in Technical Services." *Library Acquisition: Practice & Theory* 17 (1993): 85–87.

Report from the 1992 ALA Annual Conference recounting the author's experience at The Ohio State University Libraries in redesigning some of the serials acquisitions and cataloging work flows with the ultimate goal in mind of reducing organizational barriers and integrating work flowing through diverse automated systems.

AHC6. McGreer, Anne. "ALCTS Creative Ideas in Technical Services Discussion Group." *Library Acquisition: Practice & Theory* 19, no. 3 (1995): 344–45.

Discusses self-managed teams and changing functions in technical services as means of dealing with the need to reorganize and restructure responsibilities in the changing library environment. While it is merely a conference report and is, thus, very brief, it highlights the high priority areas—areas dealt with in greater depth by longer papers listed in this section.

AHC7. Russell, Gordon, and Karin den Beyker. "Technical Services/Public Services: New Wine in Old Bottles: The Organizational Structure of Libraries in the Electronic Age." Sound cassette recorded at the AALL Eighty-seventh Annual Conference, Seattle, Wash., July 9–13, 1994. Valencia, Calif.: American Association of Law Libraries, 1994.

Gordon and den Beyker planned and moderated a program with presentations by John Doyle, who looks at history of librarianship over the last several decades and "the end of librarianship as we know it"; Kathy Price, who discusses information service, the increasingly universal ability to provide that service, and the breaking down of libraries' organizational barriers; and Diane Helman, who considers the reality of structural changes and questions their significance.

AHC8. Williams, Johnette J. "Technology and Library Organizational Structure." *Community & Junior College Libraries* 8, no. 1 (1995): 93–101.

Reports on a study to determine the extent to which community college library organizational structure changed as a result of implementing automated library systems. The findings showed that about 50 percent of the library directors responding claimed to be dissatisfied with their current organizational structures but were not doing much to change them significantly. Many staffing shifts occurred in technical services, but they did not constitute true restructuring.

AHC9. Wilson, Karen A. "Reorganization of Technical Services Staff in the 90s: A Report of the ALCTS Creative Ideas in Technical Services Discussion Group, American Library Association, Midwinter Meeting, Los Angeles, February 1994." *Technical Services Quarterly* 12, no. 2 (1994): 50–55.

Reports on five roundtable discussions involving about fifty participants focusing on architectural, design, and implementation issues. Discussion leaders were Johanna Bowen of the State University of New York at Cortland, Anaclare Evans of Wayne State University, Carolyn Sherayko of Indiana University, Michael Samson of Wayne State University, and Suzanne Sweeney of Stanford University. Examines practical problems and solutions drawn from recent experiences and offers advice to those embarking on renovations or the design of new facilities. Includes brief bibliography of relevant sources.

AI
Database Management

AIA
Bibliographic Networks

AIA1. Keys, Marshall. "On the Future of the OCLC Regional Networks." *Library Administration & Management* 6 (Winter 1992): 10–14.

Provides a thoughtful review of the history and development of the bibliographic utilities and their distributors within specific geographic areas, the regional networks. Discusses current trends and their implications affecting the roles of utilities and regional networks, and concludes that the days when the relationship was exclusive are over. A new working relationship between these two entities is needed to ensure the prosperity of both.

AIA2. "OCLC Affiliated U.S. Regional Networks: A Special Partnership." *OCLC Newsletter* no. 218 (Nov./Dec. 1995).

Issue of the utility's newsletter containing news about the international scene, headquarters, regional utilities, and individual member libraries, plus a message from OCLC President K. Wayne Smith on the issue's theme, which introduces the section on the regional networks.

AIA3. Panko, W. B., et al. "Networking: An Overview for Leaders of Academic Medical Centers." *Academic Medicine* 68 (July 1993): 528–32.

Provides information about networking for decision makers by suggesting a conceptual framework around which to make informed decisions about networking in the organization. The framework covers basic connectivity, added value to build a critical mass, and the network as future integrative focus. In each area, typical problems, likely solutions, and benefits are identified and discussed.

AIB
Expert Systems

AIB1. Hawks [Diedrichs], Carol Pitts. "Expert Systems in Technical Services and Collection Management." *Information Technology and Libraries* 13 (Sept. 1994): 203–12. [Compiler's Note: In January 1996, Carol Pitts Hawks changed her name to Carol Pitts Diedrichs].

After a brief explanation of expert systems in general, the author provides a "state-of-the-art" review of applications in the technical services, that is, of published proposals and reports of expert system projects in cataloging, acquisitions, serials control, preservation, and collection management. She concludes that little has been accomplished and not much more is in progress. She says, "With downsizing, 'rightsizing,' and the emergence of contracting out technical services, particularly cataloging services, to outside vendors and utilities, it is unlikely that libraries will devote much attention to developing expert systems for technical services...."

AIB2. Jeng, Judy. "Expert System Applications in Cataloging, Acquisitions, and Collection Development: A Status Review." *Technical Services Quarterly* 12, no. 3 (1995): 17–28.

Presents a brief introduction to expert systems, reviews the history of expert systems in the library, and describes major developments of expert systems in technical services. Also discusses barriers to the development of new systems, namely, lack of requisite knowledge, inability to transport systems easily, and negative perceptions about cost and potential benefits. Concludes that, despite the fact that only a handful of applications are in use, barriers to development can be overcome and expert systems might be expected to play significant roles in the future.

AIB3. Morris, Anne, ed. *The Application of Expert Systems in Libraries and Information Centres.* London: Bowker-Saur, 1992.

Reviews the progress made (as of 1992) in applying expert systems technology to library and information work and suggests future impact of expert systems and artificial intelligence on libraries. Includes chapters on indexing, cataloging, and online information retrieval. Bibliography and glossary.

AIB4. Ridley, M. J. "An Expert System for Quality Control and Duplicate Detection in Bibliographic Databases." *Program: Automated Library and Information Systems* 26 (Jan. 1992): 1–18.

Describes QUALCAT (Quality Control in Cataloguing) project at the University of Bradford (Bradford, W. Yorks, England), which sought to apply automated quality control to databases of bibliographic records. The project used an expert system to detect putative duplicate bibliographic records and determine if they were, in fact, duplicates, and, if so, which were the best records.

AIB5. Saffady, William. "The Bibliographic Utilities in 1993: A Survey of Cataloging Support and Other Services." *Library Technology Reports* 29 (Jan./Feb. 1993): 5–141.

The latest in a series of comparative studies of the bibliographic utilities performed periodically since 1976. Examines the current range of services offered by OCLC, RLIN, Utlas, WLN, AGILE III, Interactive Access System, and DRANET. The report is followed by an extensive bibliography.

AIC
Hardware and Software

AIC1. Archer, John. "Give Me Barcodes, or Give Me Carpal Tunnel!" *Against the Grain* 7 (Feb. 1995): 36, 78.

Despite the humorous title, this paper, growing out of discussions at the 14th annual Charleston Conference, presents five serious recommendations about barcodes to local system vendors and urges librarians to review the function and capabilities of the barcodes within their local systems.

AIC2. Kaplan, Michael. "Technical Services Workstations Improve Productivity." *Library Systems Newsletter* 15, no. 7 (1995): 68–69.

Describes the features of workstations that employ graphical user interfaces and provides a brief report of a 1994 survey of users conducted among members of the Standing Committee on Automation of the Cooperative Cataloging Council.

AIC3. Matthews, Joseph R., and Mark R. Parker. "Local Area Networks and Wide Area Networks for Libraries." *Library Technology Reports* 31 (Jan./Feb. 1995): 5–110.

Provides information describing LAN and WAN technologies and illustrates how they have been used by libraries. Five sections cover LANs, WANs, linking technologies, communication options, and a call to action. Closes with a section on future developments and applications.

AIC4. Technical Services Workstations. Judith M. Brugger, Michael Kaplan, and Joseph A. Kiegel, comps. SPEC Kit, no. 213. Washington, D.C.: Association of Research Libraries, Office of Management Services, 1996.

Results of a survey by the Program for Cooperative Cataloging Standing Committee on Automation. The technical services workstation (TSW) can be defined as a personal computer that has been customized for use in technical services departments, but also includes the entire suite of standard administrative applications. It is networked, located either on a local area network or a TCP/IP campus network, and from there connected to the Internet. TSWs were in use in 65 percent of the sixty responding ARL member libraries. Provides information on hardware, software, and ergonomic considerations. The survey reveals that libraries are moving closer to business-world standards for new installation, with Pentium-based machines predominating. Sample documents include equipment upgrade proposals, training documents, and information about ergonomics. Brief bibliography.

AID
Local Systems

AID1. Boss, Richard W. "Client/Server Technology for Libraries with a Survey of Vendor Offerings." *Library Technology Reports* 30 (Nov./Dec. 1994): 681–744.

Included here primarily because of the survey of local system client/server applications. Boss begins by defining and describing client/server technology, then discusses management issues, including staffing, training, ongoing support and costs, and concludes with the report of the survey of how thirty local library systems are using or planning to use the technology. Contains extensive bibliography of sources.

AID2. ———. "Technical Services Functionality in Integrated Library Systems." *Library Technology Reports* 31 (Nov./Dec. 1995): 619–770.

Organized by type of function, covers cataloging, acquisitions, serials control, standards and interfaces, and hardware and software environments. Each section gives a brief general discussion of various functionalities, followed by an evaluation of performance for that functionality in twenty-nine local library systems.

AID3. "**Loading the GPO MARC Tapes: 1992 Preconference of ALA/Government Documents Round Table.**" *DTTP: Documents to the People* 20 (Dec. 1992): 207–32.

Proceedings of a preconference to the 1992 ALA annual conference in San Francisco, sponsored by MARCIVE, a vendor specializing in catalog conversion and tapeloading the GPO tapes. Features five papers covering the history of GPO cataloging and tape production, planning for the tape load, authority control, profiling the tapes for loading into the local system, and performing the post-load cleanup. Offers both practical details and a general approach to tapeloading issues.

AID4. Matthews, Joseph R., and Mark R. Parker. "Microcomputer-Based Automated Library Systems: New Series, 1993." *Library Technology Reports* 29 (March/April & May/June 1993): 149–456.

A comprehensive survey of the features and capabilities of twenty-eight microcomputer-based library systems that perform technical services function. Illustrated with sample screens. Although the material is somewhat dated, it is valuable as a guide to the functions and capabilities of many systems currently being offered to libraries.

AID5. McCue, Janet, with a sidebar by Dongming Zhang. "Technical Services and the Electronic Library: Defining Our Roles and Divining the Partnership." *Library Hi Tech* 12, no. 3 (1994): 63–70. [Also identified as Issue 47.]

Describes three projects implemented at Cornell University's Mann Library that illustrate the range of activities in which technical services staff have participated to provide the organizational framework for successful electronic service to users. The three projects involve numeric, full-text, and bibliographic information. In Sidebar 2, Zhang discusses the issues involved with providing access to electronic journals.

AID6. Saffady, William. "Integrated Library Systems for Minicomputers and Mainframes: A Vendor Study." *Library Technology Reports* 30 (Jan./Feb. and March/April 1994): 5–232.

A thorough, comprehensive survey of the state of the art for local systems that utilize a single database to perform cataloging, online catalog access, circulation control, and other technical services functions. Part I introduces the concepts, history, characteristics, and capabilities of available products and gives detailed reports on eight vendors (DRA, Dynix, Gaylord, Innovative, NOTIS, Sirsi, Sobeco Ernst & Young, and VTLS); Part II continues with reports on several more vendors (Best-Seller, CARL, Comstow, EliAS, Ex Libris, Gateway, GEAC, IME, Information Dimensions, ILS, MARCORP, NSC, and SLS).

AID7. Schottlaender, Brian, ed. *Retrospective Conversion: History, Approaches, Considerations.* Binghamton, N.Y.: Haworth Press, 1992. Also published as *Cataloging & Classification Quarterly* 14, nos. 3/4 (1992).

Reviews the considerable history and background that has accumulated over the last two decades on retrospective conversion. It is an important body of information for readers who may be addressing the issue for the first time. Also includes papers suggesting approaches that utilize recently developed capabilities of computer systems for input and translation from noncomputerized to computer-readable form and from one database to another.

B
Acquisitions

Karen A. Schmidt

Acquisitions literature in recent years has been marked by two main themes: the impact of technology and the place of acquisitions in the "new" library. Technology has indeed come to acquisitions, building on a long and slow history of automating various parts of the whole. Acquisitions has never been easy to automate. Unlike other processes in the library, a complete acquisitions automation system has to please the library internally and the business world externally. For some years now, acquisitions departments in many libraries throughout the country have had successful automated operations. The business world has lagged somewhat behind. The final pieces are falling into place, however, as the publishing and vending worlds embrace the X12 standards that allow for full electronic communication among all points of the triangle created by acquisitions, publishing, and selling. Articles on EDI (electronic data interchange) abound not just in the library literature now, but in the publishing field as well. Acquisitions staff can finally look forward to working in a fully automated environment within a very short time with automated systems that can be updated immediately with marketplace information.

This technological change has been accompanied by a move on the part of vendors to increase their value-added services by offering fully cataloged, shelf-ready materials. The subsequent effect on acquisitions has caused those in this area of librarianship to question and defend their role in the library and to analyze how technology has changed the nature of their work. When books can be brought into the library on approval plans ready for circulation and described in the library online catalog, both acquisitions and technical processing managers are forced to reevaluate which—and how many—services they offer to their institutions. These articles build on acquisitions literature and denote a maturation within acquisitions that is fruitful for both acquisitions and the rest of the library.

Perhaps the more interesting aspect of recent acquisitions literature is the role of acquisitions in the library organization. Today, the focus is on how acquisitions work relates to interlibrary loan, collection development, access and ownership, and cataloging of materials. It has strong ties with the legal implications of licensing agreements, and many acquisitions librarians regularly are involved in negotiating complicated networking agreements for electronic resources, not only for their own libraries but also for consortia of libraries.

This new arena, in which the acquisitions librarian in large libraries is an important player in management decisions, is a new phenomenon indeed. The gestational period to this present state has been long and the players many and vocal. Acquisitions has had an interesting evolution in librarianship. Unlike reference or cataloging, which have been standards of our profession and understood—more or less—by all, acquisitions has been something of a backwater enterprise. Although it once enjoyed prominence in the library school curriculum, by the 1940s acquisitions was lucky to be considered an optional credit course in most library schools and generally has been relegated to a one- or two-class discussion in a technical services primer course. Thus, many acquisitions librarians today, be they in charge of no staff or fifty, thousands or millions of dollars, have had to learn their craft on the fly.

This education by serendipity and association has had an important effect on the way in which acquisitions has matured and on the literature it has spawned. Like other areas in this profession, a relatively small group of acquisitions librarians have made a significant difference in the way in which acquisitions has grown. Librarians such as Christian Boissannas (Cornell), Joe Barker (University of California–Berkeley), Carol Pitts Diedrichs (The Ohio State University), and Joyce Ogburn (previously at Yale and now at Old Dominion University) set the tone for acquisitions as it is now practiced. These particular individuals and their colleagues created a synergy in acquisitions in the 1980s that was not unlike, on a much larger scale, the tone that Dewey, Bowker, Wilson, and their colleagues set for librarianship as a whole at the turn of the century. This kind of serendipitous alliance achieved two things: it built on the solid work of acquisitions librarians such as Scott Bullard, Harriet Rubedella, Jennifer Cargill, and Sharon Bonk and it recorded its new ideas for the future, thus providing an opportunity for growth in acquisitions where none had existed before. This is a new turn of events for acquisitions, where reinventing the wheel has been something of an occupational pastime, as a reading of early library literature in the field of acquisitions will tell.

Thus we see in acquisitions literature a new focus on the growth of acquisitions and its integration into the professional discourse of librarianship that seemed to have been reserved more for cataloging, bibliographic instruction, and technological and general management issues. Meta Nissley **(BA5)**, Sharon Propas and Vicky Reich **(BA6)**, Ron Ray **(BG6)**, and Ann O'Neill **(BB8, BH5)** serve as examples of this evolution in our literature with their thoughtful and thought-provoking essays and research.

Even so, the output of current acquisitions literature is slim compared to other areas of the profession. This is a phenomenon that continues to hound editors of acquisitions and technical service journals, who search for meaty, research-based work in acquisitions. What exists often *is* solid, but there is not always a lot of it. Many articles in acquisitions now focus on international aspects, including descriptions of acquisitions for foreign libraries and the severe difficulties acquisitions librarians in many countries face not only in acquiring and processing material, but also technologically and monetarily. Other topics, often covered as "Notes" in some journals, concern EDI (electronic data interchange) and its application to publishing, acquisitions, and accounting; vendor evaluation; and general management issues. Interviews with well-known figures whose work in some way affects acquisitions and collections are a new focus in acquisitions literature. Undoubtedly the best interviews now are seen in *Against the Grain* **(BAC2)**, with current editor Katina Strauch and her editorial staff ferreting out the best and brightest to share their insights. *Library Acquisitions: Practice & Theory* **(AF1)**, under the direction of Carol Pitts Diedrichs, has done similarly successful interviews. Both of these publications are creating an important body of literature for acquisitions that, while not necessarily of the standard refereed journal variety, are nonetheless extraordinarily important records of our history and our future.

Much of what is "happening" in acquisitions today is occurring at conferences. The two popular annual meetings, Feather River **(BAD3)** on the west coast in Blairsden, California, in the spring, and the Charleston Conference **(BAD2)** on the east coast in the fall, usually present the best in acquisitions and collection development each year. The presentations are not always done as publishable papers and the best parts of the conferences usually are the unscripted and lively interactions that occur at the question-and-answer sessions after presentations, during panel discussions, and over a glass of wine at conference events. These unpublishable but important acquisitions events miss the media, except as summaries in *Library Acquisitions: Practice & Theory* (which does a terrific job of keeping the profession apprised of these conferences as well as ALA committee work and presentations); more is the pity, as the real movement forward in acquisitions often occurs there. It is to both the credit and the debit of acquisitions that these important contributions are largely vocal and not written—to the credit for the spirit that acquisitions librarians are expressing in the 1990s and to the debit for often depriving our colleagues not in attendance of what we have learned.

Approval plans continue to occupy a solid portion of the literature on acquisitions. Some of this is a backward glance at how these gathering plans developed, including a renewed interest in the career of Richard Abel **(BB2, BB8)**, who pioneered the modern approval plan. Other concerns address the politics of the approval plan, and the comparative effectiveness of the U.S. and U.K. plans. The historical perspectives in particular are useful to the library community at large. Most of the inquiry and study into methods of building and maintaining approval plans published today is reiterative of earlier work that still stands as definitive text.

Where there is money, there are questions of ethics. Acquisitions has moved beyond the "I know it when I see it" phase of interpreting ethical situations into an honest analysis and definition of what constitutes ethical behavior and ethical dilemmas. The publication in 1994 of the statement on *Principles & Standards of Acquisitions Practice* **(BI1)** by the ALA ALCTS/PVLRC (Publisher/Vendor/Library Relations Committee) is a capstone to a long history of work in this area. In addition, other writings on ethics are now more rooted in a combination of business practices and philosophical concerns.

Finally, a look at this new crop of literature in acquisitions demonstrates an evolution away from purely procedural issues and toward more ideological concerns. The notion of acquisitions as part of a postmodern, chaotic condition is quite a huge leap from earlier writings on vendor selection and evaluation. This emergence into a more intellectual environment probably began with Joe Hewitt's 1989 essay "On the Nature of Acquisitions" **(BA4)**, which validated many acquisitions librarians' feelings of isolation and professional marginalization. While there continue to be many useful descriptive works of acquisitions procedure, this emergence of an intellectual approach to the work of acquisitions is a positive step indeed.

Chapter Content and Organization

This chapter begins with general works, including textbooks, guides and manuals, periodicals, and sources of expertise. Following sections address various topics within the purview of acquisitions librarians: approval plans, automation of acquisitions, out-of-print material, gifts and exchanges, administration, publishing, and ethics. Many resources described in the *GTSR* chapter on Acquisitions remain relevant and current. The reader is encouraged to consider *GTSR* and this work as complementary tools for the practice of acquisitions.

BA
General Works

BA1. Barker, Joseph W. "Acquisitions Principles and the Future of Acquisitions: Information Soup, the Soup-Hungry, and Libraries' Five Dimensions." *Library Acquisitions: Practice & Theory* 17 (1993): 23–32.

"Think readiness" is the maxim of Barker's insightful essay on the interconnections of acquisitions, publishing, library use, and the intellectual world in general. Barker always pushes the imagination, asks the difficult question, and generally spices up the world of acquisitions and collection development.

BA2. Bloss, Alex. "The Value-Added Acquisitions Librarian: Defining Our Role in a Time of Change." *Library Acquisitions: Practice & Theory* 19, no. 3 (1995): 321–42.

At a time when libraries are reevaluating their organizations for cost-effectiveness, acquisitions can find itself in the spotlight as administrators ask how outsourcing might save money and time. Bloss analyzes how acquisitions librarians add value to their organization. Ron Ray and Christian Boissannas offer their responses to Bloss's assertions, with Bloss having the final word. This is a solid revisiting of the role of the acquisitions librarian in the library.

BA3. Hawks [Diedrichs], Carol Pitts. "Building and Managing an Acquisitions Program." *Library Acquisitions: Practice & Theory* 18, no. 3 (1994): 297–308.

A presentation at the ALCTS Business of Acquisitions Institute in 1993, this thoughtful overview of the challenges and rewards of managing acquisitions is a "must-read" for both the new and the initiated acquisitions librarian. Hawks is straightforward, honest, accessible and provides a number of good ideas for the daily management of acquisitions.

BA4. Hewitt, Joe. "On the Nature of Acquisitions." *Library Resources & Technical Services* 33 (April 1989): 105–22.

Listed in GTSR, this essay is worthy of a repeat citation as a pivotal treatise on the role of the acquisitions librarian, the importance of acquisitions in the library, and the future of this work. This article continues to be referred to, read, and is probably responsible for helping move acquisitions forward in a meaningful way within the library profession.

BA5. Nissley, Meta. "Rave New World: Librarians and Electronic Acquisitions." *Library Acquisitions: Practice & Theory* 17 (1993): 165–73.

Focusing on the role of librarians in the new technological world that is emerging, Nissley discusses the opportunities and problems associated with these new advances. How the business world responds to these changes and to the demands placed on them by acquisitions librarians will be crucial to the formulation of how acquisitions will look in the future.

BA6. Propas, Sharon, and Vicky Reich. "Postmodern Acquisitions." *Library Acquisitions: Practice & Theory* 19, no. 1 (1995): 43–48.

Propas and Reich define acquisitions in the larger context of change as a postmodern condition. We may look at change as chaos or a chance to revitalize our work. The authors make a compelling and refreshing argument for the utility of change in acquisitions and libraries.

BAA
Textbooks, Guides, and Manuals

BAA1. Association for Library Collections & Technical Services, Acquisitions Section, Education Committee. *Acquisitions Course Syllabus.* Online. URL: gopher://gopher.uic.edu:70/11/library/ala/ala-xiii/ala13c/50713001/50714045

A joint effort between library practitioners and educators. Includes a bibliography and can be used whole or in part for courses on acquisitions at the graduate level, in unit within technical services, resources, and collections development courses, acquisitions institutes and workshops, and in internal staff development and training events. Also available through the ALA home page (URL: http://www.ala.org) following the links from the Organization to Divisions to ALCTS to the gopher.

BAA2. Bosch, Stephen, Patricia Promis, and Chris Sugnet. *Guide to Selecting and Acquiring CD-ROMs, Software, and Other Electronic Publications.* Acquisitions Guidelines, no. 9. Chicago: American Library Association, 1994.

Practical recommendations for selection and acquisition of electronic materials (CD-ROMs with the exception of music CDs; videotapes and videodiscs; instructional and recreational video games; bibliographic and full text databases; and software). This guide examines major acquisitions issues and presents the basic steps involved in the order process. Useful to new practitioners and experienced librarians seeking clarification on new media.

BAA3. Miller, Heather S. *Managing Acquisitions and Vendor Relations: A How-to-Do-It Manual.* New York: Neal-Schuman, 1992.

Miller covers much more than the simple business relationship between vendors and librarians. Included are chapters on bidding and contracts, pricing, vendor evaluation, the out-of-print market, ethics, and more. This is supplementary information to other sources of information and is useful mainly for bringing together a diverse discussion of library-vendor issues.

BAB
Bibliographies

BAB1. German, Lisa. "A Closer World: A Review of Acquisitions Literature, 1992." *Library Resources & Technical Services* 37 (July 1993): 255–60.

A brief review of acquisitions literature published in a single year—1992.

BAC
Periodicals

BAC1. *ACQNET*. Serial online. Boone, N.C.: Appalachian State University Library, Eleanor Cook, ed. no. 1– , 1990– . Irregular. ISSN 1057-5308. URL: http://www.library.vanderbilt.edu/law/acqs/acqnet.html

ACQNET continues to be the main voice of information for acquisitions librarians. The editorship has moved from Christian Boissannas to Eleanor Cooke and with this move has come a shift in content and commentary, with less philosophical discussions. It remains mandatory for acquisitions administrators. *ACQflashes* are time-sensitive, single-issue postings. To subscribe, send a message to LISTSERV@LISTSERV.APPSTATE.EDU, with the command: SUB ACQNET-L [your name]; messages may be sent to ACQNET-L@LISTSERV.APPSTATE.EDU. Affiliated with *AcqWeb* **(BAD4)**.

BAC2. *Against the Grain*. Charleston, S. C.: Katina Strauch. v. 1– , 1989– . Five times a year. ISSN 1043-2094. URL: http://www.spidergrahpics.com/atg/atgld.html

Subtitled "Linking Publishers, Vendors and Librarians," *ATG* is a smart, sassy, and very important journal for acquisitions and collection development issues. The movers and shakers in the field are frequently interviewed (in 1995, for example, interviews with Knut Dorn of Harrassowitz, Joseph Alen of the Copyright Clearance Center, Dan Tonkery of Readmore, and

Becky Lenzini of CARL were among those featured) and late-breaking news about publishing and libraries is available throughout. The style is quick and easily digestible, and the content just keeps getting better and better. This is a must-buy investment for acquisitions and collection development librarians.

BAD
Sources of Expertise

PROFESSIONAL ASSOCIATIONS

BAD1. American Library Association, Association for Library Collections and Technical Services, Acquisitions Section.
The Acquisitions section of the ALCTS is defining more clearly its role as the professional voice for acquisitions librarians and is actively interacting with other related sections and divisions within ALA. It also sponsors important continuing education events that inform new acquisitions librarians about important facets of their work that are often missing from library-school education. Reports of acquisitions-related discussion groups are also found in *LAPT* **(AF1)**.

CONFERENCES

BAD2. The College of Charleston Conference: Issues in Book and Serial Acquisitions. Charleston, S.C. Once a year; November.
and
BAD3. The Feather River Institute on Acquisitions and Collection Development. Feather River Conference Center, Calif. Once a year; May.
These two annual conferences continue to be pivotal meeting places for discussions of cutting edge issues in acquisitions and collection development. The reports of these meetings are regularly summarized in *Library Acquisitions: Practice & Theory* **(AF1)**. The full papers of these meetings also may be found most often in *LAPT*. For annotation, *see also* **AGB4**.

WORLD WIDE WEB SITES

Descriptions (in the form of home pages) of acquisitions departments abound on the Web. Most are intended as resources for staff of the owning library and provide information about the library, plus links to useful resources on the Internet. At present, one of the most effective search engines to find these sites is Lycos, searching "Library acquisitions." The acquisitions departments listed here are representative only.

BAD4. *AcqWeb.* Online. URL: http://www.library.vanderbilt.edu/law/acqs/acqs.html

A WWW site concentrating on acquisitions issues. Particularly useful is the directory of publishers and vendors, with hotlinks to their Internet sites. Also provides links to general reference resources: dictionaries, currency conversion, postal information, etc.

ACQUISITIONS DEPARTMENTS HOME PAGES

BAD5. Pennsylvania State University Libraries, Acquisitions Services. URL: http://www.libraries.psu.edu/iasweb/acq/hpg_acq1.htm

BAD6. University of Miami Libraries, Acquisitions Department. URL: http://www.library.miami.edu/acqui/welcome.html

BAD7. University of North Carolina–Chapel Hill, Acquisitions Department. URL: http://www.unc.edu/-acqdept/hmenubs.html

This is an excellent example of the kinds of things that can be done with an acquisitions Web site. In addition to the usual organizational information about acquisitions that helps users contact the right staff member, this department also has listed its major accomplishments, brought up vendors' and publishers' e-mail addresses, and provided hot links to publisher information for the Library.

BAD8. University of Washington Libraries, Acquisitions Division. URL: http://staffweb.lib.washington.edu/acq

BB
Approval Plans

BB1. Arnold, Amy E. "Approval Slips and Faculty Participation in Book Selection at a Small University Library." *Collection Management* 18, no. 1/2 (1993): 89–102.

A study at Auburn University at Montgomery explores the factors influencing faculty participation in book selection. Many academic libraries have already walked this path with college faculty as they worked through the role of librarians and teaching faculty in book selection. For the academic libraries that have not yet done so or that need to revisit this area, this article offers a good discussion of the relevant issues. This paper received the Blackwell North American research award.

BB2. Biblarz, Dora. "Richard Abel." *Against the Grain* 4 (June 1992): 24–27; (Sept. 1992): 20–24; (Nov. 1992): 35–39.

Biblarz interviewed Richard Abel about the history of his approval plan, the promises his ideas held, and the problems he encountered along the way. This is an important record of a man whose ideas changed the way in which libraries do business. It is a very good read, as well.

BB3. Brown, Lynne C. Branche. "An Expert System for Predicting Approval Plan Receipts." *Library Acquisitions: Practice & Theory* 17 (1993): 155–64.

Expert systems are not often applied to acquisitions procedures, but there is no reason why they might not be used to tackle some of the work in acquisitions and collection development. Brown describes the methodology in establishing an expert system to determine if titles will come on an approval plan. A selective bibliography is also included.

BB4. Eldredge, Mary. "United Kingdom Approval Plans and United States Academic Libraries: Are They Necessary and Cost Effective?" *Library Acquisitions: Practice & Theory* 18, no. 2 (1994): 165–78.

Price differentials between U.K. and U.S. editions of books continue to be an issue among acquisitions and collection development librarians. Eldredge offers a good addition to the Kruger work **(BB6)** and a chance for newer acquisitions librarians to review the issues of dual-country publications.

BB5. Franklin, Hugh L. "Sci/Tech Book Approval Plans Can Be Effective." *Collection Management* 19, no. 1/2 (1994): 135–45.

In this study of efficient ways to acquire scientific and technical books, CIP slips are compared to book approval plans with three vendors. The result shows a 95 percent hit rate for the books, with receipts occurring in an acceptable time frame. It also shows that a slip approval plan can be a workable alternative.

BB6. Kruger, Betsy. "U.K. Books and Their U.S. Imprints: A Cost and Duplication Study." *Library Acquisitions: Practice & Theory* 15, no. 3 (1991): 301–12.

Kruger's study of the cost and duplication rate between U.K. and U.S. books received on approval plans is a seminal piece of research that shows the extent to which U.K. and U.S. markets are becoming more similar. Overlap for the publishers studied was at 99 percent over a twelve-month period, with U.S. books coming in more cheaply at the time of the study.

BB7. Nardini, Robert F. "Approval Plans: Politics and Performance." *College & Research Libraries* 54 (Sept. 1993): 417–25.

Examines the inherent political nature of approval plans in academic libraries, how they affect staff involvement in selection, cooperation with campus faculty, and vendor relations. He notes that few concrete performance standards for approval plans exist. He explores the difficulty of evaluating an approval plan's performance and emphasizes the need for good communication between libraries and their vendors. Of special note is Nardini's suggestion that cooperative approval plans between two libraries may spring up as budgets tighten.

BB8. O'Neill, Ann L. "How the Richard Abel Co., Inc. Changed the Way We Work." *Library Acquisitions: Practice & Theory* 17 (1993): 41–46.

Richard Abel and his approval plan revolutionized contemporary acquisitions by bringing the vendor into full play in the selection of materials for libraries. Although his company eventually failed, his ideas have not. O'Neill covers an important part of library/vendor history in a well-researched article. The interviews with Abel in *Against the Grain* **(BB2)** are a fine complement to this.

BB9. Warzala, Martin. "The Evolution of Approval Services." *Library Trends* 42 (Winter 1994): 514–23.

Warzala examines the history and evolution of approval plans, from the Greenaway Plan to the present. Warzala is employed by Baker & Taylor, one of the major approval plan suppliers in the U.S., and so his perspective is especially interesting. This essay appeared as part of a larger *Library Trends* issue on financial changes affecting libraries.

BC
Automation of Acquisitions

BC1. Cox, John. "EDI: The Modern Way to Do Business Together." *Collection Management* 19, no. 3/4 (1995): 95–105.

This is a very practical paper on the applications of electronic data interchange for librarians, vendors, and publishers. Cox works for B. H. Blackwell and provides a good business perspective to this topic. Also included are appendixes on other types of standards (BISAC, SISAC, ISO, etc.).

BC2. Dunshire, G. "The Potential of the Internet and Networks for Library Acquisitions." *Taking Stock: Libraries and the Book Trade* 3 (Nov. 1994): 44–49.

The Internet and other networks offer a wide variety of services to acquisitions, including bibliographic information, ordering, and correspondence opportunities. This review of the current available services is a good introduction to the topic.

BC3. Heseltine, R. "Electronic Information and Acquisitions." *Taking Stock: Libraries and the Book Trade* 2, no. 2 (1993): 1–6.

This article reviews acquisitions, automation, and acquisition's sometimes uneasy relationship with new technologies. It provides a good overview of automation successes and failures, suggests ways in which acquisitions can deal with nonprint products, and looks to the future of acquisitions in a virtual library world.

BC4. Hudson, Gary A. "The MSUS/PALS Acquisitions Subsystem Vendor File." *The Acquisitions Librarian* 11 (1994): 161–86.

The acquisitions portion of MSUS/PALS (Minnesota State University System/Project for Automated Library Systems) includes wide-ranging vendor information that allows for analysis across institutions. Hudson describes the components of vendor records that bring quantification to the process of vendor selection and analysis. The basic elements of this acquisitions system will be useful in other library settings.

BC5. Kelly, Glen. "Electronic Data Interchange (EDI): The Exchange of Ordering, Claiming, and Invoice Information from a Library Perspective." *Collection Management* 19, no. 3/4 (1995): 77–94.

Kelly offers an "in the trenches" description of how EDI has been working in a Canadian academic library since 1987. He also includes a non-annotated but selective bibliography on EDI, which is a useful starting place for catching up with past developments.

BC6. Montgomery, J. G. "The Internet in Acquisitions Work: A Status Report." *Trends in Law Library Management and Technology* 6 (Feb. 1995): 1–5.

This article is listed here as an example of the ways in which specialized libraries, as well as more general academic and public libraries, are using the Internet in their acquisitions work. Along with the Dunshire article **(BC2)**, this provides a good overview of how the Internet can be incorporated into acquisitions processes.

BD
Out-of-Print Material

BD1. Eldredge, Mary and William Ludington. "Comparison of Out-of-Print Searching Methods." *Library Acquisitions: Practice & Theory* 17 (1993): 427–32.

The three main methods of O.P. searching (advertising, dealer searching, and online matches) are studied and reported on here. With O.P. work being labor intensive and not often cost-effective, this is a valuable addition to acquisitions literature. Advertising wins out as the most cost-effective.

BD2. Interloc: The Electronic Marketplace for Books. Online. P.O. Box 5, Southworth, WA 98396. Phone: 206-871-3617; fax: 206-871-5626; URL: http://host.interloc.com/

This service began in 1994 as an online service for libraries and dealers to match desiderata and dealer stock. Also included is a "Missing and Stolen

Books Database." It is easy to use, fairly inexpensive, and a good use of technology for the difficult world of O.P. searching. Interloc publishes *Interloc News*, which keeps users up to date on issues and provides information about upcoming book fairs and other book-related events.

BE
Gifts and Exchanges

BE1. Olsen, Margaret S. "The More Things Change, the More They Stay the Same: East-West Exchanges 1960-1993." *Library Resources & Technical Services* 39 (Jan. 1995): 5–21.

Exchanges are a major source of acquisitions for eastern European publications. Olsen provides a historical perspective of these important relationships, discusses the difficulties in maintaining exchanges with countries undergoing significant changes, and gives guidance in planning for future exchange work.

BF
Vendor Selection and Evaluation

BF1. Brown, Lynne C. Branche. "Vendor Evaluation." *Collection Management* 19, no. 3/4 (1995): 47–56.

Working at Pennsylvania State University (PSU) at the time this article was written, Brown looks at vendor evaluation through the eyes of a version of TQM that was instituted at PSU Libraries. Brown looks at expectations from vendors and discusses how to analyze vendor performance in the face of these expectations.

BF2. Fisher, William. "A Brief History of Library-Vendor Relations since 1950." *Library Acquisitions: Practice & Theory* 17 (1993): 61–69.

This is a historical overview of how librarians and agents developed their business relationship over the years and includes an interesting discussion on the impact of technology on acquisitions work. Fisher offers a unique addition to the literature of library-vendor relations.

BF3. Reid, Marion T. "Closing the Loop: How Did We Get Here and Where Are We Going?" *Library Resources & Technical Services* 39 (July 1995): 267–73.

Reid queried key vendors of library materials about their companies' automation, their perceptions of how acquisitions work is being transformed by automation, and their view of acquisitions librarians' capabilities over the past twenty-five years. She notes that acquisitions librarians and vendors alike are more involved in technology and need to build upon the technological skills they possess.

BF4. Rouzer, Steven M. "A Firm Order Vendor Evaluation Using a Stratified Sample." *Library Acquisitions: Practice & Theory* 17 (1993): 269–77.

Using Johns Hopkins University Library as the laboratory, Rouzer reports on a controlled study of domestic firm orders. This is an interesting example of how good research can have an immediate and important impact on daily operations.

BG
Administration

BG1. Astle, Deanna L. "Staff Involvement: The Key to the Successful Merger of Monograph and Serial Acquisitions Functions at Clemson University Library." *Library Acquisitions: Practice & Theory* 19, no. 4 (1995): 427–30.

Support staff are usually the vital portion of acquisitions work and are overlooked in the literature. Astle's case study practically stands alone as an example of how staff can be actively involved in the daily management of acquisitions.

BG2. Boakye, G. "Challenges and Frustrations of an Acquisitions Librarian in a Developing Country: The Case of Balme Library." *Taking Stock: Libraries and the Book Trade* 3 (May 1994): 26–31.

This is a useful wake-up call to U.S. acquisitions librarians who may feel at times that their lot in life is a difficult one. A number of articles appear every year on acquisitions in other countries that have fewer resources than do we. This is a particularly good example of how acquisitions librarians are coping in far-from-ideal situations.

BG3. Flowers, Kay. "Collection Development and Acquisitions in a Changing University Environment." *Library Acquisitions: Practice & Theory* 19, no. 4 (1995): 463–69.

Acquisitions and collection development are often close siblings in libraries, but there is a dearth of discussion about this in the literature. Flowers addresses how changes in collection development have affected acquisitions at one institution. Although mainly about collection development, the handling of acquisitions here is of interest.

BG4. Intner, Sheila S. "The Relationship of Acquisitions to Resource Sharing: An Informal Analysis." *Resource Sharing and Information Networks* 9, no. 2 (1994): 61–73.

Intner takes on the cutting-edge issue of how access versus ownership affects acquisitions operations, including vendor interaction, relationship of acquisitions to new resource sharing initiatives, and general role within the library. A new definition of acquisitions is suggested.

BG5. Ogburn, Joyce. "Changing Relationships in the Acquisition and Delivery of Library Materials: A Survey." *Collection Management* 19, no. 3/4 (1995): 11–27.

Seeking to understand how new technology and document delivery systems are changing the way in which libraries and vendors do business, Ogburn surveyed a variety of librarians, publishers, and vendors about their experiences. The survey was not scientific, but nevertheless is useful in its description of trends and concerns.

BG6. Ray, Ron. "The Dis-Integrating Library System: Effects of New Technologies in Acquisitions." *Library Acquisitions: Practice & Theory* 17 (1993): 127–36.

Similar in scope to the Ogburn article noted above, this article examines the effects of technology on acquisitions and how acquisitions operations are changed and enhanced by vendors' networking and telecommunications advances. Acquisitions is viewed within the context of an integrated library system that technology makes possible.

BG7. Saunders, Laverna M. "Transforming Acquisitions to Support Virtual Libraries." *Information Technology and Libraries* 14 (March 1995): 41–46.

This is a good companion piece to the Intner article **(BG4)** and considers how the new technologies have changed the face of acquisitions. This was a paper presented at the VTLS annual directors conference.

BG8. Shirk, Gary M. "Contract Acquisitions: Change, Technology, and the New Library/Vendor Partnership." *Library Acquisitions: Practice & Theory* 17 (1993): 145–53.

Shirk brings the perspective of being both a successful former acquisitions librarian and a leader in the vendor community to the issue of technology. This essay has a good philosophical bent that raises the discussion to a higher level.

BG9. Stanley, Nancy Marke, and Lynne Branche Brown. "Reorganizing Acquisitions at the Pennsylvania State University Libraries: From Work Units to Teams." *Library Acquisitions: Practice & Theory* 19, no. 4 (1995): 417–25.

TQM has been adopted extensively in some libraries. Stanley and Branche-Brown describe how the team approach works in one academic library acquisitions department. This is a straightforward look at how reorganization of a basic library procedure can succeed or fail.

BH
Publishing

BH1. Clark, Tom. "On the Cost Differences between Publishing a Book in Paper and in the Electronic Medium." *Library Resources & Technical Services* 39 (Jan. 1995): 23–28.

Clark brings an informative and rational explanation to the publishing process for both print and electronic publications. He points out why and how electronic publishing can be substantially cheaper and brings a librarian's perspective to the question of price and publication.

BH2. Edwards, F. "Licence to Kill?" *Bookseller*, no. 4613 (May 20, 1994): 22–25.

This essay examines the Net Book Agreement and the Library License in the United Kingdom, and calls for moderation in libraries' appeal for their abolition. This will be of particular interest to librarians who purchase regularly from Great Britain, where these regulations affect the cost of books.

BH3. Germain, J. Charles. "Publishing in the International Marketplace." *Library Acquisitions: Practice & Theory* 19, no. 2 (1995): 225–29.

Germain reviews the issues facing publishing today, including price increases, technological costs, and other economic issues. He follows the historical development of these issues and tells why librarians and vendors should care. This is a useful overview of the complex and rapidly changing publishing scene.

BH4. Lu, Min-Huei. "A Study of OP and OSI Cancellations in an Academic Library." *Library Acquisitions: Practice & Theory* 18, no. 3 (1994): 277–87.

This is an excellent example of good research that helps the acquisitions librarian plan operationally and advise bibliographers and selectors in the library. Lu looked at out-of-print and out-of-stock reports on orders from 1988 to 1990 placed by the library at the State University of New York–Stony Brook. The data provide specific forecasts on O.P. rates in broad disciplines, among other useful observations.

BH5. O'Neill, Ann L. "The Gordon and Breach Litigation: A Chronology and Summary." *Library Resources & Technical Services* 37 (April 1993): 127–33.

One of the more interesting bits of recent library-publisher history is the contentious relationship that sprang up between the publisher, Gordon and Breech, and academic libraries throughout the world. O'Neill tracks the history of this relationship and the litigation that ensued. This is an excellent profile of the issues.

BH6. Petit, Michael J. "The Evaluation, Selection, and Acquisition of Legal Looseleaf Publications." *Library Acquisitions: Practice & Theory* 17 (1993): 417–26.

Loose-leaf publications are the acquisitions librarian's bane. This is a unique and valuable discussion of how these publications work, how to evaluate them, and how to deal with the publishers.

BI
Ethics

BI1. American Library Association, Association for Library Collections & Technical Services, Publisher/Vendor–Library Relations Committee. *Principles & Standards of Acquisitions Practice.* Chicago: American Library Association, 1994.

This one-page listing of standards of ethics for acquisitions librarians is the culmination of the thinking of several people over a long period. It is a succinct statement of expectations that should hang on the walls of acquisitions librarians everywhere. A copy (suitable for framing) may be obtained from the ALCTS office.

BI2. Bushing, Mary C. "Acquisitions Ethics: The Evolution of Models for Hard Times." *Library Acquisitions: Practice & Theory* 17 (1993): 47–52.

Bushing departs from the usual case-study approach to ethics for acquisitions librarians. She looks at principles of conduct in developing a framework for ethical behavior, noting that in times of hunger, it is "harder not to be a thief." Contemplating ethics for hard times allows us to put ethics to a true test.

BI3. Dean, Barbara. "Toward a Code of Ethics for Acquisitions Librarians." *Against the Grain* 4 (Nov. 1992): 20–24.

Dean sets out eight tenets of ethics that acquisitions librarians should follow. Karen Schmidt, Joe Barker, and Gary Shirk then respond to her suggested outline with some modifications of their own. This is an interesting interactive discussion of ethics and policies that conveys a sense of high ethics.

BI4. Johnson, Peggy. "Ethical Considerations in Decision Making." *Technicalities* 14 (Dec. 1994): 2–4.

Ethics remains a topic of interest to acquisitions and collection development librarians and crops up often in conversations. Johnson's essay aptly captures the ethical issues and moral and legal conflicts that arise when librarians interact with the profit industry.

BI5. Presley, Roger L. "Firing an Old Friend, Painful Decisions: The Ethics between Librarians and Vendors." *Library Acquisitions: Practice & Theory* 17 (1993): 53–59.

Presley explores the changing nature of the relationships between librarians and vendors as technology increases the complexities of the services available. His description of the new demands that librarians have placed on vendors is especially interesting.

C

Descriptive Cataloging

Janet Swan Hill

Cataloging, in general, and description, in particular, are often regarded as areas within librarianship where change proceeds at a glacial pace and practitioners are determinedly resistant to any alterations to their routine. While descriptive cataloging may have started as a genteel art, it has been at least a century since it could be characterized as such. Far from being static, the field of cataloging has accommodated, embraced, and spurred significant change over the past several decades. One year viewed in isolation may give the impression that not much is happening, but when several years are taken together, considerable movement can be seen. The size of this chapter is a testament to the interest in and importance of many topics.

Chapter Content and Organization

This chapter covers the three-year span of 1993–1995. It combines an update to the chapters on descriptive cataloging and authority control in the *Guide to Technical Services Resources (GTSR)* and a review of current work and trends in those areas. The emphasis of the Descriptive Cataloging chapter in *GTSR* was on "resources that may assist catalogers in performing descriptive cataloging" (p. 46);

thus the materials included were mainly reference works, standards, and manuals. That emphasis continues to a certain extent in this chapter, but other materials and references are included to provide a snapshot of work in the field. So active were researchers and authors in the past several years and so broad was their scope that items listed in this chapter can be considered only representative of the types of work being done, and not as a comprehensive or even near-comprehensive gathering. As in *GTSR*, materials are primarily "resources for catalogers in the United States who wish to perform cataloging that conforms to prevailing national standards" (p. 46).

Topics included are descriptive cataloging and authority work and general cataloging management issues. For the most part, catalog use studies, works on catalogs in general, works on technical services overall, and works on the MARC formats are excluded as outside the scope of this chapter, and/or because they are being treated in depth by other authors. A few papers that might be considered slightly off the topic are included (e.g., Nicholson Baker's article on discarding card catalogs **[CAC2]** and Fred Kilgour's paper on search keys **[CBF7]**) because of the strength of their impact on or reception by the cataloging community. Some works specific to serials cataloging are included, but readers interested in serials cataloging should consult the chapter on Serials Management.

The chapter on Descriptive Cataloging in *GTSR* contains a lengthy section on sources for information used in catalog records, from electronically accessible name headings, to bibliographic records on microfiche, cards purchased from materials vendors, machine-readable tape subscriptions, CD-ROM databases, to printed union catalogs. The same sorts of data are still available in the same types of formats, with library automation vendors, materials vendors, and networks making especially rapid progress in offering an increasing array of options for acquiring data. So rapidly are these resources changing and growing that catalogers are advised not to rely only on printed listings. One should also seek up-to-date information directly from vendors, through perusing advertisements, and by haunting professional conference exhibits.

Print materials are by far the most common type of work included, but electronic resources such as CD-ROMs and resources electronically accessible through the Internet are also listed. Special issues of journals devoted to a single overall topic are entered under the entire issue, rather than giving citations to individual papers within the issue. Resources that consist primarily of software rather than information are generally excluded. Also, largely absent are references to vendors and other commercially available services. Ephemeral items, such as brief newsletter items, reports of meetings, workshop announcements, press releases, and discussion threads from electronic discussion forums are also excluded though such resources can do a great deal to enhance the context into which more substantial works fit. This chapter can stand on its own as a review of work, but readers interested in how materials covered relate to other available resources should refer to *GTSR*.

CA
General Works

CAA
Textbooks, Guides, and Manuals

CAA1. Chan, Lois Mai. *Cataloging and Classification: An Introduction.* 2nd ed. New York: McGraw-Hill, 1993.

One of the best basic texts on cataloging, this second edition is significantly changed from the previous edition, with updated examples, including MARC tagged examples and examples of screen displays. Covers description, subject cataloging and classification, authority control, and processing for both manual and machine-readable catalogs. For annotation, *see also* **DAB1**.

CAA2. Joachim, Martin D., ed. *Languages of the World: Cataloging Issues and Problems.* New York: Haworth, 1993. Also published as *Cataloging and Classification Quarterly* 17, nos. 1/2 (1993).

A collection of papers addressing the difficulties catalogers encounter in handling materials in foreign languages, from transliteration, to translation, to authority control. This is not a manual or a compendium of solutions, although some solutions are suggested or described in individual papers.

CAA3. Mann, Thomas. *Library Research Models: A Guide to Classification, Cataloging, and Computers.* New York: Oxford University Press, 1993.

Mann, a reference librarian at the Library of Congress, has done all catalogers a service through publication of this work. He examines the various systems used for the arrangement of information, delineating advantages and disadvantages. His observations on how cataloging and classification contribute to information access are clearly expressed, and should be both heartwarming to catalogers who sometimes wonder if anyone appreciates their work, and useful to catalogers and managers seeking a "clincher" of a quotation to justify or explain certain of their activities and aims.

CAA4. Sha, Vianne Tang. *Internet Resources for Cataloging.* Online. URL: http://www.law.missouri.edu/vianne/cat.htm

Sha, at the University of Missouri (Columbia) School of Law Library, describes and provides links to cataloging-related resources that can be found on the Internet, including home pages for national libraries, professional associations, national bibliographic utilities, and major library automation systems, as well as links to cataloging tools and training resources, and cataloging-related software. The increasing availability of cataloging resources on the Internet, accessible to catalogers as they sit at their workstations, is a recent development that holds great promise for increasing efficiency and standardization in cataloging, and Sha's compendium provides a useful gateway to many of the most needed resources.

CAB
Bibliographies

CAB1. Knutson, Gunnar. "The Year's Work in Descriptive Cataloging, 1992." *Library Resources & Technical Services* 37 (July 1993): 261–75.

Bibliographic essay reviewing the 1992 descriptive cataloging literature. Notes particular attention given to cataloging special materials, developing expert systems, and looking for ways to streamline cataloging processes.

CAC
Basic Issues and Reconsiderations

The rapid evolution of online catalogs, of the formats of information resources, and of the means of accessing them has sparked discussion of some of the basic tenets of cataloging. The following items are representative only.

CAC1. Ayres, F. H. "Bibliographic Control at the Cross Roads." *Cataloging & Classification Quarterly* 20, no. 3 (1995): 5–18.

Ayres states that "Bibliographic control is at a stage of development that could lead to a dramatic improvement in its scope and effectiveness," and posits that the best results will require more than merely revising existing rules or rethinking individual principles. Describes traditional approaches and principles (for example, the main entry as a descriptive linchpin), noting where changes in operating context and technological developments offer an opportunity—even a requirement—to change historical approaches (for example, adopting a "manifestation" entry).

CAC2. Baker, Nicholson. "Discards." *New Yorker*, April 4, 1994: 64–86.

Although written by a nonlibrarian, addressed to a general audience, and published outside the professional literature, this song of praise for card catalogs and lament for their demise, combined with a critique of the kinds of access provided by their online replacements, was probably more widely read and heatedly discussed than any other article related to cataloging in the past three years. Baker's arguments contain some inaccuracies and misconceptions, but hearing the views of an articulate library user is invaluable to catalogers and others in librarianship as they strive to create catalogs that are effective, useful, and attractive to those who must use them.

CAC3. Caplan, Priscilla. "You Call It Corn, We Call It Syntax-Independent Metadata for Document-Like Objects." *The Public Access Computer Systems Review* 6, no. 4 (1995): 19–23.

A brief summary of the OCLC Metadata Workshop whose report is listed below **(CAC9)**. For those not used to considering cataloging information as "metadata," the full report may be a bit daunting, and this capsule description may provide the sort of introduction that makes sense of the longer item. Also available online at URL http://www.nlc-bnc.ca/documents/libraries/cataloging/caplan3.txt

CAC4. Fattahi, Rahmatollah. "A Comparison between the Online Catalog and the Card Catalog: Some Considerations for Redesigning Bibliographic Standards." *OCLC Systems & Services* 11, no. 3 (1995): 28–38.

Fattahi explores the essential differences between online and card catalogs, from the processes of creation (input) through output, and concludes that online catalogs represent such a significant departure from card catalogs that the bibliographic standards that govern content of the catalogs, including cataloging codes and the ISBDs, need to be revised to reflect the "conceptual as well as the practical differences" between the two types of catalogs.

CAC5. Gorman, Michael. "The Corruption of Cataloging." *Library Journal* 120, no. 15 (1995): 32–34.

An impassioned statement decrying the damage being done to cataloging departments and thereby to library service through application of such management tactics as modernizing, restructuring, downsizing, reengineering, rightsizing, and outsourcing. Gorman notes how easy it sometimes seems to cut costs by cutting a cataloging operation, and describes the centrality of the cataloging operation and of catalogers to the mission of libraries. He notes that the rift between catalogers and reference librarians "has been one of the more damaging aspects of modern librarianship," and urges mutual respect and recognition of interdependence.

CAC6. LeBlanc, James D. "Cataloging in the 1990s: Managing the Crisis (Mentality)." *Library Resources & Technical Services* 37, no. 4 (1993): 423–33.

A consideration of various approaches to addressing the "crisis in cataloging," which in LeBlanc's terms equals receiving more materials than you can process fully. Includes proposals for rethinking current attitudes concerning quality versus quantity.

CAC7. Intner, Sheila S. "Ethics in Cataloging." *Technicalities* 13, no. 11 (1993): 5–8.

An exploration of matters facing catalogers and cataloging managers, from copy cataloging to downsizing, priorities, services, etc. that include an ethical component.

CAC8. Mangan, Elizabeth U. "The Making of a Standard." *Information Technology and Libraries* 14, no. 2 (1995): 99–101.

A history of the creation of the Content Standard for Digital Geospace Metadata. This is one example of the increasing degree to which cataloging data are being treated as just one type and system of data that may be needed to describe and assure the retrieval of information resources.

CAC9. Weibel, Stuart, Jean Godby, Eric Miller, and Rodney Daniel. "OCLC/NCSA Metadata Workshop Report." Dublin, Ohio: OCLC, Inc. Online. URL: http://www.oclc.org:5047/oclc/research/publications/weibel/metadata/dublin_core_report.html

The official report of a three-day workshop at OCLC aimed at "reaching a consensus on a core set of metadata elements to describe networked resources." The list of elements is being referred to as the Dublin Core and includes those elements deemed necessary to describe any type of information resource or "document-like object." The impetus for the meeting was the explosive growth of the Internet and World Wide Web, and the increasing complexity of finding things there, coupled with the realization that catalogers and indexers will not be able to cope with everything online, and that the assistance of the information providers themselves is required. Librarians were only one group represented. The report includes the list of core elements, along with definitions, rationales, and examples. It seems likely that the term "metadata" will become a buzzword for "data about data," or "cataloging."

CAD
Periodicals, Including Electronic Journals

Periodicals that deal with cataloging are largely the same as they were when *GTSR* was published. Two of the items below are described in *GTSR*, but have experienced changes worth noting. Electronic journals are included in this section; electronic discussion lists and forums are included in the sections to which they are more closely related.

CAD1. *Cataloging & Classification Quarterly.* New York: Haworth Press, v. 1– , Fall 1980– . Quarterly. ISSN 0163-9374. URL: http://stirner.library.pitt/~haworth/ccq.html

CCQ remains one of the central journals for the field. It recently created a Web home page at URL: http://stirner.library.pitt.edu/~haworth/ccq.html.

51

The home page includes announcements and special editorials. An especially attractive feature is that as soon as the content of an issue is settled, often more than a year in advance of publication, it is posted on the home page, complete with abstracts of articles.

CAD2. *CONSERline: Newsletter of the CONSER (Cooperative Online Serials) Program, Library of Congress and OCLC, Inc.* Online. Washington, D.C.: Library of Congress, Serial Record Division. no. 1– , Jan. 1994– . Semi-annual (irregular). ISSN 1072-611X. Continues the newsletter *CONSER*. URL: http://lcweb.loc.gov/acq/conser/consrlin.html

Edited by Jean Hirons, this periodical contains news of the CONSER Program and information of interest to the serials cataloging community. To subscribe, send a message to LISTSERV@LOC.GOV with the command: SUBSCRIBE CONSERLINE [your name]. Archived at Web site.

CAD3. *LCCN: LC Cataloging Newsline.* Online. Washington, D.C.: Library of Congress, Cataloging Directorate, v. 1, no. 1–, Jan. 15, 1993– . Quarterly (irregular). ISSN 1066-8829.

A complement to, not a replacement for, the *Cataloging Services Bulletin*; *LCCN* is edited by Robert M. Hiatt and contains announcements, press releases, explanations of policy decisions, meeting reports, new publications, job listings, etc. To subscribe, send a message to LISTPROC@LOC.GOV, with the command: SUBSCRIBE LCCN [your name]. Archives for volumes 1–3 available at URL gopher://marvel.lc.gov:70/11/services/cataloging/lccn. Volume 4 is available at URL http://www.lcweb.loc.gov/catdir/lccn and earlier issues will be converted as time permits.

CAE
Sources of Expertise

With the development and increased use of the Internet, of gophers, and of the World Wide Web, catalogers find tapping into the expertise of colleagues, both through electronic newsgroups, through publication of e-mail addresses, and through the proliferation of home pages easier than ever.

TECHNICAL SERVICES HOME PAGES

There has been a virtual explosion in the production of home pages by libraries and library departments with access to the World Wide Web. Many of these home pages appear intended to (or will eventually, regardless of intent) replace local policy manuals, memoranda, routing lists, and routed copies of items of interest.

While this is certainly of benefit to local catalogers, the fact that these pages are available to anyone with access to the World Wide Web will have a considerable impact well beyond local libraries. It is already possible to see that certain home pages are developing individual "personalities" and the architects of the home pages are discovering that they can provide access to a wide array of useful resources simply by linking to resources that may already have been mounted elsewhere. Because of this, the "content" of many home pages may be redundant because they contain links to each other and many provide links to the same resources. The redundancy cannot be relied on; some pages may contain links to some materials and not others. The number of home pages available is growing seemingly daily and the pages themselves are being modified and augmented constantly. In addition, some pages are changing their addresses or are dropping off the Web. No list can be considered either complete or totally accurate, but the following items are representative of the resources available.

CAE1. Auburn University Libraries Cataloging Department. URL: http://www.lib.auburn.edu/catalog/docs/tools.html
   Includes a cataloging FAQ, departmental directory, the local preprocessing procedure manual, and links to various Internet cataloging resources, such as network documentation, other Web sites, etc.

CAE2. Cataloguer's Toolbox. Memorial University of Newfoundland Queen Elizabeth II Library Cataloguing Homepage. URL: http://www.mun.ca/library/cat/
   Includes local documents, links to remote library OPACs, the Library of Congress gopher and home page, Cutter tables, MARC documentation, and a list of recently acquired cataloging tools.

CAE3. *Cornell University Library Technical Services Manual.* URL: http://www.library.cornell.edu/tsmanual/
   Includes a fairly detailed manual for local processing, covering topics such as the bibliographic record, item records, holdings records, shelflisting and authority records. Also includes links to code lists and resources concerning format integration and cataloging electronic resources.

CAE4. Library and Information Science. Mansfield University. URL: http://www.clark.net/pub/lschank/web/library.html
   Maintained by Larry Schankman, this page is rich in resources. It includes an annotated list of library and information science resources, plus links to such diverse resources as technical services documents, policy statements pertaining to library science and services, home pages and gophers of library-related companies and associations, and links to OPACs.

CAE5. MIT Cataloging Oasis. Massachusetts Institute of Technology Libraries. URL: http://macfadden.mit.edu:9500/colser/cat/
This page is "designed to provide many of the 'little bits' of information catalogers need on a daily basis." Maintained by Eric Celeste as part of ongoing research, it includes links to local Cutter tables, OCLC and MARC documentation, MARC code lists, and local policies and procedures.

CAE6. Northwestern University Library Technical Services. URL: http://www.library.nwu.edu/tech
This page is still largely under construction. Eventually, it will provide local documentation for all of technical services, but at present has mainly materials of interest to serials librarians, including a link to the Interactive Electronic Serials Cataloging Aid (IESCA).

CAE7. Princeton University Libraries Technical Services Department. URL: http://infoshare1.princeton.edu/tech/hptsd.html
This page is still under active construction. It includes organizational and staff information and local documentation.

CAE8. QTECH Web. Queens University Libraries. URL: http://stauffer.queensu.ca/techserv/qtech.html
Page maintained by Sam Kalb at Queens University Libraries, Queens University, Kingston, Ontario. This is a well-developed page with links to cataloging tools such as Cutter tables, local procedures, LC and NLM resources, MARC documentation, and recent articles of interest. Also includes links to commercial, network, organizational, and other technical services home pages, as well as to various electronic discussion lists.

CAE9. Technical Processing Online Tools (TPOT). University of California at San Diego Libraries. URL: http://tpot.ucsd.edu/
Includes local policies and procedures, news of internal and external meetings, local and system documentation, annual reports, and lists of cataloging tools recently received.

CAE10. University of Virginia Cataloging Services Department. URL: http://www.lib.virginia.edu/cataloging
Includes articles and papers, local policies and procedures, links to other Web sites, including technical services home pages, newsgroups, and special project home pages.

CAE11. Wayne State Gopher Menu. URL: gopher://gopher.libraries.wayne.edu:70/11/techserv
Maintained by B. Lessin, this page is aimed primarily at Wayne State's own staff. It includes a technical services directory and the local manual of technical services procedures.

OTHER WORLD WIDE WEB SITES AND HOME PAGES

In addition to the home pages of individual libraries, a number of organizations of interest to cataloging librarians maintain home pages or gopher sites through which one can gain access to a vast array of information about the organization, as well as documentation, order blanks, calendars of events, and project reports. The following are the names and addresses of some of those that may be of greatest interest.

National Libraries

CAE12. CONSER. URL: http://lcweb.loc.gov/acq/conser
The home page for CONSER, the cooperative program for serials cataloging. Originally standing for CONversion of SERials Project, CONSER now stands for the Cooperative ONline SERials Program. Links to resources of use to CONSER participants and other serials catalogers.

CAE13. Library of Congress. URL: http://www.lcweb.loc.gov
From here you can access MARVEL, the Library's gopher-based campus wide information system. Also available from this site are CONSER **(CAE12)** and the LC Cataloging Directorate **(CAE14)**.

CAE14. Library of Congress Cataloging Directorate. URL: http://lcweb.loc.gov/catdir
Home page providing information of use to all catalogers, not just those working at the Library of Congress.

CAE15. National Library of Medicine [U.S.]. URL: gopher://gopher.nlm.nih.gov

CAE16. National Library of Canada. URL: http://www.nlc-bnc.ca/ehome.htm

Professional Associations

CAE17. American Library Association. URL: http://www.ala.org/ or URL: gopher://gopher.ala.org
Lists of American Library Association officers and staff, news releases, policies, and general information can be found on the ALA home page and gopher.

CAE18. Association for Library Collections and Technical Services. URL: http://www.ala.org/alcts.html/ or URL: gopher://gopher.ala.org:70/11/alagophxiii/alagophxiiialcst

Networks and Vendors

CAE19. OCLC Online Computer Library Center, Inc. URL: http://www.oclc.org or URL: gopher://oclc.org.

CAE20. Research Libraries Group, Inc. URL: http://www.rlg.org

CAE21. WLN. URL: http://wln.com or URL: gopher://rsloa.wln.com

CB
Management Issues

While cataloging is, at its heart, an intellectual activity, its performance within an organization is a production process requiring effective organization and utilization of personnel, equipment, and other resources. Balancing the intellectual needs against the realities of such things as budgets and time is a constant challenge for cataloging managers. As the pace of technological evolution has quickened, the choices to be made about just how a particular operation will be handled have increased in number and complexity and the imperative to "keep up" and "not be left behind" has become stronger. Certain management issues are always represented in the literature (e.g., cooperative projects, organization of cataloging) while other subjects spring up essentially anew (e.g., outsourcing, workstations), and still others come and go at irregular intervals (e.g., authority work, training), giving testimony to the opportunities that may develop during a given period as well as to the particular matters that are posing special problems and absorbing attention.

CBA
Catalogers

In addition to items touching on the nature of catalogers, their retention, recruitment, morale, etc., this section includes papers on education for cataloging, since these are considerations of how and why people may become catalogers in the first place. Works about on-the-job training, however, are contained in the section on Training **(CBB)**.

CBA1. Benemann, William E. "The Cathedral Factor: Excellence and the Motivation of Cataloging Staff." *Technical Services Quarterly* 10, no. 3 (1993): 17–25.

An examination of the relationship between quality control and productivity and how appreciation of the purpose for which cataloging is done enhances both job satisfaction and performance.

CBA2. Callahan, Daren, and Judy MacLeod. "Recruiting and Retention Revisited: A Study of Entry Level Catalogers." *Technical Services Quarterly* 11, no. 4 (1994): 27–43.

Seventy-nine entry-level academic catalog librarians were asked about their prior job experience, education, current work environment, and career goals in an effort to identify reasons why librarians may or may not remain in cataloging. Attitudes expressed by library educators and colleagues are important, as were what the catalogers perceived as deficiencies in formal preparation for cataloging.

CBA3. Clack, Doris H. "Education for Cataloging: A Symposium Paper." *Cataloging & Classification Quarterly* 16, no. 3 (1993): 27–37.

A talk originally given at the Third Annette Phinazee Symposium at the School of Library & Information Sciences, North Carolina Central University. Gives a brief history and explores some of the current issues that have an impact on how cataloging is taught and how its teaching is perceived, and ends with recommendations for the future including recognition of the importance of bibliographic control to the field. Clack notes the difficulty of either adequately conveying the importance of the cataloging or teaching it effectively in the time currently allotted and given current attitudes.

CBA4. Evans, Anaclare F. "The Education of Catalogers: The View of the Practitioner/Educator." *Cataloging & Classification Quarterly* 16, no. 3 (1993): 49–57.

Educators and practitioners often differ on the subject of what is most important to teach prospective catalogers. As a practitioner who also teaches cataloging in a library school, the author presents her opinions about the "ideal" curriculum (which would consist of a beginning class, an internship, and an advanced class for all) and gives details about the preferred content of such courses.

CBA5. Horny, Karen L. "Taking the Lead: Catalogers Can't Be Wallflowers!" *Technicalities* 15, no. 5 (1995): 9–12.

Observations about the need for catalogers to be actively engaged in keeping up with technological developments and in making decisions about how best to handle and take advantage of information sources that are becoming available in an increasing number of formats, and through an increasing variety of vehicles.

CBA6. Intner, Sheila S. "The Re-Professionalization of Cataloging." *Technicalities* 13, no. 5 (1993): 6–8.

The author muses on the recent history of cataloging, in which departments once rife with professionals were altered in response to what was considered to be a significant decrease in the extent to which professional training and perspective were essential to the cataloging operation. The change has been to departments primarily staffed by paraprofessionals and clerical staff. Intner considers the more recent changes in cataloging, in technology, and in organizational structures that make "re-professionalization" desirable.

CBA7. MacLeod, Judy, and Daren Callahan. "Educators and Practitioners Reply: An Assessment of Cataloging Education." *Library Resources & Technical Services* 39, no. 2 (1995): 153–65.

The results of a survey that asked both practitioners and cataloging educators to rank the importance of items of course content, the chief objectives of cataloging education, cataloging practicums, what constitutes adequate preparation, on-the-job training, etc. The differing views of the two groups provide much food for thought.

CBA8. McAllister-Harper, Desretta. "An Analysis of Courses in Cataloging and Classification and Related Areas Offered in Sixteen Graduate Library Schools and Their Relationship to Present and Future Trends in Cataloging and Classification and to Cognitive Needs of Professional Academic Catalogers." *Cataloging & Classification Quarterly* 19, no. 3 (1993): 99–123.

The trends and needs of cataloging professionals are compared to course descriptions. Although discerning course content based on a course description is often difficult, the author concludes that the curricula surveyed are lagging in preparing catalogers for the future. She recommends that more effort be put into projecting future needs, courses be more standardized among schools, a specialty in cataloging be increasingly recognized, and certificates be awarded to those who complete a cataloging-intensive curriculum.

CBA9. Palmer, Joseph W. *Cataloging and the Small Special Library.* SLA Research Series, no. 7. Washington, D.C.: Special Libraries Association, 1992.

A report, based on surveys, of practices and trends in small special libraries. This is not a how-to-do-it book, but to a librarian facing management of such a library, it may provide useful information about how other similar libraries are structured and how they handle their work.

CBA10. Reimer, John J. "A Practitioner's View of the Education of Catalogers." *Cataloging & Classification Quarterly* 16, no. 1 (1993): 39–48.

The author urges cooperation between educators and practitioners to find common ground in preparing prospective librarians to become competent in cataloging. He discusses the value of laboratories, internships, and curriculum enhancement in order to assure a useful mix of theory and practice.

CBB
Training

CBB1. *CatSkill, An Interactive Multimedia Training Package to Teach AACR2 and MARC.* CD-ROM. Canberra, Australia: Learning Curve Pty. Ltd. and Doc Matrix Pty. Ltd., distributed by InfoTrain, 1994.

A software package available in both Windows and Macintosh versions. The program has beautiful graphics and arresting animation, and gives the user a choice of learning and applying ABNMARC, CANMARC, UKMARC, and USMARC tagging. Covers *AACR2R* (with 1993 amendments). Title pages and other information from materials being used as cataloging examples are reproduced in facsimile. The program provides detailed instruction, includes tests, and allows results to be stored online for use by an instructor. As with any training resource, currency is always a concern since, although principles embodied in what is covered may be unchanged, specific rules and codes are being altered constantly. Experienced catalogers and copy catalogers may find that the lessons move rather slowly but, for a student without cataloging experience, the pace and detail of instruction may not be a problem.

CBB2. Clack, Mary Elizabeth. "The Role of Training in the Reorganization of Cataloging Services." *Library Acquisitions: Practice & Theory* 19, no. 4 (1995): 439–44.

A description of the process by which the Cataloging Services Department of Harvard's Widener Library was restructured to produce a team environment, with special emphasis on the training that was done to orient personnel to the new organizational culture and to new roles.

CBB3. Cundiff, Margaret Welk. *Cataloging Concepts: Descriptive Cataloging.* Washington, D.C.: Library of Congress, Cataloging Distribution Service, 1993.

A training program for catalogers and others, developed at the Library of Congress. The program was designed to be used in a classroom setting, but it also can be used successfully for individual self-directed study. Consists of *Instructor's Manual* and *Trainee's Manual* (each two volumes), which can be purchased separately; it is assumed that each student will have an individual copy of the trainee manual. Includes tests, examples, exercises. Although the stated purpose "to teach library staff members how to understand, interpret and use current . . . catalog records" might seem to imply that this is primarily useful to those who will not actually have to catalog, in fact, *Cataloging Concepts* provides a highly valuable and structured beginning to a cataloger's training.

CBB4. *Training Catalogers in the Electronic Era: Essential Elements of a Training Program for Entry-Level Professional Catalogers.* Online. URL: gopher://ala1.ala.org:70/00/alagophxiii/alagophxiialcts/alagophxiiialctseduc/50714034.document

Developed by ALA/ALCTS/CCS/Committee on Education, Training and Recruitment for Cataloging. A training outline of topics that should be included in a training program beginning with job orientation, moving through cataloging and classification, authority control, and catalog maintenance through management issues. This is not a how-to-do-it document, nor does it indicate what to include or how to convey it under the topics listed, but it still provides a useful starting point for those creating or revising an approach to training new catalogers. Includes a brief bibliography.

CBC
Workstations

As more and more cataloging operations are automated, as general-purpose computers are replacing dumb or vendor-specific terminals in technical services, and as an increasing number of cataloging tools are being made available either through the Internet or via CD-ROM or other locally loaded resources, cataloging managers are seeking efficiency, comfort, and ease of operation through use of "cataloger's workstations." The term is as yet ill-defined and easily confused with other legitimate uses of the word "workstation," which can mean nothing more glamorous than a desk, so sometimes discussions can be held at cross-purposes. The following items give an idea of the ways in which the term is being used in cataloging, of what is already available, and of what may become available in the future.

CBC1. Brisson, Roger. "The Cataloger's Workstation and the Continuing Transformation of Cataloging." Parts I and II. *Cataloging & Classification Quarterly* 20, no. 1 (1995): 3–24; 20, no. 2 (1995): 89–104.

Part I covers the conceptual foundations and development of cataloger's workstations, with reference to the technological developments and context in which development is taking place, and the types of features that have been thought to be desirable for workstations. In Part II, Brisson explores technical requirements and configurations for workstations, touching on LANs, use of the Internet, and integration of software running under Windows.

CBC2. Gomez, Joni. "A Cataloger's Workstation: Using a NeXT Computer and Digital Librarian Software to Access the *Anglo-American Cataloguing Rules*." *Library Resources & Technical Services* 37, no. 1 (1993): 87–95.

An experiment in which the NeXT computer was determined to have potential as a cataloger's workstation, but additional programming will be necessary to provide the flexible and sophisticated searching needed to access *AACR2R* and other tools.

CBC3. Lange, Holley R. "Catalogers and Workstations: A Retrospective and Future View." *Cataloging & Classification Quarterly* 16, no. 1 (1993): 39–52.

Includes a historical survey of catalogers' tools and equipment, from manual tools to typewriters through the early days of automation, then through early attempts at creating multipurpose cataloging workstations, to visions for the future. This piece provides interesting context for discussions of what workstations should eventually encompass, but contains little specific detail.

CBC4. The Library Corporation. *ITS for Windows: Integrated Technical Services Workstation.* CD-ROM. Inwood, W.Va.: The Library Corporation, n.d.

A "descendant" of Bibliofile, the ITS workstation provides cataloging software, including software to retrieve copy as well as the "Cataloger's Reference Shelf," a collection of cataloging tools such as MARC formats and code lists, and the *Library of Congress Rule Interpretations*. The workstation can be used in conjunction with a library's overall automation system. Increasingly, library automation vendors and service organizations are offering a workstation, which includes software used to perform cataloging as well as information resources such as MARC formats and code lists, rules, etc. The number of companies offering such services and the content of the package can be expected to change constantly.

CBC5. Library of Congress. *Cataloger's Desktop.* CD-ROM. Washington, D.C.: Library of Congress, 1994– . Quarterly.

A CD-ROM resource containing a wide variety of cataloging tools published by the Library of Congress, including subject headings and MARC code lists. For descriptive cataloging, the *Desktop* includes the *Cataloging Concepts: Descriptive Cataloging* manual **(CBB3)**, official rule interpretations and decisions, manuals for map cataloging, loose-leaf cataloging, etc. More resources, such as the *NACO Participants' Manual* **(CBF11)**, are added as they become available. Includes search software and the ability to create personal files, notes, bookmarks, links, etc. Requires a powerful and well-equipped personal computer. The convenience of having all of these authoritative cataloging resources available in one place and utilizing the same search software might well make the purchase price and equipment requirements worthwhile.

CBD
Organization of Cataloging

CBD1. Black, Leah, and Colleen Hyslop. "Telecommuting for Original Cataloging at the Michigan State University Libraries." *College & Research Libraries* 56, no. 4 (1995): 319–23.

Telecommuting often is mentioned as one opportunity offered by automated library operations, but logistical concerns for materials have stopped many from pursuing it for cataloging. This paper reports on a six-month experiment. Advantages and disadvantages, including issues of workflow, morale, physical setting, feelings of isolation, access to needed resources, etc. are considered.

CBD2. Ruschoff, Carlen. "Perspectives on the Cataloging Literature: Cataloging's Prospects: Responding to Austerity with Innovation." *Journal of Academic Librarianship* 21, no. 1 (1995): 51–57.

Considers the changes made in cataloging operations in response to limited funding, as reflected in the literature. As many catalogers know, though others may doubt it, many changes have been made. Pressure to cut back while increasing speed and output continues. Ruschoff's overview may acquaint cataloging managers with other avenues that they might pursue in their own attempts to respond to funding difficulties and also provide useful ammunition to managers needing to point out how they have already "given at the office."

CBD3. Schuneman, Anita, and Deborah A. Mohr. "Team Cataloging in Academic Libraries: An Exploratory Survey." *Library Resources & Technical Services* 38, no. 3 (1994): 257–66.

A survey of eight libraries in which cataloging is accomplished by teams of personnel working on materials belonging to particular subjects, languages, or formats. Issues of productivity and morale are examined, revealing a slight increase in productivity, but views regarding impact on morale were mixed.

CBE
Levels and Types of Cataloging

Although many noncatalogers are under the impression that the cataloging rules are monolithic and inflexible, many options exist for how quickly, in how much detail, by whom, and in what manner particular items or types of items may be cataloged. In addition to the options offered by standards that have been in use for some time, recent years have seen significant work done to define new levels and types of cataloging for particular purposes. The challenge to catalogers is to ascertain what approach is best for what materials, given limitations in budget, time, nature of the catalog's users, availability of preexisting data, sophistication of the catalog apparatus, contractual obligations, cooperative arrangements, and the uses to which the materials will be put. The items cited below explore these issues from various perspectives.

CBE1. Braun, Janice, and Lola Raykovic Hopkins. "Collection-Level Cataloging, Indexing, and Preservation of the Hoover Institution Pamphlet Collection on Revolutionary Change in Twentieth Century Europe." *Technical Services Quarterly* 12, no. 4 (1995): 1–8.

This description of a preservation project includes information about the level of cataloging, workflow, decision process, etc. Although information about the cataloging *per se* is sparse, the description of a project into which collection-level cataloging is fit may be useful to libraries considering or undertaking similar projects.

CBE2. Cromwell, Willy. "The Core Record: A New Bibliographic Standard." *Library Resources & Technical Services* 38, no. 4 (1994): 415–24.

The standard for Minimal Level Records has been in existence for some years, but changes in technology, in cooperative projects, and experience sparked an exploration of a more useful standard. Cromwell describes the development of a new standard for less-than-full-level cataloging that is more useful and more conducive to record sharing than the older minimal level standard, but less expensive to follow than the standard for full level records.

CBE3. Erbolato-Ramsey, Christiane, and Mark L. Grover. "Spanish and Portuguese Online Cataloging: Where Do You Start from Scratch?" *Cataloging & Classification Quarterly* 19, no. 1 (1994): 75–87.

Describes a process used to examine patterns of copy availability for a particular category of materials, how this information can be used to determine when to catalog originally, and when to wait for copy. Although the project was aimed at Spanish and Portuguese materials, the process might be adopted for investigations of copy availability for other types of works.

CBE4. Haynes, Kathleen J. M., Jerry D. Saye, and Lynda Lee Kaid. "Cataloging Collection-Level Records for Archival Video and Audio Recordings." *Cataloging & Classification Quarterly* 18, no. 2 (1993): 19–32.

Description of a project to create a local database for these materials. The project created collection-level records suitable for contribution to OCLC that would alert scholars to the existence of the collection and guide them to the more powerful item-level searching provided by the local database. Discusses cataloging problems encountered, such as choice of cataloging tools, name authority, etc. An interesting description of a project that uses mutilple technologies and approaches to achieve its end.

CBE5. Intner, Sheila S. "The Floating Standard: One Answer to Cataloging Schizophrenia." *Technicalities* 15, no. 6 (1995): 7–9.

Discusses the difficulties of making all cataloging standards apply equally to all situations. Considers the problems of failing to identify and deal successfully with those things over which we have little or no control, such as Internet structure and the design of existing library automation systems.

CBE6. ———. "Taking Another Look at Minimal Level Cataloging." *Technicalities* 14, no. 1 (1994): 3–5, 11.

A discussion of issues surrounding minimal level cataloging, including those characteristics that make catalogers reluctant to utilize it and those that have given it a "bad name."

CBE7. LeBlanc, James D. "Towards Finding More Catalog Copy: The Possibility of Using OCLC and the Internet to Supplement RLIN Searching." *Cataloging & Classification Quarterly* 16, no. 1 (1993): 71–83.

A sample of French and Italian titles from the Cornell University cataloging arrearage was searched in OCLC, RLIN, and the Internet-available OPACs of six large libraries in an effort to determine the usefulness of incorporating an Internet search into local routines. Hit rates in individual catalogs were less than one third those on either network at best. Only 4 percent of the sample found copy in an OPAC that was not also available on either OCLC or RLIN.

CBE8. Saunders, Richard. "Collection- or Archival-Level Description for Monographic Collections." *Library Resources & Technical Services* 38, no. 2 (1994): 139–47.

Proposals for several methods of providing access to monographic materials so that materials with related provenance can be linked in the catalog even if they cannot be housed together all the time. Many libraries deal repeatedly with requests to provide such information and access. One or more of the methods described may be just what a particular library needs to accede to such requests within the rules for standard practice and staffing limitations.

CBE9. Schultz, Lois Massengale. *A Beginner's Guide to Copy Cataloging on OCLC/Prism.* Englewood, Colo.: Libraries Unlimited, 1995.

Includes a definition of copy cataloging, discusses training and workflow issues, and includes many examples with discussion. Although focused on processes as performed on OCLC, this manual has much useful information that can be generalized to other systems.

CBE10. Smith, Stephen J. "Cataloging with Copy: Methods for Increasing Productivity." *Technical Services Quarterly* 11, no. 4 (1994): 1–11.

A description of practices introduced at the University of Illinois at Urbana-Champaign (UIUC) for the purposes of enhancing productivity in terms of numbers and timeliness, in order to strengthen the service function of cataloging and to reinforce the team cataloging environment. This paper is not about copy cataloging, *per se*, but rather about management of a significant cataloging operation. The UIUC Libraries have an extremely large copy-cataloging operation; while other institutions may not be able to adopt their practices as is, many libraries may be able to adapt them to local needs.

CBE11. Tsui, Susan L. "Strategies for Reducing Billable OCLC Searches Used in Cataloging." *Cataloging & Classification Quarterly* 16, no. 2 (1993): 93–106.

This description of the strategies introduced at the University of Dayton Libraries in response to changes in the OCLC searching pricing structure should be of interest to all cataloging managers interested in saving money—which is to say, all cataloging managers.

CBF
Authority Work

In the years prior to the introduction of online catalogs, the process of authority work received scant attention. Many libraries that acquired catalog cards from outside sources for most of the materials they added to their collections de-emphasized or even discontinued local authority work, relying on the work performed by the Library of Congress and other card suppliers to make their catalogs consistent and coherent. In the first flush of excitement at the potential power of online catalogs, many were ready to believe that authority control would no longer be necessary. As libraries gained experience with online catalogs, however, and as local catalogs and bibliographic network databases grew larger, it became increasingly clear that authority control was not dispensable. While some time-honored authority processes might indeed be "overkill" in an automated environment and other processes might be made significantly more efficient through application of new technologies, some activities were more critical than ever. The belief that machines are capable of everything dies hard, however. Along with papers about advances in authority control, works continue to explore and explain the need for authority control in an effort to refute the papers that continue to posit a need for less than we are currently trying to provide.

Authority Control occupied its own chapter in *GTSR*. In its present position as a section of another chapter, the topic is treated with considerably less detail. This is not to be taken as an indication of a change in its importance.

CBF1. Aliprand, Joan M. "Linking of Alternate Graphic Representation in USMARC Authority Records." *Cataloging & Classification Quarterly* 18, no. 1 (1993): 27–62.

A description of the capability in the USMARC format to link a Latin script field to the 880 field, which contains the graphic representation of the field content in non-Roman vernacular form, and a discussion of the need for authority control in a multi-script environment.

CBF2. Bangalore, Nirmala S. "Authority Files in Online Catalogs Revisited." *Cataloging & Classification Quarterly* 20, no. 3 (1995): 75–94.

Replicates Arlene Taylor's influential 1983 study, in which user requests resulting in zero hits in online catalogs were analyzed to determine the extent to which online authority files might improve search results. Taylor's conclusion that authority work to effect collocation is necessary is confirmed, but Bangalore takes issue with earlier conclusions about aspects of authority control that may be unnecessary or not cost-effective. She concludes instead that online linkage of authority records to bibliographic files is a justifiable expense in terms of improved search results.

CBF3. Bourdon, Francoise. *International Cooperation in the Field of Authority Data: An Analytical Study with Recommendations.* Translated by Ruth Webb. UBCIM Publications—New Series, vol. 11. New Providence, R.I.: Saur, 1993.

Examines problems encountered in international sharing of name and title authority data. Considers the underlying principles and possible reasons for difficulties, analyzes the possible uses of shared authorities data, including what elements are necessary to such sharing, and makes suggestions for some next steps.

CBF4. Byrd, Jacqueline and Kathryn Sorury. "Cost Analysis of NACO Participation at Indiana University." *Cataloging & Classification Quarterly* 16, no. 2 (1993): 107–23.

Participation in any cooperative project involves costs as well as benefits, and participation in elite projects such as NACO can be viewed as an honor. A study of work required for contributing NACO records as compared to work that would be done by the library if it were not a NACO member revealed that NACO participation doubled the library's name authority costs.

CBF5. Edelbute, Thomas. "A Pro-Cite Authority File on a Network." *Technical Services Quarterly* 12, no. 3 (1995): 29–40.

Description of an authority file developed locally for the USDA Agricultural Research Service and Agricultural Engineering faculty at the University of Missouri–Columbia. The difficulties encountered and issues addressed have general relevance.

CBF6. Irwin, Dale. "Local Systems and Authority Control." *Cataloging & Classification Quarterly* 16, no. 2 (1993): 55–69.

Outlines the information that a library needs about authority control in general and suggests ways in which authority control may be handled in an automated system to help evaluate the particular authority capabilities of an automation system under consideration.

CBF7. Kilgour, Frederick G. "Effectiveness of Surname-Title-Words Searches by Scholars." *Journal of the American Society for Information Science* 46, no. 2 (1995): 146–51.

An experimental database was mounted to simulate scholarly use of an online catalog to determine the frequency of one-screen displays when a scholar searches for known items using keywords and implied Boolean operators. A high success rate (92.8 percent) was experienced when using a search formula of surname plus first and last title words. Kilgour's conclusion that simple title-page transcription may be sufficient for known-item searches led to this article's being hotly debated on AUTOCAT (the Library and Authorities Discussion Group) and rebutted by the author of one of the papers he cited as supporting some of his arguments. Kilgour's reputation and the attractive simplicity of the solution it apparently offers make this an article that catalogers cannot afford to miss.

CBF8. Lin, Joseph C. "Undifferentiated Names: A Cataloging Rule Overlooked by Catalogers, Reference Librarians, and Library Users." *Cataloging & Classification Quarterly* 19, no. 2 (1994): 23–48.

The concept of an undifferentiated name, i.e., a name that is used in cataloging records in identical form without qualifying elements to represent more than one person or entity, is relatively new. This paper discusses the problems that undifferentiated names may cause in bibliographic access and proposes a modification of the current rule.

CBF9. McCurley, Henry H., Jr. "The Benefits of Online Series Authority Control." *Technical Services Quarterly* 11, no. 3 (1994): 33–50.

A companion piece to **CBF10** in which the emphasis is on the ways in which having the series authority file online benefits technical and public services personnel and patrons.

CBF10. ———. "Implementation of an Online Series Authority File at Auburn University." *Cataloging & Classification Quarterly* 18, no. 2 (1993): 41–57.

A description of a two-part project to convert retrospectively an existing manual authority file and to begin establishing all new series authorities in the local system. Covers the rationale for undertaking the project, benefits realized, and problems encountered.

CBF11. *NACO Participants' Manual.* Amy M. McColl, comp. Washington, D.C.: Library of Congress, 1994.

The manual of procedural guidelines for libraries participating in NACO. Contains detailed, often field-by-field instructions for the content and content designation of authority records created or revised for incorporation in the Library of Congress authority file. Includes examples of full authority records formatted for the major bibliographic utilities. Although intended for NACO participants, this manual has much value in any library that creates MARC-format authority records for its own system or that must use and interpret such records obtained from a network or other resource.

CBF12. Preece, Barbara G., and Barbara Henigman. "Shared Authority Control: Governance and Training." *Technical Services Quarterly* 11, no. 3 (1994): 19–31.

A description and discussion of the process and underlying policies that control the governance of authority control for the 800 Illinois libraries sharing the ILLINET online catalog. To libraries grappling with the difficulties of sharing responsibility for authority control among a few libraries or library departments, this paper may be both useful and comforting.

CBF13. Reimer, John J. "National Series Authority File Derailed?" *Cataloging & Classification Quarterly* 19, no. 1 (1994): 5–8.

This guest editorial is in reaction to a Library of Congress proposal to abandon significant aspects of series authority work. It offers reasons, based on benefits derived, for retaining detailed series authority control.

CBF14. Weiss, Paul. "The Expert Cataloging Assistant Project at the National Library of Medicine." *Information Technology and Libraries* 13, no. 4 (1994): 267–71.

A description of NLM's project to apply expert systems to authority work, including a description of system design and evaluation. The reasons this project was abandoned, despite NLM's considerable resources and in spite of conventional wisdom about the applicability of expert systems to cataloging operations, provide food for thought.

CBF15. Younger, Jennifer. "After Cutter: Authority Control in the Twenty-first Century." *Library Resources & Technical Services* 39, no. 2 (1995): 133–41.

The author proposes that the concept of "utility" be introduced to considerations of authority control, to guide librarians in determining where to focus their efforts for the greatest good and to address the greatest need.

CBG
Outsourcing

Running a cataloging operation is expensive. It requires significant numbers of highly trained personnel, who utilize expensive and complicated equipment and refer to expensive, extensive, and complicated standards and information resources. Providing the necessary training and supervision for catalogers and finding catalogers and others with the necessary education to handle the variety of materials acquired for a collection often can seem beyond the resources of an individual library. In some cases, the amount and types of materials to be handled do not allow for economy of scale. These and other reasons, combined with the library community's years of experience in contracting with outside vendors for retrospective conversion services, have caused many cataloging managers and library administrators to consider whether they might be better served by contracting with an outside vendor to perform cataloging for them. "Outsourcing," as the activity is called, is viewed variously as the only sensible thing to do, as a hallmark of ignorance, or as an indicator of a failure to fulfill societal obligations. No doubt the truth lies somewhere between the extremes, but it may be some years until it will be determined just where.

CBG1. Duranceau, Ellen. "Vendors and Librarians Speak on Outsourcing, Cataloging and Acquisitions." *Serials Review* 20, no. 3 (1994): 69–83. For annotation, *see also* **FBB2.**

Various authors, including those working in libraries as well as some vendors of services, discuss outsourcing of various technical services functions.

CBG2. Hirshon, Arnold, and Barbara Winters. *Outsourcing Technical Services: A How-to-Do-It Manual for Librarians.* New York: Neal-Schuman, 1995.

Provides detailed and practical advice on investigating, contracting for, and carrying out various technical services operations, including cataloging, through outsourcing.

CBG3. Hirshon, Arnold, Barbara Winters, and Karen Wilhoit. "A Response to 'Outsourcing Cataloging: The Wright State Experience.'" *ALCTS Newsletter* 6, no. 2 (1995): 26–28.

A rebuttal and clarification of David Miller's summary **(CBG4).**

CBG4. Miller, David. "Outsourcing Cataloging: The Wright Experience." *ALCTS Newsletter* 6, no. 1 (1995): 7–8.

A summary and commentary on a presentation given by Barbara Winters, Associate University Librarian for Central Services at Wright State University, at the 1994 meeting of the New England Technical Services Librarians. Wright State was the first academic library of significant size to outsource virtually all of its cataloging operations. Winters' presentation covered Wright State's reasons for the action, as well as its subsequent experience.

CBG5. *Outsourcing Cataloging, Authority Work and Physical Processing: A Checklist of Considerations.* Prepared by the Commercial Technical Services Committee of the Association for Library Collections & Technical Services. Marie A. Kascus and Dan Hale, eds. Chicago: American Library Association, 1995.

Covers many facets of the question of outsourcing major components of cataloging and technical services operations, including identifying the need for outsourcing, weighing pros and cons, identifying work that may need to be retained locally, anticipating and implementing changes in workflow, selecting a vendor, financial considerations, and more.

CBG6. Rider, Mary M. "PromptCat: A Projected Service for Automatic Cataloging—Results of a Study at the Ohio State University Libraries." *Cataloging & Classification Quarterly* 20, no. 4 (1995): 23–44.

Reports a test of a projected OCLC service in which libraries will receive full catalog records for approval or firm-order books at the same time the books are shipped by the vendor. In the test, Baker & Taylor supplied vendor records to OCLC for matching to Ohio University Libraries catalog records and forwarding to the libraries. A high degree of quality in record selection and match was found.

CBG7. Waite, Ellen J. "Reinvent Catalogers!" *Library Journal* 120, no. 18 (1995): 36–37.

A diatribe against current organization of cataloging as "futile, wasteful efforts to perfectly describe the physical book." Highly supportive of outsourcing.

CBG8. Wilson, Karen A. "Outsourcing Copy Cataloging and Physical Processing: A Review of Blackwell's Outsourcing Services for the J. Hugh Jackson Library at Stanford University." *Library Resources & Technical Services* 38, no. 4 (1995): 359–83.

Describes the history and rationale of the project and the quality of services received during the first year of the outsourcing project. The rate at which records required local modification was tracked and, by the end of the project, the vendors had achieved an error rate of less than 1 percent.

CBH
Backlogs

CBH1. Camden, Beth Picknally, and Jean L. Cooper. "Controlling a Cataloging Backlog; or Taming the Bibliographical Zoo." *Library Resources & Technical Services* 38, no. 1 (1994): 64–71.

A great many cataloging operations have some sort of cataloging backlog, but few admit it openly and fewer still write about it. Camden and Cooper draw conclusions from their experience and suggest that providing online catalog access to materials in a cataloging backlog may reduce the need for special projects or policies aimed at eliminating backlogs.

CBI
Cooperative Projects and Databases, Networks

Effective and affordable bibliographic control is critically dependent on a high degree of cooperation among libraries and organizations of all types. Starting with the Library of Congress' public sale of catalog cards, advancing through the creation and development of bibliographic utilities, and continuing through the evolution of NACO to the development of the Cooperative Cataloging Council (CCC) and the Program for Cooperative Cataloging (PCC), cooperative endeavors to provide and improve bibliographic control are an inseparable part of cataloging operations. The number, type, scope, and sophistication of such endeavors are increasing constantly. News of the activities of emerging programs often appears in the form of press releases posted electronically to discussion forums and home pages and as brief items in newsletters. As projects gain more maturity, they are increasingly reported in less ephemeral forums.

CBI1. Baker, Barry B., ed. *Cooperative Cataloging: Past, Present and Future.* New York: Haworth, 1993. Also published as *Cataloging & Classification Quarterly* 17, nos. 3/4 (1993).

A collection of papers, covering the "colorful past, controversial present, and exciting and challenging future" of cooperative cataloging including the major national cooperative cataloging programs such as the National Coordinated Cataloging Program (NCCP), CONSER, the United States Newspaper Program, and cooperative projects to catalog microform sets. Papers touch on issues such as strengths and weaknesses of the programs, authority control, and the conflict between a desire to maximize use of copy and the need for contribution of original cataloging. Emphasis is strongly on the United States and North America, but there is some coverage of programs being carried on elsewhere.

CBI2. Barnett, Judith B. "OCLC Cataloging Peer Committees: An Overview." *Cataloging & Classification Quarterly* 16, no. 4 (1993): 67–76.

The history, development, structure, present role, and impact on database quality of cataloging peer committees in OCLC.

CBI3. Bryant, Philip. "Quality of a National Bibliographic Service: In the Steps of John Whytefeld—An Admirable Cataloguer." *International Cataloguing and Bibliographic Control* 24, no. 2 (1995): 29–31.

A discussion of what constitutes quality in the cataloging produced by a national bibliographic agency as one aspect of the path toward universal bibliographic control. Written from the point of view of the British Library Bibliographic Service, the discussion is nevertheless applicable to the product of any national agency.

CBI4. *COOPCAT: Cooperative Cataloging*. Electronic discussion group. List owner: Carol Wlaton Hixson, Indiana University, Bloomington.

The semi-official list of the Cooperative Cataloging Council, most postings are announcements and press releases, as well as some questions and discussions related to various cooperative initiatives. *COOPCAT* is often the most rapid and reliable source for information about publication of task force reports. To subscribe, send a message to LISTSERV@IUBVM.UCS.INDIAN.EDU with the command SUBSCRIBE COOPCAT [your name]. Messages can be sent to: COOPCAT@IUBVM.UCS.INDIANA.EDU

CBI5. Kaneko, Hideo. "RLIN CJK and the East Asian Library Community." *Information Technology and Libraries* 12, no. 4 (1993): 423–26.

A brief history of cooperative cataloging projects for East Asian materials, with an emphasis on RLIN.

CBI6. Kiegel, Joseph, and Merry Schellinger. "A Cooperative Cataloging Project between Two Large Academic Libraries." *Library Resources & Technical Services* 37, no. 2 (1993): 221–25.

Describes the development, implementation, and evaluation of a cooperative cataloging arrangement between the University of Minnesota Libraries and the University of Washington Libraries. Touches on the challenges of working with different automated systems and bibliographic utilities and of accommodating differences in policies such as authority control.

CBI7. O'Neill, Edward T., Sally Rogers, and W. Michael Oskins. "Characteristics of Duplicate Records in OCLC's Online Union Catalog." *Library Resources & Technical Services* 37, no. 1 (1993): 59–71.

A study to determine what data elements may cause the creation of duplicate records. Findings provide information to help catalogers and searchers identify and accept existing matching records as an aid to more efficient copy cataloging. Also may help system designers create more useful matching algorithms for record loaders.

CBI8. *Program for Cooperative Cataloging*. Online. URL: http://lcweb.loc.gov/catdir/pcc/pcc.html

Includes news and announcements, reports of task groups, including matters relating to descriptive and subject cataloging as well as authority control, information about PCC, its governance and committees, as well as links to standards such as the core record for JACKPHY (Japanese, Arabic, Chinese, Korean, Persian, Hebrew, Yiddish) materials, electronic newsletters related to PCC interests, and various other types of documentation. Most links that refer to PCC specifically are to the LC MARVEL gopher.

CBI9. Romero, Lisa. "Original Cataloging Errors: A Comparison of Errors Found in Entry-Level Cataloging with Errors Found in OCLC and RLIN." *Technical Services Quarterly* 12, no. 2 (1994): 13–27.

Compares errors created by entry-level catalogers with errors found on two bibliographic databases in an effort to identify particular "areas of cataloging difficulty" in monographic cataloging, so that educators and trainers can take special care when covering them.

CBI10. Ross, Rosemary E. "A Comparison of OCLC and WLN Hit Rates for Monographs and an Analysis of the Types of Records Retrieved." *Information Technology and Libraries* 12, no. 3 (1993): 353–60.

A study conducted at Central Washington University discovered that for the materials it acquires, there is not a significant difference in matching record hit rates between OCLC and WLN. The study examined such variables as type and level of cataloging, age of materials, format, classification, etc. The project description may be useful to others needing to determine hit rates, assess relative copy quality, etc.

CBI11. Saffady, William. "The Bibliographic Utilities in 1993: A Survey of Cataloging Support and Other Services." *Library Technology Reports* 29, no. 1 (1993).

This special issue includes an overview, then separate chapters on OCLC, RLIN, Utlas, and WLN. Each chapter has a background section, then covers Equipment and Communications, Data Base Characteristics, Record Retrieval, Cataloging Support, Output Products, and Other Products and Services. An excellent resource for those wishing to choose a bibliographic network or needing to reconsider their present provider.

CBI12. Sercan, Cecilia S. "Where Has All the Copy Gone: Latin American Imprints in the RLIN Database." *Library Resources & Technical Services* 38, no. 1 (1994): 56–63.

As libraries begin performing their cataloging on local databases rather than directly on utilities and as RLIN libraries cease inputting acquisitions records onto RLIN, one side effect may be delayed or decreased availability of copy for other users of the database.

CBI13. *Towards a New Beginning in Cooperative Cataloging: The History, Progress, and Future of the Cooperative Cataloging Council.* David W. Reser, comp. Washington, D.C.: Library of Congress, 1994.

Includes the history and strategic plan of the Cooperative Cataloging Council, a group formed in 1993 and convened by the Library of Congress, plus the final reports of its six task groups: More, Better, Faster, Cheaper; Availability and Distribution; Authorities; Standards; Cataloger Training; Foreign MARC. An excellent source for anyone wishing to identify and understand the issues related to cooperative and shared cataloging.

CBI14. Tsao, Jai-hsya. "The Quality and Timeliness of Chinese and Japanese Monographic Records in the RLIN Database." *Library Resources & Technical Services* 38, no. 1 (1994): 60–63.

Copy availability for Chinese and Japanese materials is still a problem, as is quality of records found.

CBI15. Wu, Ai-Hwa. "With Characters: Retrospective Conversion of East Asian Cataloging Records." *Information Technology and Libraries* 12, no. 4 (1993): 427–31.

A brief history and examination of options and problems for retrospective conversion of East Asian materials.

CC
Standards and Their Application

Works concerning the standards for description, in whole or in part, are considered first, followed by works concerning the standards for content designation.

CCA
Standards for Descriptive Content

CCA1. *Anglo-American Cataloguing Rules, Second Edition, 1988 Revision Amendments 1993.* Chicago: American Library Association, 1993.

Includes official rule revisions, typographical and index corrections. Revisions are formatted for insertion into the looseleaf edition of the *Code*, or for tipping into the bound versions.

CCA2. Duke, John K. "Slow Revolution: The Electronic AACR2." *Library Resources & Technical Services* 38, no. 2 (1994): 190–94.

A history and description of the project by the publishers of the *Anglo-American Cataloguing Rules* and its revisions to encode the text of *AACR2R* in SGML (Standard General Markup Language) as a first step in making the code available to end users in machine-readable form. Includes a discussion of problems encountered and principles applied.

CCA3. Heaney, Michael. "Object-Oriented Cataloging." *Information Technology and Libraries* 14, no. 3 (1995): 135–153.

Considers description, content designation, and authority control as inextricable parts of the enterprise of cataloging. The author concludes that *AACR2* is "not so much a plateau of maturity as an evolutionary blind alley." Proposes a different approach and a process for moving toward it.

CCA4. Howarth, Lynne C., and Jean Weihs. "*AACR2R:* Dissemination and Use in Canadian Libraries." *Library Resources & Technical Services* 38, no. 2 (1994): 179–89.

A survey of the extent and frequency of use of *AACR2R* in Canadian libraries and of the preferred publication format for the rules. A very high level of use was found, even among smaller libraries, and a distinct preference for loose-leaf format.

CCA5. ———. "*AACR2R* Use in Canadian Libraries and Implications for Bibliographic Databases." *Library Resources & Technical Services* 39, no. 1 (1995): 85–99.

Report of a study to establish a baseline profile of Canadian libraries contributing to shared databases and to determine the extent to which *AACR2R* is used in original contributions.

CCA6. Jacobowitz, Neil A. "A Comparison of *AACR2R* and French Cataloging Rules." *Cataloging & Classification Quarterly* 20, no. 1 (1995): 47–60.

As cataloging records from foreign sources become increasingly available, librarians may find it useful to understand the ways in which records prepared according to other important cataloging codes differ from those prepared according to Anglo-American rules. Using the rules for printed monographs, the author compares *AACR2R* to the French cataloging code, pointing out significant differences in the areas of choice of access points, uniform titles, and extent of bibliographic description.

CCB
Descriptive Content Standards: Special Topics

CCB1. Bierbaum, Esther Green. "A Modest Proposal: No More Main Entry." *American Libraries* 25, no. 1 (1994): 81–84.

Although the title tends to imply a more sweeping proposal than is made, this is a genuinely modest proposal for a change to terminology more appropriate to a non-card-catalog environment.

CCB2.   Fattahi, Rahmatollah. "Anglo-American Cataloguing Rules in the Online Environment: A Literature Review." *Cataloging & Classification Quarterly* 20, no. 3 (1995): 25–50.

A detailed bibliographic essay that includes consideration of the problems with continuing to develop and apply a standard originally created for a manual system, as well as the difficulties of making radical changes, considering the MARC formats and the inertia represented by so many large catalogs, databases, and automation systems based on the current standards.

CCB3.   Hu, Qianli. "How to Distinguish and Catalog Chinese Personal Names." *Cataloging & Classification Quarterly* 19, no. 1 (1994): 29–60.

Although the cataloging code provides rules for handling Chinese personal names, the process can be bewildering to those unfamiliar with such names, and the resulting cataloging decisions often can be questionable. Hu describes simple methods for handling Chinese names and explains how Chinese personal names may be differently formatted in different countries. Also includes lists of common Chinese last names in various formats and transliterations.

CCB4.   Marker, Rhonda, and Melinda Ann Reagor. "Variation in Place of Publication: A Model for Cataloging Simplification." *Library Resources & Technical Services* 38, no. 1 (1994): 17–26.

A proposal for a change in policy and of a Library of Congress rule interpretation to reduce the number of instances in which a new bibliographic record is required.

CCB5.   Nelson, David, and Jonathan Marner. "The Concept of Inadequacy in Uniform Titles." *Library Resources & Technical Services* 39, no. 3 (1995): 238–46.

A proposal for simplification of the decision process in establishing uniform titles.

CCB6.   ———. "Dates in Added Entries: An Analysis of an AUTOCAT Discussion." *Cataloging & Classification Quarterly* 20, no. 2 (1995): 71–88.

A summary of a discussion thread concerning the practice of adding dates to analytical added entries. Although the primary purpose of the article is to illustrate the usefulness of AUTOCAT (the Library and Authorities Discussion Group) as a means of cataloging problem solving, the detailed discussion of the problem is itself enlightening.

CCB7.   Schimizzi, Anthony J. "The Library of Congress and the 'Bibliography' Note." *Cataloging & Classification Quarterly* 18, no. 2 (1993): 59–70.

Describes the Library of Congress' various interpretations of *AACR2* rule 2.7B18 concerning representation in the catalog record of bibliographies and bibliographical references in works being described. Schimizzi takes issue with current practice and recommends a return to the interpretation in effect from 1984 to 1989. Although this issue may seem a small one to some, the paper provides ample evidence of the importance of even small decisions and the difficulties encountered by libraries attempting to find any aspect of cataloging to simplify.

CCB8. Weintraub, Tamara S., and Wayne Shimoguchi. "Catalog Record Contents Enhancement." *Library Resources & Technical Services* 37, no. 2 (1993): 167–80.

A study to see the extent to which enhancement or inclusion of a contents note for monographic materials could increase citation-based and subject access to materials in an online catalog. Twenty-three percent of books were found to benefit from such enhancement, with approximately equal improvement to subject and citation-based access.

CCB9. Winke, R. Conrad. "Discarding the Main Entry in an Online Cataloging Environment." *Cataloging & Classification Quarterly* 16, no. 1 (1993): 53–70.

Winke posits that the concept of main entry has little meaningful purpose in today's cataloging environment and proposes ways in which *AACR2R* and the MARC formats might be altered to accomplish its abandonment.

CCB10. Yee, Martha M. "What Is a Work? Part 1: The User and the Objects of the Catalog." *Cataloging & Classification Quarterly* 19, no. 1 (1994): 9–27; Part 2: "The Anglo-American Cataloging Codes." 19, no. 2 (1994): 5–22; Part 3: "The Anglo-American Cataloging Codes." 20, no. 1 (1995): 25–46; Part 4: "Cataloging Theorists and a Definition Abstract." 20, no. 2 (1995):3–24.

A series of articles exploring one of the most basic concepts in the Anglo-American cataloging codes, considering how it relates to the objects of the catalog and to the users and how it is incorporated in and applied in the codes.

CCC
Independent Cumulations of Rule Interpretations

While the Library of Congress rule interpretations are available from many sources, including LC's own *Cataloging Service Bulletin* (ISSN 0160-8029), cumulated *Library of Congress Rule Interpretations,* and *Cataloger's Desktop* **(CBC5),** catalogers may prefer separate stand-alone cumulations. It is interesting to note that the need to interpret the cataloging rules is felt beyond the Anglo-American community. A recent query on AUTOCAT elicited the information that catalogers in Chile use translations of both *AACR2* and the *LC Rule Interpretations.* Similarly, the Swedish Cataloguing and Classification

Committee of the Swedish Library Association produces rule interpretations for the Swedish version of *AACR*.

CCC1. **Howarth, Lynne C., comp.** *AACR2 Decisions & Rule Interpretations.* 6th ed. Loose-leaf. Chicago: American Library Association, 1995.

Consolidates the official interpretations and application decisions "which the Library of Congress, the National Library of Canada, the British Library, and the National Library of Australia have made to govern their use" of *AACR2R*, 1988 Revision, and including the 1993 amendments. Indicates "ownership" of decisions. Updated approximately every two years.

CCC2. *Oberlin LCRI Cumulated: Prototype Edition.* Online. URL: http://ocaxp1.cc.oberlin.edu~library/lcri/Prototype.HTML

Oberlin has published cumulations of Library of Congress rule interpretations in a loose-leaf format with updates since 1981. An experimental prototype version, covering Chapters 5 and 23 and the Index, was made available in 1995, with the note "If there is enough support out there for this type of Web publication we would hope to complete the mark up of the full text of all LCRIs by the end of 1995. Subscriptions could be set up for an entire campus network or Internet domain." The prototype version is very easy to navigate; each rule interpretation is individually accessible by clicking or the user can browse.

CCD
Application of the Standards to Special Types of Materials

RARE BOOKS

CCD1. *Examples to Accompany "Descriptive Cataloging of Rare Books."* Chicago: Association of College & Research Libraries, American Library Association, 1993.

Prepared by the Bibliographic Standards Committee of ACRL's Rare Books and Manuscripts Section to accompany the Library of Congress' *Descriptive Cataloging of Rare Books*, which is in turn a manual to supplement *AACR2R*. Includes fifty-three examples, derived from fifty titles whose title pages, with other pertinent information, are reproduced in the manual. Relevant rules are cited with each example.

CCD2. *EXLIBRIS: Rare Books and Special Collections Forum.* Electronic discussion group. Moderator and list owner: Everett C. Wilkie, Connecticut Historical Society.

Replaces *NOTRBCAT*, a list originally tied to the NOTIS system. This list covers all aspects of rare books and special collections librarianship, but a significant number of postings are related to cataloging. Established in 1990 by Peter Graham at Rutgers University; migrated to the University of California/Berkeley, where it is managed by Wilkie of the Connecticut Historical Society. To subscribe, send a message to LISTPROC@LIBRARY. BERKELEY.EDU, with the command: SUBSCRIBE EXLIBRIS [your name]. Messages may be sent to EXLIBRIS@LIBRARY.BERKELEY.EDU.

MANUSCRIPTS, ARCHIVES

CCD3. *Standards for Archival Description, A Handbook: Information Systems, Data Exchange, Cataloging, Finding Aids, Authority Control, Editing and Publishing, Statistics.* Compiled by Victoria Irons Walch for the Working Group on Standards for Archival Description. Chicago: Society of American Archivists, 1994.

Not a manual or set of standards itself, this is a sourcebook to standards and guidelines for describing and providing access to archival collections.

MUSIC AND SOUND RECORDINGS

CCD4. Hartsock, Ralph. *Notes for Music Catalogers: Examples Illustrating AACR2 in the Online Bibliographic Record.* Soldier Creek Music Series, no. 3. Lake Crystal, Minn.: Soldier Creek Press, 1994.

Provides copious real examples from Library of Congress cataloging records, shown alongside the relevant *AACR2R* rule and Library of Congress and music rule interpretations.

MOTION PICTURES AND VIDEORECORDINGS

CCD5. Yee, Martha. "The Concept of *Work* for Moving Image Materials." *Cataloging & Classification Quarterly* 18, no. 2 (1993): 33–40.

An exploration of the concepts of "work" and "related works" as they apply to this special category of material. This paper forms a companion to Yee's four-part discussion of the concept of work for all types of materials, "What Is a Work?" **(CCB10).**

CCD6. ———. "Manifestations and Near-Equivalents: Theory, with Special Attention to Moving-Image Materials." *Library Resources & Technical Services* 38, no. 3 (1994): 227–55.

Examines *AACR2R's* treatment of editions, versions, etc., in order to aid in definition of such "states" for moving-image materials. Provides the underlying discussion for the follow-up paper listed next **(CCD7).**

CCD7. ———. "Manifestations and Near-Equivalents of Moving-Image Works: A Research Project." *Library Resources & Technical Services* 38, no. 4 (1994): 355–72.

Explores, through sampling over a hundred representative titles, various ways in which films can be identified as editions (manifestations) or only slight variations from an original. Although the paper deals specifically with films, similar situations arise and similar questions need to be asked and answered in connection with other types of materials.

ELECTRONIC AND ONLINE RESOURCES

CCD8. Caplan, Priscilla. "Providing Access to Online Information Resources: A Paper for Discussion." Online. URL: http://www.nlc-bnc.ca/documents/libraries/cataloging/caplan2.txt

This paper was written in response to MARBI Discussion Paper No. 54 and distributed through the online discussion groups *USMARC-L, PACS-L,* and *CNI-DIR*. Caplan suggests that "online information resources" fall into at least two categories: electronic data resources and online systems/services, and their description requires somewhat similar, though not identical, treatment. Although this paper is written in response to and in terms of the MARC formats, it may help catalogers struggling to understand these new types of information resources to organize their thoughts and better handle the materials with which they increasingly are dealing.

CCD9. Cataloguing and Indexing of Electronic Resources. Online. URL: http://www.nlc-bnc.ca/ifla/II/catalog.htm

Not a manual or standard, this page provides links to a variety of resources, including papers, journals, tools, and standards.

CCD10. *Catriona: Cataloguing and Retrieval of Information over Networks.* Online. URL: gopher://ukoln.bath.ac.uk:7070/00/Link/Catriona/catriona.pub

A feasibility study of the British Library Research and Development Division for a project that is to last six months. Only the proposal is posted so far, but progress reports are promised.

CCD11. *EMEDIA: Electronic Cataloging Issues in Libraries.* Electronic discussion group. List owner: Eric Childress, Elon College.

A discussion forum for those responsible for creating, adapting, and revising cataloging rules and practices for electronic media. Postings include announcements, meeting reports, questions and answers about specific problems, discussion of existing standards for cataloging, etc. To subscribe, send a message to LISTSERV@VAX1.ELON.EDU, with the command, SUBSCRIBE EMEDIA. Messages may be sent to EMEDIA@VAX1.ELON.EDU

CCD12. Gaynor, Edward. "Cataloging Electronic Texts: The University of Virginia Experience." *Library Resources & Technical Services* 38, no. 4 (1994): 403–13.

Description of a project to provide full MARC records and headers using Text Encoding Initiative (TEI) specifications for electronic texts and to integrate the resulting records with other OCLC records.

CCD13. *Guidelines for Bibliographic Description of Interactive Multimedia.* Prepared by the ALCTS Interactive Multimedia Guidelines Review Task Force. Laurel Jizba, ed. Chicago: American Library Association, 1994.

Created in response to perceived inadequacies in *AACR2* for handling this type of material, these guidelines provide practical advice for determining whether an item is in fact multimedia, and for cataloging material, applying *AACR2R* whenever possible. Includes many examples. Other more formal standards may eventually develop, but for those in need of guidance now, these may be valuable.

CCD14. Hoogscarspel, Annelies. *Guidelines for Cataloging Monographic Electronic Texts at the Center for Electronic Texts in the Humanities.* Technical Report, no. 1. New Brunswick, N.J.: Rutgers, the State University, 1994. Also available via anonymous FTP to ceth.princeton.edu as catguid.ps

Guidelines developed for catalogers working with the Rutgers Inventory of Machine-Readable Texts in the Humanities. These guidelines were developed independently in part because of inadequacy of *AACR2R* for these materials. More formal standards may eventually develop, but these may be valuable for those in need of guidance now.

CCD15. *Interactive Electronic Serials Cataloging Aid (IESCA).* Online. URL: http://www.library.nwu.edu/iesca

Designed by Wei Zhang and John Blosser and supported by Northwestern University Library. Provides "ready access to cataloging rules, interpretations, examples of MARC bibliographic records in serial and computer file formats linked to instructional annotations, and a glossary of cataloging and computer terminology." Examples are given for electronic bulletins, digests, journals, and newsletters, and there are also "hypothetical examples of multiple versions."

CCD16. National Library of Canada Electronic Publications Pilot Project. Online. URL: http://www.nlc-bnc.ca/eppp/e3pe.htm

The home page for a small pilot project to acquire, catalog, and permanently store a small number of formally published Canadian online journals as the first step in planning the management of electronic publications at NLC. This page is a showcase for the project because it provides links to the journals themselves, rather than a detailed project description or presentation of standards used.

CCD17. Olson, Nancy B., ed. *Cataloging Internet Resources: A Manual and Practical Guide.* Dublin, Ohio: OCLC, Inc., 1995. Also available at URL: http://www.oclc.org/oclc/man/9256 cat/toc.htm or by anonymous FTP at: ftp.rsch.oclc.org/pub/internet_cataloging_project/Manual.txt

This manual was prepared specifically for use by participants in the OCLC project "Building a Catalog of Internet Resources." Based on *AACR2* and intended to be used in conjunction with it. Examples are coded for OCLC.

Most of the manual is concerned with description and access, though there are sections on classification and coding.

CCD18. *Proceedings of the Seminar on Cataloging Digital Documents, October 12–14, 1994.* University of Virginia Library, Charlottesville, and the Library of Congress. Online. URL: http://lcweb.loc.gov/catdir/semdigdocs/seminar.html
   Includes the full text of eight papers on topics such as the Text Encoding Initiative, organizing digital archives, and cataloging interactive multimedia, digital images, and Internet resources.

SERIALS, INCLUDING NEWSPAPERS

CCD19. *CONSER Cataloging Manual.* Washington, D.C.: Library of Congress, 1993. Loose-leaf. Base text, plus updates.
   A highly detailed and authoritative resource, used as the official manual for CONSER participants. The base text is divided into three sections: Original Cataloging; Adapting Records for Online Cataloging; Special Types of Serials and Special Problems. Update No. 1 (1993) includes "Interpreting Pre-AACR2 Serial Cataloging Records and Direct Access Computer File Serials." Update No. 2 (1995) includes "Microform Serials," plus additions and changes to previously issued text. The manual is useful far beyond its main target audience.

CCD20. *CONSER Editing Guide, 1994 Edition.* Washington, D.C.: Library of Congress, 1994. Loose-leaf. Base text, plus update service.
   Significantly revised from previous editions to encompass format integration. This is the official guide to content designation for CONSER participants. Includes a history of the CONSER project, a field-by-field guide to the USMARC format as applied to serials, and numerous examples. Essential to CONSER participants and highly useful to all serials catalogers.

CCD21. *Tools for Serials Catalogers: A Collection of Useful Sites and Sources.* Vanderbilt University Library. Online. URL: http://www.library.vanderbilt.edu/ercelawn/serials.html
   Maintained by Ann Ercelawn, this home page offers links to information of particular interest to serials catalogers. Includes cataloging documentation, reference sources, national and international serials programs and organizations, e-mail addresses for cataloging questions, minutes for the ALCTS Committee to Study Serials Cataloging, etc.

NONBOOK MATERIALS

CCD22. Fecko, Mary Beth. *Cataloging Nonbook Resources: A How-to-Do-It Manual for Librarians.* New York: Neal-Schuman, 1993.

Covers visual materials including motion pictures and videorecordings, projected and non-projected graphics, sound recordings, cartographic materials, computer files, microforms, electronic resources (serials, catalogs, and online databases and electronic services), and kits. The primary focus is description, although examples of full bibliographic records are given both in "card-image" and in MARC-coded versions. Separate chapter on classification. *AACR2R* rules are mainly paraphrased or interpreted, so it is not always possible to tell whether an instruction comes from the rules or from the author.

CCD23. Weihs, Jean, and Lynne C. Howarth. "Nonbook Materials: Their Occurrence and Bibliographic Description in Canadian Libraries." *Library Resources & Technical Services* 39, no. 2 (1995): 184–97.

A survey of Canadian libraries indicates that nine types of nonbook materials are common in library collections, but that while 75.1 percent of responding libraries use *AACR2R* for their book cataloging, fewer use *AACR2R* for cataloging nonbook materials, and 11.5 percent report not cataloging nonbook materials at all.

AUDIOVISUAL MATERIALS

CCD24. Varughese, Lola, and Gayle Poirier. "A Brief Survey of ARL Libraries' Cataloging of Instructional Materials." *Cataloging & Classification Quarterly* 18, no. 11 (1993): 125–35.

Seventy-five libraries responded to a survey investigating practices and policies followed for the cataloging of audiovisual materials including kits, teacher manuals, workbooks, and other instructional materials. Survey results reveal a wide variety of practice, with libraries often devising systems of their own.

REALIA

CCD25. Urbanski, Verna, with B. C. Chang and B. L. Karon. *Cataloging Unpublished Nonprint Materials: A Manual of Suggestions, Comments and Examples.* Edited by E. Swanson. Lake Crystal, Minn.: Soldier Creek Press, 1992.

Coverage is limited to description. Covers sound recording, motion pictures and videorecordings, graphic materials, computer files, realia, and kits; not intended for materials being added to an archival or manuscript collection. Relevant *AACR2R* rules are quoted, followed by examples of their application, and "discussion." This should be regarded only as an adjunct to *AACR2R* and may be useful primarily to catalogers with relatively little experience or those in need of a little "hand-holding" as they tackle unfamiliar materials.

SPECIAL CATEGORIES NOT TIED TO PHYSICAL FORMAT

Government Publications and Legal Materials

CCD26. Maben, Michael. "The Cataloging of Primary State Legal Material." *Cataloging & Classification Quarterly* 18, no. 1 (1993): 103–15.

An examination of the provisions and application of *AACR2* and *Library of Congress Rule Interpretations* as they apply to legal material contained in state reports, court rules, session laws, codes of law, administrative codes, and attorney general opinions.

CCD27. Sherayko, Carolyn C., ed. *Cataloging Government Publications Online.* New York: Haworth, 1994. Also published as *Cataloging & Classification Quarterly* 18, nos. 3/4 (1994).

Fourteen papers, most describing the experiences of various libraries in adding government publications records to online catalogs. Several articles deal with specific issues in cataloging government publications, such as cataloging titles on CD-ROM and availability of copy. Not a "how-to-do-it" publication, but some of the observations on others' experience may be useful to documents librarians contemplating including their cataloging in an online catalog.

Reproductions

CCD28. *Guidelines for Bibliographic Description of Reproductions.* Prepared by the Committee on Cataloging: Description and Access of the Association for Library Collections & Technical Services, Bruce Chr. Johnson, Principal Editor. Chicago: American Library Association, 1993.

Designed to be used in conjunction with *AACR2R*, includes practical guidance on cataloging various types of reproductions, including microforms, photocopies, reprints, etc. Some provisions of the guidelines are experimental; if they prove to be useful and readily applicable, they may form the basis for future proposed rule revisions.

CCE
Standards for Content Designation

CCE1. Coyle, Karen, ed. *Format Integration and Its Effect on Cataloging, Training, and Systems: Papers Presented at the ALCTS Preconference, American Library Association Annual Conference, June 26, 1992.* ALCTS Papers on Library Technical Services and Collections, no. 4. Chicago: American Library Association, 1993.

Includes revised versions of papers presented at an ALCTS preconference in 1992, touching on a wide variety of issues and including ample background and examples, treating all types of materials from the simplest monographic record to such materials as archivally controlled serials. Assumes some knowledge of MARC formats. Provides an "overview of Format Integration and its history, discusses its effect on cataloging and systems, and explores training and documentation."

CCE2. Library of Congress, MARC Editorial Division. *MARC Conversion Manual–Authorities (Names). Content Designation Conventions and Online Procedures.* 2nd ed. Loose-leaf. Washington, D.C.: Library of Congress, 1993.

Field-by-field instructions, including all changes. Examples reflect MARBI decisions made as of January 1993.

CCE3. *USMARC Format for Bibliographic Data, Update No. 1.* Washington, D.C.: Library of Congress, Cataloging Distribution Service, 1995.

This first update to the 1994 edition completes the Format Integration (FI) process, incorporating into the regular format specifications those FI and related changes that were documented previously in "Future" sections. Changes falling into this category are generally limited to the Leader and fields 006 to 008.

D
Subject Analysis Systems

Nancy J. Williamson

As stated in the *Guide to Technical Services Resources,* "the literature of subject analysis encompasses any resources that aid in the analysis, representation, and retrieval of recorded information *about* persons, places, events, and topical subjects" (p. 68). More specifically, it includes all of the resources that support the three basic processes of information handling—the content analysis of documents and information, the categorization and representation of that content, and the access to and retrieval of documents and information from both manual and automated systems. This situation has not changed. Over the years, however, the resources that belong under the topic "subject analysis systems" have become more and more diverse, embracing new theories, new technologies, and combinations of these. Distinctions between modes of subject access have blurred and intertwined. An examination of such resources over the past three years suggests three major trends. These are the continuation of the traditional approach to subject work, a rapidly growing interest in both theoretical and practical research, and the tremendous impact of technology on how bibliographic and information systems are designed and accessed.

Inevitably, changes in the terminology of bibliographic classification and subject headings will be continuous and will require constant and frequent updating of these systems. While the intellectual content of these systems reflects their origins and background, the tools that support them and control their manipulation in online catalogs and bibliographic databases are having a dynamic impact on the way the data are accessed, displayed, and used. Bibliographic classification has experienced a great deal of recent activity. At the time of writing, the publication of the *Dewey Decimal Classification*, Edition 21 **(DBB3)** is imminent, with Abridged Edition 13 **(DBB2)** to follow within a year. OCLC Forest Press also is becoming more deeply involved in electronic products with the production of *Dewey for Windows* **(DBB5)** and its exciting venture into its *NetFirst* database **(DD3)**, through which OCLC is accepting the challenge of organizing access to the Internet. The *Library of Congress Classification* is also moving forward electronically. At long last, the LCC schedules are about to become available in machine-readable form and LC's *Classification Plus* **(DBC2)** will provide an electronic product that will integrate the use of LCC and LCSH. Such a move to integration promises to aid in a more efficient holistic approach to subject access in online catalogs. All this activity suggests that the traditional systems will not die—they will adapt.

In a second trend, there is ever increasing evidence of a plethora of research seminars and workshops in classification, or "knowledge organization"—a term now coming into frequent use. The International Society for Knowledge Organization (ISKO) held its biennial conference in Copenhagen in June 1994, and in October 1994 an Allerton Institute on "New Roles for Classification in Libraries" was held at Allerton Park, University of Illinois. Its proceedings will be published in an upcoming issue of *Cataloging & Classification Quarterly*. The ASIS Special Interest Group on Classification Research (ASIS SIG/CR) holds "classification workshops" annually **(DAA2)**. Plans also are being formulated for the Sixth International Study Conference on Classification Research to be held in London in 1997. Collectively, these conferences focus on a broad spectrum of research activities concerned with the derivation, representation, and organization of recorded knowledge. The Allerton Institute focused on classification in libraries, while most of the other conferences may include the traditional classification and subject descriptor systems. They are much more broadly based, drawing on research activities in communication, linguistics, cognitive psychology, and systems outside the traditional library setting. Participants are a cross section of librarians, educators, theoreticians, and information management and systems personnel. The conferences themselves are playing fields for the dynamic interchange of ideas on principles, practices, and design of information systems in a variety of domains and at an interdisciplinary level. It is the diversity and the third trend—the increasing application of the technologies—that bind these activities together.

Chapter Content and Organization

The chapter on Subject Analysis Systems is organized in a manner similar to that in *GTSR* with the addition of two subsections. A section on Conference Proceedings has been included under General Works. Internet Subject Access has been added at the end of the chapter to suggest a few resources that look at the complex problem of subject access to the Internet. It is inevitable that more attention has been given to specialized problems of subject analysis systems. More electronic products and software are included here. When appropriate, the technological aspects are dispersed among the various sections, a decision made on the grounds that technologies are tools and, in this kind of list, it is better to group by topical content than format.

In selecting items to be included in this update, an attempt has been made to respond to some of the deficiencies pointed out in reviews of *GTSR* and in direct contact with readers. Also included are some important items that were omitted from the 1993 publication. A number of major textbooks have been updated and revised; they are included here. In selecting the electronic products and software, the author recognizes that she has touched only the "tip of the iceberg." An objective for future supplements should be more emphasis on electronic resources. Finally, with so many conference proceedings, there might have been more individual entries for individual papers. However, number, diversity, and nature of the papers make it impossible, and not necessarily desirable, to provide separate entries and annotations for each paper. As a compromise, an effort has been made to identify major areas of content in the annotations for the complete items.

DA
General Works

DA1. Aluri, Rao, Alasdair Kemp, and John J. Boll. *Subject Analysis in Online Catalogs.* Englewood, Colo.: Libraries Unlimited, 1991.

The authors examine the nature of subject access in online catalogs as systems of interrelated components, depending on the number, specificity and accuracy of subject headings, and classification numbers assigned to bibliographic records. However, these are considered in the light of other factors in the bibliographic record, such as comprehensiveness of data, the richness of uncontrolled vocabulary appearing in fields such as titles, the user interface module, and the qualities users bring to their searches in online catalogs. All these areas are identified and examined in terms of overall usefulness in subject access in online catalogs. An important and helpful guide intended for the use of library and information science students and educators, as well as indexers and practitioners.

DA2. Beatty, Sue. "Subject Enrichment Using Contents or Index Terms: The Australian Defense Force Academy Experience." In *Advances in Online Public Access Catalogs,* Vol. 1, edited by Marsha Ra, 93–104. Westport, Conn.: Meckler, 1992.

As the title indicates, the author describes a project that experiments with enrichment of catalog records. The library uses LCSH and the project is based on procedures used by Pauline Atherton Cochrane in the Subject Access Project at Syracuse University in the late 1970s. The procedures are described together with results of a study on the project by Mary Micco, Indiana University of Pennsylvania. Micco looks at the impact of the project on information retrieval and the relevance of materials retrieved. Includes examples of contents pages showing term selection.

DA3. Cherry, Joan M. "Improving Subject Access in OPACs: An Exploratory Study on Users' Queries." *Journal of Academic Librarianship* 18, no. 2 (1992): 95–99.

In a study at the University of Toronto Libraries, observers recorded protocols for 100 OPAC search sessions. This article reports on the analysis of those protocols and examines zero-hit subject searches to explore the effectiveness of various conversions of users' queries to improve recall. Findings of this exploratory study merit consideration in efforts to improve subject access in OPACs. They demonstrated that "keyword subject, keyword title or title searches using the original query from the user's zero-hit subject search were as fruitful or more fruitful than new searches constructed from cross references provided by LCSH." Nevertheless, data revealed that the keyword subject/keyword title/title approaches were seldom used. Conclusions pointed to the need to encourage users receiving zero-hits to try alternative types of searches or to provide OPAC software to automatically convert zero-hit subject queries to other types of searches. A carefully researched project of interest to designers and users of online catalogs.

DA4. Curl, Margo Warner. "Enhancing Subject and Keyword Access to Periodical Abstracts and Indexes." *Cataloging & Classification Quarterly* 20, no. 4 (1995): 45–55.

Curl addresses subject access in an online catalog of bibliographic records for periodical articles and indexes (as opposed to books and other kinds of materials). Considers the addition of Library of Congress Subject Headings and expanded content notes. The questions addressed in the research are "Can subject and keyword access to bibliographic records be improved within the existing cataloging standards and practices?" and "Are there other ways of providing additional access to subject content?" The answers given are "yes" and "no." "Yes" because the notes area of the MARC record could be adapted; "no" because of the amount of time and effort involved in reexamining the items to decide what additions should be made. An interesting addition to a growing discussion of this topic.

DA5. Down, Nancy. "Subject Access to Individual Works of Fiction: Participating in the OCLC/Fiction Project." *Cataloging & Classification Quarterly* 20, no. 2 (1995): 61–70.

Describes the OCLC Fiction Project from the point of view of a practicing cataloger involved in the project. Four types of access to individual works of fiction are discussed: form/genre, characters, geographic setting, and topical. A topic of increasing interest to both public service and cataloging professionals.

DA6. "Enhancing USMARC Records with Tables of Contents." In *Advances in Online Public Access Catalogs*, Vol. 1, edited by Marsha Ra, 105–13. Westport, Conn.: Meckler, 1992.

This MARBI Discussion Paper (no. 46) responds to library and information science specialists interested in enhancing subject access to library materials. With advances in technology, resource sharing, and increasing use of databases, there is an ever growing need for improved information retrieval through the enrichment of catalog records by various methods, such as supplying more than the usual number of subject headings and adding tables of contents and indexing terms. Explores methods of providing enhanced access using the current USMARC bibliographic format and identifies additional content designators being considered to enhance the format's capabilities for carrying table of contents information. The background for such an initiative is outlined briefly and possible options for inclusion of contents information in MARC records are discussed. The paper is concerned with the technicalities of enhancement of records and is designed as a basis for further discussion and possible further action by the USMARC Advisor Group and others.

DA7. LeBlanc, Jim. "Classification and Shelflisting as Value Added: Some Remarks on the Relative Worth and Price of Predictability, Serendipity, and Depth of Access." *Library Resources & Technical Services* 39 (July 1995): 294–302.

A thoughtful consideration of classification and shelflisting and the intrinsic value of browsing. Using some rough data from a test at Cornell University, the author estimates the cost of maintaining the collocative and alphabetical integrity of shelflist files by or about literary authors. The author ponders "optimal browsability."

DA8. Quinn, Brian. "Recent Theoretical Approaches in Classification and Indexing." *Knowledge Organization* 21, no. 3 (1994): 140–47.

Selective review of recent studies in classification and indexing theory. Discusses a wide variety of important issues including subjectivity versus objectivity, attempts to develop a theory of indexing, the theoretical role of automation in classification, theoretical approaches to a universal classification scheme, concept theory as an approach to universal classification, concept systems as definition systems, classification theory in the humanities, and a theoretical approach to the classification of fiction. As the result of his study, the author notes that much work in classification theory comes from sources outside the United States and explores some of the reasons for the non-American origin. An excellent overview of the subject for students, educators, and other information professionals.

DA9. Taylor, Arlene G. "On the Subject of Subjects." *Journal of Academic Librarianship* 21 (Nov. 1995): 484–91.

An overview article on the need for support for effective subject retrieval in information systems now and in the future. Using as her starting point the evidence that "subject cataloging continues to be disparaged by many librarians" in spite of evidence that "the subject index . . . is still one of the most commonly used search access points in the online catalog," the author examines five fundamental issues involved in the debate: subject access and the Internet (including standards), controlled vocabulary versus keywords (including specific entry), use of classification for enhanced subject access, online public access catalog, and LCSH. She concludes that subject analysis needs "innovation" and suggests four possible approaches: identification of new ways for determining which types of information need more subject analysis and which need less; better education of reference librarians as to ways in which classification and subject headings can be used to improve searches; more national-level cooperation in creating subject access; and the implementation of techniques research has shown will reduce the frustration of users who retrieve either nothing or too much. An important contribution to the debate, bringing together a significant amount of research on the topic well supported with citations.

DA10. Wittenbach, Stefanie A. "Building a Better Mousetrap: Enhanced Cataloging and Access for the Online Catalog." In *Advances in Online Public Access Catalogs,* Vol. 1, edited by Martha Ra, 74–92. Westport, Conn.: Meckler, 1992.

The author defines an enhanced record and explores options for embellishing the basic USMARC record. Options discussed in detail are additional subject headings, tables of contents, summaries, abstracts or contents notes, classification, back-of-the-book indexes, and natural-language subject headings. In the belief that enrichment of the records with additional data should not be the sole means of improving the system, improved systems features also are discussed. The latter suggests more advanced search capabilities.

Among such features are the improvement of displays relating subject headings and other content-bearing terms, establishment of links between keywords and authority records, and monitoring of transaction logs to identify frequently used search terms and phrases that result in no matches. Possible impact of enhanced records on library users, collections, and the cataloging operation are discussed. Wittenbach challenges librarians to rethink current rules, methods, and tools with the goal of improved OPACs. Discusses many studies that have investigated online catalogs and that contribute to this topic. Well documented with citations; an excellent overview of the issues and possibilities.

DAA
Conference Proceedings

DAA1. Albrechtsen, Hanne, and Susanne Oernager, eds. *Knowledge Organization and Quality Management: Proceedings of the Third International ISKO Conference 20–24 June 1994, Copenhagen, Denmark.* Advances in Knowledge Organization, vol. 4. Frankfurt am Main, Germany: INDEKS Verlag, 1994.

Another in the series of international conferences held biennially by the International Society for Knowledge Organization (ISKO). Contains refereed papers, many by well-known experts, that are the product of worldwide research efforts in the broad area of knowledge organization. The fifty-four papers are organized under twelve topics reflecting the organization of the conference itself. The titles of these topics indicate the diversity and breadth of coverage: Quality in Knowledge Organization, Theory of Knowledge, Linguistics in Knowledge Organization, Concept Representation in Systems Design, Knowledge Organization in Specific Domains, Communications and Knowledge Organization, Online Public Access Catalogues, Knowledge-Based Systems, Tools and Techniques for Organization of Knowledge, Thesauri Facing New Technologies, Restructuring Classification Schemes and Thesauri, and Future Prospects. Contains a list of contributors with names, addresses, telephone and fax numbers and, where available, e-mail addresses. Name and subject index. Of interest to those involved in classification research. Practitioners will have an interest in investigations carried out in a variety of information-centered institutions.

DAA2. *ASIS SIG/CR Classification Research Workshops.* Medford, N.J.: Published by Information Today, Inc. for the American Society of Information Science, 1990–

A series of one-day preconference workshops, or seminars, held annually since 1990, in conjunction with ASIS annual meetings. These meetings of the ASIS Special Interest Group on Classification Research bring together researchers and practitioners from a broad range of disciplines and areas of application for purposes of exchanging ideas, discussing problems, and sharing research in classification (i.e., knowledge organization). For each workshop, a prepublication volume of papers is prepared for the participants and later produced in a fully published form. Six conferences have been held and, to date, four have appeared in fully published form. A perusal of the output indicates a wide diversity of topics ranging through such topics as subject indexing principles, the development of the *Art & Architecture Thesaurus (AAT)*, object-oriented classification, images and classification, classification in special fields (e.g., patents, building theory, nursing interventions), the relevance of classification theory to database design, frame-based systems, thesaurus development, and many other related topics. This series provides an excellent overview of the activities, both theoretical and practical, taking place in the field of knowledge organization in North America.

DAA3. Williamson, Nancy J., and Michèle Hudon, eds. *Classification Research for Knowledge Representation and Organization: Proceedings of the Fifth International Study Conference on Classification Research, Toronto, Canada, June 24–28, 1991.* FID Publication, no. 698. Amsterdam, Netherlands: Elsevier, 1992.

Proceeding of the fifth in a series of conferences held at irregular intervals since 1957. Conference objectives are to present and discuss the results of research, to identify problems, and to consider future prospects. Over the years, the intellectual content has evolved from the narrow definition of bibliographic classification to a broader and more interdisciplinary view of knowledge organization. The forty papers presented at the 1991 conference cover a spectrum of topics from classification theory and cognitive science (in general) through traditional classification, thesaurus design and development, and graphics display to expert systems. An important collection for all who are interested not only in research, but also in new developments in the field.

DAB
Textbooks, Guides, and Manuals

DAB1. Chan, Lois Mai. *Cataloging and Classification: An Introduction.* 2nd ed. New York: McGraw Hill, 1994.

A much needed new edition of a work originally published in 1981. One of the best basic textbooks on all aspects of descriptive and subject cataloging. Part 3 on subject cataloging, Part 4 on classification, and sections of Part 5 on USMARC formats are particularly relevant to subject analysis. All chapters have been edited and rewritten to reflect recent developments, in particular reflecting the impact of technology on cataloging and classification. Emphasis is given to computer processing of records and the characteristics and application of the USMARC formats. Each section is introduced by a bibliography that includes the basic tools of the operation, background reading, and further reading. Includes examples and exercises on the application of various tools with suggested answers in an appendix. Unfortunately, the answers to the DDC exercise may become outdated with the publication of *DDC 21* **(DBB3)**. However, the topics themselves should remain useful. Glossary of terms and examples of MARC records. An essential text for beginning students and a tool for practicing cataloging personnel. For annotation, *see also* **CAA1**.

DAB2. Kao, Mary L. *Cataloging and Classification for Library Technicians.* New York: Haworth Press, 1995.

A textbook on the fundamentals of cataloging and classification for technicians. Useful in library technicians educational programs and in training library technicians in libraries.

DAB3. Sha, Vianne Tang. *Internet Resources for Cataloging.* Online. URL: http://www.law.missouri.edu/vianne/cat.htm

Sha provides hot links to cataloging-related resources on the Internet, including a section called "Internet Resources for Cataloging: Subject Cataloging and Classification." For annotation, *see also* **CAA4**.

DAC
Periodicals

Periodicals listed in section **CAD** of the chapter "Descriptive Cataloging" are also of value to subject analysis specialists.

DAC1. *Classification Issues for Knowledge Organization.* The Hague, Netherlands: International Federation for Information and Documentation (FID), v. 1, 1995– . Quarterly.

Official publication of FID/CR, the Committee on Classification Research for Knowledge Organization of FID, and issued quarterly as an insert in the *FID News Bulletin.* Its objective is to record activities and developments in research and practice in knowledge organization worldwide. Provides reports of FID/CR activities, including UDC developments; reports of conferences at both the international and national levels; reviews of standards, electronic products, and other important publications of interest in knowledge organization; announcements; and calls for papers. As a new publication, *Classification Issues* is still developing its policy on coverage. Its ultimate goal is to be as international and cosmopolitan as possible and to include short, invited contributions on major developments in the field. The editor also makes use of the LIS-FID discussion group **(DAE1)** for timely announcements.

DAC2. *Knowledge Organization.* Frankfurt am Main, Germany: International Society for Knowledge Organization, v. 20, no. 1, 1993– . Quarterly. ISSN 0943-7444. Continues *International Classification.*

With the first issue of 1993, *International Classification* changed its title to *Knowledge Organization.* The change reflects changing terminology, not a change in objectives or content, and resulted when the editorial board wanted to make clear the journal is interested in "all questions of knowledge organization," not just the narrower concept of "classification." It is the organ of ISKO and its scholarly articles cover a wide range of topics on all aspects of knowledge organization and representation, e.g., conceptology, classification, thesauri, indexing, relevant linguistic problems, and terminology. Also contains book reviews, reports on conferences and activities in knowledge organization worldwide, announcements of conference, and calls for papers. An annotated bibliography of knowledge organization literature, which appears once a year, is an excellent and comprehensive overview of research and publication in the discipline. Of primary interest to researchers, theorists, and educators.

DAD
Bibliographies

DAD1. Gupta, Sushima. *Decimal Classification System: A Bibliography for the Period 1876–1994.* March 1995. URL: http://www.oclc.org/oclc/fp/bibl/front.htm

Bibliography (arranged chronologically) of over 700 items with subject and author indexes.

DAD2. Stone, Alva T. "That Elusive Concept of 'Aboutness': The Year's Work in Subject Analysis, 1992." *Library Resources & Technical Services* 37 (July 1993): 277–98.

A comprehensive bibliographic essay that references over 130 publications on subject analysis. The essay is organized into the categories of theoretical foundations, cataloging, practices, subject access in online environments, and specialized materials topics.

DAE
Sources of Expertise

The sources of expertise listed in section **CAE** of the chapter "Descriptive Cataloging" are of equal value to subject analysis specialists.

DAE1. LIS-FID. Electronic discussion group. International Federation for Information and Documentation (FID).

The aim of LIS-FID is to promote discussion of information management and library and information science in an international context. Home page (under development) available at URL: http://nisp.ncl.ac.uk/hypermial/lists-k-o/lis.fid.html. To subscribe, send a message with the command JOIN LIS-FID [your name] to MAILBASE@MAILBASE.AC.UK. Send messages to

LIS-FID-REQUEST@MAILBASE.AC.UK. Archived at URL: gopher://mailbase.ac.uk:70/77/lists/lis-fid/.waisindex/files.

DAE2. OCLC Forest Press DDC Web Site. URL: http://www.oclc.org/cgi-oclc/imagemap01/fphome3.htm

This WWW home page provides links to information of interest and value to users of DDC. These include DDC Summaries (English, French, and Spanish), hot classification topics, list of libraries using the DDC, list of current editions of DDC that have been or are being translated, and more.

DB
Classification Systems

DBA
General Works

DBA1. Langridge, Derek W. *Classification: Its Kinds, Systems, Elements and Applications.* London: Bowker-Saur, 1992.

A general discussion of the principles of classification in the light of current trends in knowledge organization. Despite its title, this work focuses primarily on theory and principles. Of particular interest is the section on the *Bliss Bibliographic Classification*, which is being transformed into a fully faceted system in its second edition. An important publication for library and information science educators and students of classification theory, it will also be of interest, but of less practical value, to practicing classifiers.

DBA2. "New Roles for Classification in Libraries and Information Networks: Presentations and Reports from the Thirty-Sixth Allerton Institute, October 23–25, 1994." *Cataloging & Classification Quarterly* 21, no. 2 (1995): 3–142.

A special issue of *CCQ* containing the papers and reports of panel discussions from the Institute in which the theme centered around "new roles for library classification in the electronic age." Three keynote addresses focus on the future of traditional classification in library and information networks: considering the present and the future (Lois Mai Chan), the qualities needed for classification schemes for the information superhighway (Pat Molholt), and presenting a theoretical point of view on the future of classification (Ingetraut Dahlberg). Another trio of papers focus on the future of major universal classification schemes including: *DDC* 21 (Joan Mitchell), *UDC* (Ia McIlwaine), and *BC2* and *BSO* (E. J. Coates). Papers on LCC and the NLM classification schemes were also presented and will be included in a later issue of *CCQ*. A paper by Janet Swan Hill addressed classification from an administrator's perspective. A final paper (not presented at the institute) addresses classification access in the online catalog (Elaine Broadbent). Also includes reports from three panel discussions held during the conference. These address the question of what lies ahead for classification in information networks, classification in library and information networks abroad, and nontraditional uses of classification. Includes a twenty-seven-page annotated bibliography of books, journal articles, and Internet resources related to the conference theme. A very good overview of ideas and concerns of experts about the importance of traditional classification and its use into the next century.

DBA3. Thomas, Alan. *Classification: Options and Opportunities.* New York: Haworth Press, 1995. Also published as *Cataloging & Classification Quarterly* 19, nos. 3/4, 1995.

Sixteen articles on various classification issues of current interest. Articles are grouped under five broad topics: basic design considerations, options within standard classification systems (LCC and DDC), alternative classification systems (including Bliss, UDC, schemes of the Research Libraries of the New York Public Library, and reader-interest classification), combination platters and reclassification, and classification and the new technology. The latter includes options in classification and the *Electronic Dewey*. Of general interest to all practitioners, educators, and students.

TEXTBOOKS, GUIDES, AND MANUALS

DBA4. *Canadian MARC Communications Format: Classification Data.* Ottawa: National Library of Canada, Canadian MARC Office, 1992– . Looseleaf.

A format for organizing classification schedules into machine-readable form. Parallels the USMARC format with variations to meet Canadian needs.

DBA5. Marcella, Rita, and Robert Newton. *A New Manual of Classification.* Aldershot, Hampshire, England: Gower, 1994.

A textbook on classification with a focus somewhat different than most North American texts. The authors link their work with Berwick Sayers' *Manual of Classification,* first published in 1926 and subsequently revised, twice by Sayers himself, with the responsibility for the fourth and fifth editions being shouldered by Arthur Maltby, who maintained the tradition of Sayers. The authors of the present work claim that their work "builds upon the work of its predecessors." Their objective is to provide a clear and comprehensible discussion of classification theory, policy, and practice, together with a description of the major general classification schemes, to support students of classification in their studies. Part 1 deals with "principles and systems" briefly addressing the topics of the theory of classification and the five major general schemes (DDC, LCC, UDC, CC, and BC) together with special schemes, and a final section on indexes, thesauri, and classification. Part 2 addresses "management of classification and classification policy in the library," including a discussion of some aspects of management and application not discussed in many other texts of this kind. Part 3 centers on information technology and classification, including discussions of classification and the

OPAC and automatic classification. There appears to be some skepticism on the subject of "theory" by the authors and it is not clear that this new work is really a "new Sayers." However, times have changed. Theory still has its place, and the development and management of systems is a much needed consideration. Regardless, this work provides a useful complement to other texts on the subject and is one of the few devoted entirely to the subject. It will broaden the horizons of students in classification and be a useful tool for library and information science educators. Because this is a British text, North American users will find that it takes a somewhat different approach than similar texts produced on this continent.

DBA6. Scott, Mona L., and Christine E. Alvey. *Conversion Tables: LCC-Dewey, Dewey-LCC.* Englewood, Colo.: Libraries Unlimited, 1993.

This publication consists of two sets of tables, one ordered by Library of Congress classification number, the other by Dewey decimal classification number. Each entry is accompanied by a descriptor or brief explanatory phrase. It represents a great deal of work on the part of the authors. While there is no substitute for careful matching of the class number with the item to which it is being assigned, this tool can be helpful in some specific situations—when a whole library is being converted from one system to the other, when the subject cataloger needs a class number while editing MARC records, and when the library uses DDC but only the LCC class number is given on a record. A useful tool for any library that does original cataloging and/or tailors classification to local needs.

CLASSIFICATION OF SPECIAL TYPES OF LITERATURE

DBA7. Beghtol, Clare. *The Classification of Fiction: The Development of a System Based on Theoretical Principles.* Metuchen, N.J.: Scarecrow Press, 1994.

Adapted from the author's thesis, this work addresses the use of the theory and principles of facet analysis to develop a classification for fiction. Goals of the research were to inquire critically into various kinds of theortical and methodological issues that arise in designing a fiction analysis system and to develop an operable preliminary prototype system using both established and experimental semantic and syntactic classificatory techniques. The author feels that fiction scholarship may benefit from this work. She describes the problem of trying to classify fiction and presents historical background and the theory derived upon which to base the actual process. An important contribution to classification theory.

CLASSIFICATION IN ONLINE CATALOGS

DBA8. Kneisner, Dan, and Carrie Willman. "But Is It an Online Shelflist? Classification Access on Eight OPACs." *Cataloging & Classification Quarterly* 20, no. 4 (1995): 5–21.

This paper recognizes the fact that, in general, designers of online catalogs have neglected to pay enough attention to the search potential of catalog access through shelf order via classification. The article compares the call number searching capabilities of eight systems in which the catalogs use the Library of Congress Classification: VTLS, Inlex, Innovative Interfaces, Inc., Data Research Associates (DRA), Geac Advance, Dynix, SIRSI, and NOTIS. Five questions were addressed in the study: Are the catalog records arranged in shelflist order? Are long call numbers fully displayed? Are main entries displayed? Is free browsing, backward and forward, allowed? Do "no exact matches" still result in a screen of approximate matches? Ultimately, the eight systems were judged by how well they performed through the Internet. In looking to the future, the authors anticipate that future characteristics of online catalogs will include greater orientation to the Internet and client-server interfaces. An important study for all information professionals with concerns for the evolution of library catalogs.

STANDARDS

DBA9. *USMARC Format for Classification Data, Update No. 1.* Washington, D.C.: Library of Congress, Cataloging Distribution Service, 1995.

Includes all changes to the USMARC classification format since its approval and publication in 1990. This first update contains new and changed USMARC data elements resulting from format development work since December 1990.

DBB
Dewey Decimal Classification

GENERAL WORKS

DBB1. Sweeney, Russell. "The International Use of the *Dewey Decimal Classification*." *International Cataloguing & Bibliographic Control* 24 (Oct./Dec. 1995): 61–64.

Summarizes the use of the Dewey Decimal Classification throughout the world. Describes the translations available; the responses of those responsible for the maintenance of the DDC to the need to internationalize the English language editions; the activities of the profession in many countries in providing a focus for discussion and input into the editorial process; and the different ways that DDC is used. Among the author's suggestions are the undertaking of a comprehensive survey of use in order to provide evidence that would support the development of DDC as a more international scheme. Sweeney suggests the possibility of providing variant publications to meet the needs of the wider international community. An excellent bibliography of fifty items organized under the names of twenty-one countries is included. Updates an article published in *Dewey: An International Perspective*, edited by Robert P.

Holley (London: KG Saur, 1991). An excellent starting place for anyone wishing to study the nature and impact of DDC on library and information science worldwide.

SCHEDULES

DBB2. Dewey, Melvil. *Abridged Dewey Decimal Classification and Relative Index.* 13th ed. Edited by Joan S. Mitchell, Julianne Beall, Winton E. Matthews, Jr., and Gregory R. New. Albany, N.Y.: OCLC Forest Press, [forthcoming].

Following the usual pattern of publication, *DDC 13 Abridged* will be published soon after the full edition comes off the press. It will be based on *DDC 21* and will reflect the changes described below. Publication is expected by mid-1997.

DBB3. ———. *Dewey Decimal Classification and Relative Index.* 21st ed. Edited by Joan S. Mitchell, Julianne Beall, Winton E. Mathews, Jr., and Gregory R. New. 4 vols. Albany, N.Y.: OCLC Forest Press, 1996.

The major classification system in use in public and school libraries in North America and in libraries worldwide. This edition is known as *DDC 21*. It follows in the tradition of earlier editions. *DDC 21* includes three major revisions to *DDC 20*: 350–354 Public Administration (a complete revision), 370 Education, and 560–590 Life Sciences. Only two parts of 560–590 Life Sciences are completely revised: 570 (Biology in general) and 583 (Dicotyledons). Other notable revisions include 296 Judaism, 297 Islam, 368 Insurance, and Table 2 area numbers—47 for the former Soviet Union and—499 for Bulgaria. Terminology throughout the *Classification* has been updated to achieve currency, ensure sensitivity, and reflect international topics. Many new topics that have gained literary warrant since the publication of *DDC 20* are now mentioned in the *Classification*, e.g., rap music, virtual reality, in-line skating, snow-boarding. This is an essential publication for libraries that classify their collections in DDC and for library and information science programs. As with previous new editions, the use of *DDC 20* in catalog records produced by the Library of Congress and in national bibliographies such as *Canadiana, British National Bibliography,* and the *Australian National Bibliography* will be discontinued and superseded by the application of *DDC 21*.

DBB4. ———. *Sistema de Clasificacion Decimal Dewey.* Edition 20. 4 vols. Bogota, Colombia: Rojas Eberhard Editores LTDA (Colombia) and Information Handling Services de Mexico S.A. de C.V. under license from OCLC Forest Press, 1996.

A new Spanish edition of the *Dewey Decimal Classification*. This Spanish translation contains the full text and all of the features of *DDC 20* and replaces the 1980 translation, which was based on Edition 18. An important tool for libraries and information centers in Spanish-speaking areas of the United States and in libraries in Spanish-speaking countries abroad. Available from OCLC Forest Press in Albany, New York.

DBB5. *Dewey for Windows: DDC 21.* CD-ROM. Albany, N.Y.: OCLC Forest Press, 1996.

A successor to *Electronic Dewey*, in a Windows mode. It is a Windows-based version of the *Electronic Dewey* and includes all the features of the DOS version plus multiple user views, LAN capability, and local annotation. The initial version is based on the *DDC 21* database.

TEXTBOOKS, GUIDES, AND MANUALS

DBB6. **Chan, Lois Mai, John P. Comaromi, Joan S. Mitchell, and Mohinder P. Satija.** *Classification Decimale de Dewey: Guide practique.* A translation prepared by the Association pour l'advancement des sciences et des techniques de la documentation, Canada. Montréal: ASTED, 1996.

A French translation of the English text described below. An essential manual for DDC libraries in French-speaking and bilingual communities.

DBB7. ———. *Dewey Decimal Classification: A Practical Guide.* Albany, N.Y.: OCLC Forest Press, 1994.

Organized according to the structure of DDC itself, this work analyzes all aspects of the classification system focusing on *DDC 20*. A brief introduction provides a historical context. Includes bibliography. Exercises at the ends of relevant chapters provide a basis for practice in applying the schedules. The emphasis here is on practice and it is truly a textbook for courses in classification in library and information science, as opposed to practicing catalogers.

DBC
Library of Congress Classification

GENERAL WORKS

DBC1. **Williamson, Nancy J., Suliang Feng, and Tracy Tennant.** *The Library of Congress Classification: A Content Analysis of the Schedules in Preparation for Their Conversion into Machine-Readable Form.* Washington, D.C.: Library of Congress, Cataloging Distribution Service, 1995.

Describes the research carried out in preparation for the conversion of the Library of Congress Classification schedules into machine-readable form. A detailed content analysis of more than 3,900 pages of the printed LCC schedules was conducted to identify characteristics of the data that would need to be taken into account in the conversion. Outlines research methodology and describes general characteristics, hierarchical arrangement, tables and subarrangements, notes, and references in depth. Identifies need to make all implicit data explicit. Other recommendations include the need for editing specific kinds of data, elimination of "divide-like" instructions, consolidation and expert manipulation of tables and sub-arrangements, and the development of an end-user version of the system. A suggestion for the future is the elimination of most, or all, tables incorporating the topics into the main tables. Of value to all who are interested in the development and manipulation of LCC in machine-readable form.

SCHEDULES

DBC2. *Classification Plus.* CD-ROM. Washington, D.C.: Library of Congress, Cataloging Distribution Service, 1996– . Quarterly.

As one product of the conversion of its classification systems to machine-readable form, the Library of Congress has produced a full-text Windows-based CD-ROM product to be available "early in 1996." It is to be sold on an annual subscription with quarterly issues. Includes the Library of Congress Classification schedules and the Library of Congress Subject Headings. Uses "Folio" software and permits users to follow hypertext links between files, to view headings in an expanded hierarchical display, and to conduct complex searches. The initial version contains the entire LCSH to September 1995 together with Library of Congress Classification schedules E–F, H, J, L, R, T, and Z. Additional LCC schedules will be added through the updates and as they become available. The use of this product should increase efficiency in the use of the LCC schedules and should be of interest to practitioners, educators, and students alike.

DBC3. *Classification Plus & Cataloger's Desktop.* CD-ROM. Washington, D.C.: Library of Congress, Cataloging Distribution Service, 1996.

At the time of writing, the Library of Congress is making available a demonstration package which combines these two CD-ROM products. The demo may be acquired in two ways. A free demo disk can be obtained from the Cataloging Distribution Service (e-mail: cdsinfo@mail.loc.gov) or users of the Internet may access it via anonymous FTP at URL ftp://ftp.loc.gov and moving to directory pub/CDs/deskclas. Download all of the files in the directory and print the WordPerfect file CDINFO.WP5 for further instructions. An excellent introduction to these two systems and their working interrelationships for all technical services personnel, educators, and students.

SPECIAL SCHEDULES

DBC4. *Class FC: A Classification for Canadian History.* 2nd ed. Ottawa: National Library of Canada, 1994.

A much needed revision of a special "LCC" schedule for Canadian History. It is a revision and expansion of the first edition published in 1976. This special scheme has replaced many LC-like classification schemes formerly used in Canadian libraries. In this new edition, changes have been made in three areas. The historical periods have been brought up to date and, based on experience, changes have been made in the time periods under the provinces where the time periods had been too narrowly defined. Secondly, this edition contains many more examples under biography and special subjects than previously. Finally, the names included in the schedules have been established using the *Anglo-American Cataloguing Rules,* 2nd ed., 1988. The schedule was created from machine-readable data making it possible for NLC to do frequent updates as needed. NLC has made plans to include the *Canadian MARC Communication Format: Classification Data* in a future phase of its online bibliographic system. This will enable NLC to maintain Class FC efficiently and to distribute the next version of the schedule in both machine-readable and printed form.

DBC5. *Class PS8000: A Classification for Canadian Literature.* 2nd ed. Ottawa: National Library of Canada, 1978.

A special "LCC schedule" for Canadian literature. Developed in response to the need for a more specific treatment of the topic than provided in LC's own schedules. Widely used in Canadian academic libraries. The preface states that "it differs in principle from Library of Congress practice, both in its assembling of Canadian literature regardless of language, and in keeping novels with the rest of Canadian literature instead of placing them in PZ." It is an essential tool for Canadian academic institutions and also may be of interest to library and information science professionals in other countries to observe how one country outside of the United States has coped with its literature. Published in 1978, it is now quite ancient. However, "Individual Authors" table is constructed in such a way that class numbers can be created for contemporary authors. The schedule is applied in *Canadiana,* Canada's national bibliography, and in the National Library of Canada's own collection.

DBD
Other Classification Systems

BOOK NUMBERS

DBD1. Dick, Gerald K. *LC'S Author Numbers.* Castle Rock, Colo.: Hi Willow Research and Publishing, 1992–. Loose-leaf.

A loose-leaf publication listing the author (Cutter) numbers for literary authors whose names are found in the Library of Congress Classification schedules based on a file kept at Brigham Young University. It contains two lists, one arranged by LC's classification number and the second arranged alphabetically by author. Additional data include the classification number used by Brigham Young University, special tables used to expand the classification, the record number from LC's name authority file, the source of the class number or name, and the indicator for the form of name. Covers primarily authors writing in the English language. Content requires constant updating, hence the loose-leaf format. Useful for catalogers and authority file units in libraries that use the Library of Congress Classification. An excellent basis for classifying literature.

UNIVERSAL DECIMAL CLASSIFICATION

DBD2. *Extensions and Corrections to the UDC.* The Hague, Netherlands: UDC Consortium, v. 1, 1988– . Annual. ISSN 0014-5424.

Popularly referred to as "E & C," this publication appears annually and gives notice of the officially authorized changes to the Universal Decimal Classification. In 1995, it reached its seventeenth volume. While the principal objective is the publication of the revisions and other enhancements whereby UDC adapts itself to new developments in the realm of science and culture, recent editions (e.g., 1993, 1994, and 1995) have included articles by users and others involved in the development of the UDC system. For example, v. 17 includes articles on the use of UDC in Estonia, the Czech Republic, Portugal, and Sweden and describes projects such as the "Revision Project Class 52 - Astronomy" and the research being carried out using Class 61 - Medical Sciences to study the feasibility of restructuring UDC into a fully faceted system. Recent changes in UDC tables included here are a new schedule Computer Science, a list of countries with their names in six languages (English, French, Spanish, Dutch, Italian, and German), and a new geographic table for the former Union of Socialist Republics (USSR) reflecting its breakup into separate political entities. A bibliography of publications about UDC during the year is also included. This is the single most authoritative source on the changes and developments taking place in the UDC schedules. Of particular interest to UDC users, library educators and, to some extent, DDC users. As reported by the editor-in-chief, "plans are . . . in the formative stage for some limited cooperation with the *Dewey Decimal Classification*." Meetings have been taking place between the two editors.

FACETED CLASSIFICATION SYSTEMS

DBD3. Beghtol, Clare. " 'Facets' as Interdisciplinary Undiscovered Public Knowledge: S. R. Ranganathan in India and L. Guttman in Israel." *Journal of Documentation* 51 (Sept. 1995): 194–224.

This paper discusses the concept of "facet" as an example of interdisciplinary public knowledge. As described by the author, "facets were central to the bibliographic classification theory of Ranganathan, mathematician and librarian, and to the behavioral research of Louis Guttmann, sociologist." The term "facet" has the same meaning in both fields and the common concept was developed and exploited in both at about the same time. Yet two separate unconnected bodies of literature are grouped around the term and its associated concepts. This paper examines the origins and parallel uses of the concept as a case study in interdisciplinary knowledge. The significance of the coincidental discovery of facet analysis by Ranganathan and Guttman lies in the use of the same term for the concept and in the evolution of major theoretical structures in separate research literatures, each without reference to the other. The discovery of facet concepts has served similar functions in both disciplines and corresponds to work by others in the discovery or creation of new knowledge by information retrieval and classification. Recent research has begun to look at the recognition of the concept of facets in other disciplines. This research has shown that the development of classification systems is being discovered independently of bibliographic classification theory. While the social sciences, in general, were receptive to conceptual borrowing, library and information science was relatively isolated as a discipline. In a period when interdisciplinarity is gaining momentum, this new and refreshing approach to the study of faceted classification has further implications for the study of the structure of knowledge and the design of information systems. Of primary interest to classification theorists, systems designers, students, and teachers of classification.

SPECIAL CLASSIFICATION SYSTEMS

DBD4. Grund, Angelika. "*ICONCLASS:* On Subject Analysis of Iconographic Representations of Works of Art." *Knowledge Organization* 20, no. 1 (1993): 20–29.

The special classification system *ICONCLASS*, created by Henry de Waal for the description of occidental art, is described against the background of art-historical iconography. By means of examples, the structure and use of *ICONCLASS* and its importance for art-historical documentation are explained. A very good basic article on the subject.

DC
Subject Headings and Related Systems

DCA
General Works

DCA1. Berman, Sanford, ed. *Prejudices and Antipathies: A Tract on the LC's Subject Headings Concerning People.* 2nd ed. Jefferson, N.C.: McFarland, 1993.

While described as a new edition, the text of this work is virtually unchanged from the first edition published in 1971. There is a new foreword by

Eric Moon and a 1993 preface by Berman. Some corrections and added editorial matter; the major change is in physical format—this edition being in paperback form. This book has its critics, but will be of continued interest to catalogers and other librarians who feel strongly about what they believe is a slow pace of change in the Library of Congress Subject Headings. Of limited value in day-to-day cataloging, it has made a significant, if controversial, contribution to the development and growth of subject heading terminology.

DCA2. Drabenstott, Karen M., and Diane Vizine-Goetz. *Using Subject Headings for Online Retrieval: Theory and Practice and Potential.* San Diego, Calif.: Academic Press, 1994.

A major contribution to the use of Library of Congress Subject Headings in online retrieval. The authors describe subject analysis and subject searching in online catalogs, discuss their limitations in retrieval, and demonstrate how such limitations could be overcome through systems design and programming to enable a system to respond to users' queries with the subject search method most likely to succeed. The introductory chapter provides a general overview of subject analysis with particular refererence to the characteristics of the LCSH system. Chapters 2 and 3 focus on subject information in authority files and bibliographic records, followed by Chapter 4 which deals with the relationships between these files. The next five chapters focus on the subject queries users enter into online bibliographic systems, in terms of the kinds of queries that are likely or unlikely to succeed and queries that are exact matches to subject headings, as well as approaches available for searching and their appropriateness for users. Chapter 13 focuses on the use of search trees to provide defined paths that follow routes most likely to succeed. The final chapter is a "think piece" on the future of subject headings in online information retrieval. Each chapter has an introductory overview and concludes with a "synthesis and summary" and supporting references. Many illustrative examples. This definitive work is an important resource for subject catalogers, online searchers, library educators, and students on the whys and wherefores of subject searching in online catalogs.

DCA3. Yee, Martha M., ed. *Headings for Tomorrow: Public Access Display of Subject Headings.* Edited by Martha Yee. Chicago: American Library Association, 1992.

A short guide to aid librarians and OPAC designers in making decisions about the design and display of lists of subject headings. The authors offer guidance rather than standards or prescription; includes recommendations. Distinguishes between the structural approach and the strictly alphabetical approach to organization of headings with examples and points out arguments for each. The major portion of the contents focuses on organization and display of various kinds of headings including the display of: subdivisions, inverted headings, qualified headings, and topical subjects interfiled with names and titles. Also addresses punctuation of headings and information for users. Within each chapter, the two basic approaches are addressed and brief statements of current practice are given based on existing OPACs. Includes extensive examples demonstrating the choices together with arguments for and against each. The authors state that the best subject heading display on the OPAC screen depends on the setting and the user's understanding of the system.

STANDARDS

DCA4. Bloomfield, Masse. "A Look at Subject Headings: A Plea for Standardization." *Cataloging & Classification Quarterly* 16, no. 1 (1993): 119–23.

This article arose out of the author's previous experiences in using subject headings and work for his book, *How to Use a Library: A Guide to Literature Searching* (2nd ed. Canoga Park, Calif.: Wasefield Books, 1991). In carrying out actual searches to support the discussion in his book, he observed that subject heading systems and their application have not improved greatly over the past twenty-five years. Examples from library catalogs, periodical indexes, and abstracting services suggest topics not adequately served, variant treatment of similar topics (in the same tool), and variations in the same subject across tools. The author had hoped, with the advent of computer technology, for standardization of terminology across tools—an event that has yet to happen. He identifies, correctly, a lack of communication among various responsible bodies and readily admits his desire for standardization may only be a dream.

DCA5. Studwell, William E. "Ten Years after the Question: Has There Been an Answer?" *Cataloging & Classification Quarterly* 20, no. 3 (1995): 95–98.

A very brief article that brings up to date the progress toward improvements in the principles and practices of subject heading systems (in particular LCSH) since 1985 when the article "Why Not An AACR for Subject Headings?" was published. Studwell identifies briefly the views of various experts in the field who responded to the challenge between 1985 and 1994. From his investigation, the author concludes that while no code has emerged, "considerable progress has been made towards the goals of a code." However, he feels much is still to be done if complete standardization of subject headings is to be achieved—or can it be achieved? (See Bloomfield's article **DCA4**). The bibliography provides a fast method of locating articles on this topic for anyone interested in following developments over a period of years.

DCA6. *USMARC Format for Authority Data, Update No. 1.* Washington, D.C.: Library of Congress, Cataloging Distribution Service, 1995.

Update to the 1993 edition, containing new and changed USMARC data elements resulting from format development work since June 1993. The most notable additions to the format are several field-level data elements for genre-form terms and subdivisions. The new fields make it possible to differentiate within a USMARC authority record between genre-form and other types of terms. Other newly defined data elements include fields for government classification/call numbers and indicators in two classification number fields that allow the identification of the source of a number.

DCB
Library of Congress Subject Headings

GENERAL WORKS

DCB1. Thurlow, H. *"LCSH* Online at the State Library of Queensland." *Cataloguing Australia* 19, nos. 3–4 (1993): 253–58.

This paper outlines the processes undertaken to incorporate data from the weekly LCSH tapes into the State Library's existing authority system. A brief overview of the system is given, followed by a discussion of the technical and practical procedures in processing the tapes. The library has found it easier to maintain control of the reference structure and the system is flexible enough to permit changes in headings to reflect "Australian" terminology. A normalizing key matching process is used to avoid duplications. The use of LCSH in this manner has advantages in that it automatically provides validation for new headings, a reference structure, and permits the identification of nonstandard subjects. Among the disadvantages are the need for programming expertise and the need to be able to "massage" LCSH headings into the existing system. A very brief, but interesting insight into the problems of adapting terminology from one country to another, even if the terminologies are in the same language.

TEXTBOOKS, GUIDES, AND MANUALS

DCB2. Chan, Lois Mai. *Library of Congress Subject Headings: Principles and Application.* 3rd ed. Englewood, Colo.: Libraries Unlimited, 1995.

A much needed update of an essential textbook and support tool on Library of Congress Subject Headings. Changes reflect developments since the second edition was published in 1986. Takes into consideration the expanded use of LCSH in North America and abroad, not only in library catalogs and bibliographies, but also in online databases. Focuses on the current system. Part I deals with historical background and basic principles; Part II covers the applications of LCSH to MARC records including the assignment of subject headings in general and the treatment of certain types of materials in particular. Part III looks at future prospects for LCSH as an online retrieval tool—addressing an agenda for change, application policies, and online catalog design. A major change is the move from the "see," "see also," "x," and "xx" to the symbols USE, UF, BT, NT, RT, and SA in the examples of cross-references.

Extensive appendixes include examples of USMARC authority records, several of free-floating subdivisions, and the USMARC coding for subject heading information. Free-floating subdivisions are updated frequently in LC's *Subject Cataloging Manual: Subject Headings,* so the *SCM* should be consulted for the most up-to-date information. An essential tool for catalogers, library, and information science educators and students.

LISTS AND FILES

DCB3. Library of Congress, Office for Subject Cataloging Policy. *LC Subject Headings: Weekly Lists.* Library of Congress, Washington, D.C.: Cataloging Distribution Services, 1993– . URL: gopher://marvel.loc.gov:70/11/services/cataloging/weekly/

The weekly list of new and altered subject headings reviewed and approved by the Office of Subject Cataloging Policy is now available electronically via weekly posting on LC Marvel.

PROBLEMS IN LCSH SPECIAL SUBJECT AREAS

DCB4. Clack, Doris H. "Subject Access to African American Studies Resources." *Cataloging & Classification Quarterly* 19, no. 2 (1994): 49–66.

The author investigates access to African studies resources in light of research carried out in 1973, in which she found subject analysis of materials in this specialized subject area to be woefully inadequate. Access to these materials is examined through Library of Congress Subject Headings in online catalogs, considering such aspects as relevance of terms to the literature, specificity of terminology, and predictability of structure. Findings suggest progress has been made, but problems still remain. Clack recommends more attention to the problems of ethnicity in catalogs through the provision of additional free-floating subdivisions, the use of parenthetical expressions, and general references relative to ethnicity. An important contribution to the literature.

DCB5. Franz, Lori, John Powell, Suzann Jude, and Karen M. Drabenstott. "End-User Understanding of Subdivided Subject Headings." *Library Resources & Technical Services* 38 (July 1994): 213–26.

Report of a study that considered end-user understanding of subdivided subject headings in their current format and in the form proposed by the first recommendation of the LC Subject Subdivisions Conference. Catalog users interpreted the meanings of subject headings in the same manner as catalogers about 40 percent of the time for current forms and about 32 percent of the time for proposed forms. Authors conclude with recommendations regarding the importance of considering the implications of "simplification."

DCB6. **Gitisetan, Darrin.** "Subjects of Concern: Selected Examples Illustrating Problems Affecting Information Retrieval on Iran and Related Subjects Using *LCSH*." *Cataloging & Classification Quarterly* 20, no 3 (1995): 43–56.

Discusses the specific problem of United States bias in LCSH, using "Iran" as a particular case. The author identifies the fact that LCSH problems "stem from its not being a true thesaurus." The lack of a subject heading code and the suitability of LCSH for computerized catalogs also are cited as issues. Gitisetan considers four important aspects related to Iran—the land, people, language, and literature. Among the conclusions drawn are the needs for faster response to multicultural and ethnolinguistic topics, improvements in these areas, and improved syndetic structure and a better understanding of the idiocyncrasies of LCSH. Problems discussed are typical of problems faced by library and information science professionals in countries other than the U.S. when they elect to adopt LCSH as a subject cataloging tool. These problems are also important to American libraries with substantial collections of materials from other countries.

DCB7. **Miller, David.** "Ambiguities in the Use of Certain Library of Congress Subject Headings for Form and Genre Access to Moving Image Materials." *Cataloging & Classification Quarterly* 20, no. 1 (1995): 83–104.

Some Library of Congress Subject Headings have the potential to function as subjects and as form/genre in catalogs. This dual role is described as "confusing and inadequate." LCSH terms that were cognates with terms from *Moving Image Materials* were searched in OCLC's Online Union Catalog to determine how often these terms were used for each of the two types of access. Subject terms were used as form-genre headings approximately 53 percent of the time. The result would be a large number of cases in which users would be likely to retrieve form-genre examples as well as books about them. The author foresees the need for local solutions, but calls for an international approach to the problems. This is a topic that is discussed increasingly among those interested in improving subject access to systems that more and more contain not only books and serials, but also various other media.

DCB8. **Rogers, Margaret N.** "Are We on Equal Terms Yet? Subject Headings Concerning Women in *LCSH*, 1975–1991." *Library Resources & Technical Services* 37, no. 2 (April 1993): 181–96.

Rogers traces the evolution of LC subject headings concerning women from *LCSH 8* through *LCSH 14*. She identifies improvement and problem areas (including areas in which gender biases remain), and suggests solutions.

DCC
Other Subject Heading Lists

DCC1. **Cimino, James J.** "Vocabulary and Health Care Information Technology: State of the Art." *Journal of the American Society for Information Science* 46 (Dec. 1995): 777–82.

Controlled medical vocabularies are an essential component of almost all health care computing applications. This article reviews the vocabularies available today and some reasons why they have failed to meet the needs of application developers. Among the vocabularies discussed are the *International Classification of Diseases*, published by the World Health Organization, *Medical Subject Headings (MeSH)*, and the *Systematized Nomenclature of Human and Veterinary Medicine—SNOMED International*, along with nomenclatures in the areas of pathology, diseases and operations, and medicine. There is a brief overview of current research on vocabulary problems in the discipline and the remaining challenges yet to be addressed. Cimino calls for continued efforts to improve the design, construction, and maintenance of health care vocabularies. Of particular interest to library and information science professionals in the health sciences.

DCC2. Fountain, Joanna F. *Headings for Children's Materials: An LCSH/Sears Companion.* Englewood, Colo.: Libraries Unlimited, 1993.

The source for Library of Congress juvenile subject headings, designed "to ease the complex and often difficult process of automating the library catalog or changing from *Sears List of Subject Headings* to *Library of Congress Subject Headings (LCSH)*" as the mode of subject authority control. The list is gathered from a database of more than 20,000 subject headings used for children's literature in the libraries of a large school district processing center over ten years. It has been supplemented with names from the LC's Authority File and incorporates all modifications of LCSH up to the time of publication. Terms come from three sources of supply: LCSH itself, the LC's Annotated Card program (AC), and *Sears*. Includes a complete list of subdivisions and guidelines for using the headings. Employs extensive cross-referencing for ease of use. Explanations are clear and intelligible to the non-expert subject cataloger with a minimum of training. Of primary use for public, school, and academic librarians and catalogers and anyone using or migrating to *Library of Congress Subject Headings for Children's Materials* and in courses in cataloging and organization of materials.

DCC3. Greenberg, Jane. "Intellectual Control of Visual Archives: A Comparison between the *Art & Architecture Thesaurus* and the *Library of Congress Thesaurus for Graphic Materials.*" *Cataloging & Classification Quarterly* 16, no. 1 (1993): 85–117.

As background, Greenberg begins with a discussion of the nature of visual archives and the application of archival control theory to graphic materials. The major difference between the two tools was found to be the difference in structure—the faceted approach used in *AAT* versus the LCSH-like structure of the Library of Congress *TGM* **(EE12)**. Identifies advantages and disadvantages of the two tools. The conclusion recognizes the need to understand the differences between thesauri and subject heading lists and the need for further investigation and understanding of visual archives in the computerized environment. Extensive bibliographical notes. An important analysis for information professionals dealing with this type of material.

DCC4. Pelzer, N. L. "Veterinary Subject Headings and Classification: A Critical Analysis." *Cataloging & Classification Quarterly* 18, no. 2 (1993): 3–18.

In this study, monographic titles in the Iowa State University Library with the classification range of Veterinary Medicine were analyzed for the appropriateness of assigning subject headings that would fit in with the existing LCC schedule S (Agriculture). Suggestions are made for the improvement of access to monographic titles in the field of Veterinary Medicine which might be provided in LCSH and the LC classification scheme.

DCC5. *Repertoire de vedettes-matiere (RVM).* 10th microfiche ed. Quebec: Bibliotheque de L'universite Laval, 1993.

French language equivalent of *Library of Congress Subject Headings* developed over a period of more than forty years by L'universite Laval. Used by French language and bilingual libraries throughout Canada and Europe. Also forms part of the authorities database of France, RAMEAU (Repertoire d'authorite matiere encyclopedique et alphabe unifie). This essential authority list of French language headings was inadvertently omitted from *GTSR*.

DCC6. Winkel, Lois, ed. *Subject Headings for Children: A List of Subject Headings Used by the Library of Congress with Dewey Numbers Added.* 2 vols. Albany, N.Y.: OCLC Forest Press, 1994.

Intended to fill a long-standing need to reconcile the language of subject headings with the language of classification. Makes it possible for catalogers of children's material to consult, in a single list, all LC's juvenile subject headings paired with the Dewey numbers that go with them. Volume 1 is a *List of Headings* as they appear on the LC's bibliographic records (e.g., headings with subheadings) with recommended Dewey numbers identified. Volume 2 is a *Keyword Index* that can be understood and used by children, parents, teachers, and librarians to locate quickly headings for materials sought. Listing includes all significant words in main headings and the first word of each subdivision. Recommended for use in school and public libraries that use LC's juvenile headings.

DCD
Subject Authority Control

DCD1. Wilkes, Adeline, and Antoinette Nelson. "Subject Searching in Two Online Catalogs: Authority Control vs. Non-Authority Control." *Cataloging & Classification Quarterly* 20, no. 4 (1995): 57–79.

This study addresses the much debated question, "Is authority control necessary?" Using transaction logs from a catalog with no authority control to analyze search patterns of users, the authors found that 40 percent of the subject searches were unsuccessful. When "see" references from the authority-controlled catalog were applied to the unsuccessful searches in the catalog with no authority control, they discovered that 73.9 percent of the searches that appeared to fail would have retrieved "at least one and usually many records" if a link had been provided between users' terms and system terms. Wilkes and Nelson conclude that authority control is a very important tool in online catalogs and should be given serious consideration for inclusion in the design of any online catalog. Of interest to all information professionals who are designers of online catalogs.

DCE
Thesauri

GENERAL WORKS

DCE1. Milstead, Jessica L. "Invisible Thesauri: The Year 2000." *Online & CDROM Review* 19, no. 2 (1995): 93–94.

Brief, interesting "think piece" on the possible characteristics of thesauri of the future. Author predicts that the thesaurus will remain as a subject access tool in the year 2000, but that it may be a very different kind of tool than the one familiar today. While the design of the thesaurus has remained relatively stable, with the advent of full text databases, there is still a need for semantic assistance for effective retrieval. Small databases still may have human indexers, but large database designers will be looking at more machine-aided indexing (MAI). Today, these systems use an equivalence translation dictionary and a separate thesaurus. Milstead sees these two operations being brought together in future. Text analysis techniques to aid indexing will be used to compile thesauri and thesauri may be more likely to become retrieval tools as opposed to indexing tools. They would function as tools for automatic analysis invisible to users and serve to aid users in stating searches more precisely. Relationships among terms in thesauri may become more complex in machine systems, but this is doubtful. Milstead believes that broad categories are more likely and that cost will be a factor. How will they look in 2000+? Print and electronic versions will still be quite common. It is the use rather than the look that will change—blending machine-aided indexing and text retrieval systems, with boundaries between thesauri and other semantic tools becoming more vague. Defining the search will receive more attention in the future. Milstead concludes that, with luck, "the thesaurus may become almost invisible to most users."

DCE2. Molholt, Pat, and Toni Petersen. "The Role of the *Art & Architecture Thesaurus* in Communicating about Visual Art." *Knowledge Organization* 20, no. 1 (1993): 30–34.

Addresses the ways in which computerization and a thesaurus such as the *AAT* influence organization, description, and understanding of the visual arts. The authors discuss the ways in which the structure and content of the *AAT* serve as a binding mechanism between the many manifestations of the visual arts and the different types of organizations serving viewers. They suggest that a knowledge base such as the *AAT* enables various users to develop approaches to accessing visual art and that special nature of images influences the way in which the thesaurus is structured. The authors have recognized that a two-step process is involved: providing the means of communication (i.e., the thesaurus) and facilitating the understanding of it. An excellent overview of this topic for those interested in the nature and development of *AAT*.

STANDARDS

DCE3. American National Standards Institute. *Guidelines for the Construction, Format, and Management of Monolingual Thesauri: An American National Standard.* Bethesda, Md.: NISO Press, 1994. ANSI/NISO Z39.19-1993.

Bears little resemblance to the 1980 *Guidelines for Thesaurus Structure, Construction and Use* which it replaces. The first edition was a modest item of ten pages, while the expanded new edition is an impressive document supplemented by an extensive glossary, a large number of useful non-ambiguous examples, and several informative appendixes. The word "use" which appeared in the title of the previous version has been removed from the current title as the scope of the standard does not provide recommendations for "use" after a thesaurus has been created—although it occasionally makes reference to "indexing." While there is little formal resemblance to the previous edition, this standard does operate within the status quo. The primary difference from the 1980 edition is that it brings the American standard closer in contents to the widely known and used ISO 2788-1986 *(Guidelines for the Establishment and Development of Monolingual Thesauri)*. The main sections of the new standard, for the most part, follow the ISO standard with some internal reorganization. Important in this standard is its attention to the potential role of machines in thesaurus development. Sections 7 (Screen Display) and 20 (Thesaurus Management System) are original with this standard. As standards go, this one is well and clearly written and contains a plethora of excellent examples. An essential tool for thesaurus and database designers, library and information science educators, students, and practitioners.

LISTS AND FILES

DCE4. Milstead, Jessica, ed. *ASIS Thesaurus of Information Science and Librarianship.* Medford, N.J.: Published for the American Society for Information Science by Learned Information, Inc.,1994.

In this thesaurus, the editor "seeks to cover the fields of library and information science." Related fields are also covered to the extent that the editor feels they are warranted. Includes such fields as computer science, linguistics, and the behavioral and cognitive sciences. However, it cannot be expected to be as strong in these peripheral fields. Provides guidelines for use and gives sources used in compiling the list. Contents include a "facet list" indicating on one page the areas covered in the field of knowledge, an alphabetical display with relationships among terms, a hierarchical display, and a rotated display. Use of the thesaurus indicates that it has some limitations which are largely a result of the limitations of the *Lui-Palmer Thesaurus Construction System* **(DCE8)** software used to compile it. An important tool for library and information science programs, as well as being a useful compilation of the terminology for the library and information science professions.

TEXTBOOKS, GUIDES, AND MANUALS

DCE5. Petersen, Toni, and Patricia J. Barnett, eds. *Guide to Indexing and Cataloging with the* Art & Architecture Thesaurus. New York: Oxford University Press, published on behalf of the Getty History Information Program, 1994.

A guide and companion to a major subject thesaurus—the *Art & Architecture Thesaurus*. Part I describes the methodology and conventions employed in the construction of the thesaurus as necessary background for application of *AAT* terms in indexing and cataloging. In this section, four chapters cover *AAT* editorial principles and conventions, the process of indexing with *AAT*, application protocol related to geographic names and dates, USMARC coding and MARC/*AAT* fields, and local policy and procedural issues. Part II focuses on special format considerations with particular reference to archives, books, objects, and visual resources. Part III contains cataloging examples for a variety of types of materials. Glossary of terms, bibliography of supporting documents, and subject index. Especially valuable comparisons of *AAT* and the Library of Congress Subject Headings and the "*AAT* Field Guide to USMARC Coding." An essential tool for libraries, museums, and information centers using the *AAT*; also of value to other institutions in the process of developing thesauri for their collections. If the guide is to have heavy use, it may have to be rebound since the ring binding may not stand constant handling. References and index.

THESAURUS DESIGN AND CONSTRUCTION

DCE6. Chen, Hsinchun, et al. "Automatic Thesaurus Generation for an Electronic Community System." *Journal of the American Society for Information Science* 46, no. 3 (1995): 175–93.

Research on the automatic generation of a thesaurus using term filtering, automatic indexing, and cluster analysis techniques. Test data are derived from a comprehensive library of specialized community data and literature currently in use by molecular biologists. The resulting thesaurus contains numerous researchers' names, gene names, experimental methods, and 4,302 topical subject descriptors. The thesaurus is used as an online search aid. Tests demonstrated that the resulting thesaurus is an excellent "memory jogging" device supporting learning and serendipitous browsing. A highly technical article, but indicates that important developments are taking place. References. For annotation, *see also* **EE5**.

DCE7. Hudon, Michèle. *Le Thésaurus: Conception, elaboration, gestion.* Clé en main. Montréal: Les Éditions ASTED, 1994.

A new practical manual on the thesaurus, its characteristics, and its construction. In ten chapters, Hudon defines and characterizes the thesaurus and compares it with other sources of terminology. Separate chapters on semantic structure of thesauri, decisions related to the creation of a new thesaurus, needs analysis and specifications, and actual construction and validation of the thesaurus. One chapter looks as multilingual thesauri. Glossary, index, and numerous examples throughout the text. Very useful bibliographies of both English and French sources. Most useful to thesaurus designers and library and information science educators, students, and practitioners with a reading knowledge of the French language, but this is easy reading and has much to offer English-speaking professionals.

DCE8. *Lui-Palmer Thesaurus Construction System.* Professional Ed. Release 2.1. Software. Los Angeles: Lui-Palmer, 1994.

This well-received thesaurus construction software package replaces a "basic edition" that was first produced in 1989. Menu-driven and easy to use. Accompanied by a manual. Output provides for three types of display: an alphabetical display that lists each term with its scope note, Broader Terms, Narrower Terms, and Use For relationships; a hierarchical display showing terms and facet indicators in an indentational format just as they are displayed on the screen; and a rotated display (KWIC format) providing access on every word in each significant word in a term string. The system provides for an unlimited number of hierarchies and references and permits the establishment of polyhierarchies. Reciprocals of references are created automatically. A major weakness is the hierarchical structure built into the construction process. Nevertheless, one of the best software packages of its kind, superior to manual methods of construction, and ensures greater accuracy in terms of reciprocal relationships. The *ASIS Thesaurus of Information Science and Librarianship* **(DCE4)** is an example of a thesaurus created using this software.

DCE9. *MultiTes 5.2.* Software. Miami, Fla.: Multisystems, 1994.

This software package is similar to the *Lui-Palmer Thesaurus Construction System* **(DCE8)**, but is newer and appears to be somewhat more flexible than *Lui-Palmer.* Input of terms is still hierarchical (a limitation), but navigation and the construction of polyhierarchies are easier. Screen displays are an alphabetical list of terms, display of the record of each term and its relationships, and a hierarchical display. Facet indicators are not built into this system. Printed output is of two types—the alphabetical thesaurus and a hierarchical index. Reciprocals are created automatically.

DD
Internet Subject Access

DD1. Micco, Mary. "Subject Authority Control on the World of Internet" Part 1. *ALCTS Newsletter* 6, no. 5 (1995): A–D (supplement).

Micco focuses on authority control as support for searching across the Internet—to aid users in moving from the natural language of a query to the language of the system. She identifies the need for automatic generation of displays that will act as navigation tools to provide access to "information objects" by broad subject area. Suggests that authors of electronic documents should be encouraged to supply subject keywords and "most importantly, broad class numbers" with the help of expert systems software. These data would be refined by professionals. Problems encountered "on a macro scale" in controlling subject access to the Internet would include the information explosion on the Internet, lack of cooperative planning and governance, lack of descriptive standards, lack of a unifying information structure, and cultural clashes among technocrats, scholars, and salespeople. Problems on the micro scale include the inadequacy of existing Internet search tools (described in terms of specific systems) and the restrictions of existing library tools, such as OPAC software and keyword search engines. Effectiveness is also hampered by inadequacies of existing vocabulary control mechanisms to help the user phrase his/her queries and to determine which terms to use, and the lack of navigation tools. In conclusion, Micco calls for rethinking of the whole approach to subject access.

DD2. ———. "Subject Authority Control in the World of the Internet" Part 2. *ALCTS Newsletter* 6, no. 6 (1995): A–D (supplement)

Building on Part 1 **(DD1)**, this article addresses steps that might be taken to improve searching the Internet, including use of classification. As a first step, Micco suggests that "an effective filtering system is needed that will restrict the domain of search relative to general areas while increasing the depth and variety of index terms within that group." In this context, she discusses determining the "aboutness" of documents, identification of the information package, naming of the intended audience, extraction of concepts, assignment of classification numbers, and the weighting terms based on SGML tags. The second half of the paper addresses the development of information management tools (i.e., expert systems) to carry out the process. In this context, discussion focuses on classification systems as a means for "mapping" a subject area and the basis for authority control across the Internet, the desired characteristics for a desirable classification scheme, and the development of the system itself. A fresh approach to the problems of navigating the Internet.

DD3. *NetFirst.* Online database. URL: http//www.oclc.org/oclc/press/960219.htm

As of mid-February 1996, *NetFirst* services, an authoritative directory of Internet-accessible resources, became available on FirstSearch and EPIC,

OCLC's online reference services. The database provides coverage of resources on World Wide Web sites, listservs, and other types of discussion lists, Usenet newsgroups, and anonymous FTP sites. It initially contained 40,000 records, is updated weekly, and is expected to grow by 10,000 records per month. As the database grows, other Internet resources such as electronic journals, newsletters, gopher sites, and library catalogs will be added. The database is selective and includes only significant resources selected by a staff of editors. Each resource is described in a bibliographic-like record which includes an original abstract describing the resource. Each record has both Library of Congress subject headings and Dewey decimal classification numbers, which may be used as access points. A special feature is the inclusion of a DDC number for *each* subject heading. The presence of LCSH allows users to search the database by subject area in addition to keyword searching. DDC numbers organize the resources using a hierarchical structure ideal for browsing of resources by subject area. The database is comprehensive and current; it is produced by people using automated techniques to identify the resources which are then evaluated by the editorial staff. A number of similar databases are being developed. This is, however, the largest and most comprehensive to date and is an excellent example of enhanced subject access to resources. A valuable tool for anyone wishing to search the Internet constructively.

E
Filing and Indexing

Susan Morris

Filing and indexing are complementary activities: filing is the orderly arrangement of records in a catalog, while indexing is the generation of "a systematic guide to the items contained in, or the concepts derived from a collection or database" (*Organizing Knowledge* by Jennifer Rowley, **EAB18**) or "a list of bibliographic information or citations to a body of literature, usually arranged in alphabetical order and based on some specified datum, such as author, subject, or keyword" (*Introduction to Indexing and Abstracting* by Donald and Ann Cleveland, Englewood, Colo.: Libraries Unlimited, 1990, p. 17). In this chapter, "filing" includes screen display for online public access catalogs as well as filing of catalog cards, documents, and so forth. With the growing prevalence of automated library catalogs, the concept of filing has expanded to encompass online catalog displays and the machine filing rules and content designation conventions that support them.

The notion of "indexing" likewise is seemingly infinitely elastic; the term may denote the assignment of controlled vocabulary to facilitate future retrieval of documents, highly exhaustive analysis in conjunction with preparation of abstracts, and Internet "search engines," which aim to retrieve documents using uncontrolled terms found in the text. The works on indexing in this chapter are concerned mostly with indexing in a relatively traditional sense: either the establishment and selection of controlled vocabulary—the "specified data" on which the index is based—or the process of analyzing documents to determine what terms should be assigned to permit the retrieval of the document in the future.

As Ann O'Brien **(EC8)**, echoed by Jessica Milstead **(EAB13)**, has observed, the amount of published research in indexing has dropped precipitously over the past twenty years. This decline is due, in large part, to limited research funding in an uncertain economy, but may signal a redefinition of the concept of indexing under the stimulus of so-called *full-text indexing* (more traditionally called *keyword searching*) and Internet retrieval engines.

Much recent research in indexing and thesaurus construction has been conducted in Great Britain and Europe. The most salient publications are listed in this chapter whether the work they report originated in Europe or in North America, provided the work is relevant to state-of-the-art indexing on the North American continent. Interest in automatic generation of thesauri and computer-aided analysis of documents remains high. The literature also shows renewed research interest in Library of Congress Subject Headings, which has coincided with the British Library's 1993 decision to resume use of LCSH in *British National Bibliography* records.

Standards continued to be an area of intense activity from 1993 and 1995, with the National Information Standards Organization proposing two standards in thesaurus generation and indexing. ANSI/NISO A39.19-1993, *Guidelines for the Construction, Format, and Management of Monolingual Thesauri* **(EB1)**, revises a standard first adopted in 1980. The standard for indexes, Z39.4-199X **(EB4)**, proved controversial and had a difficult journey toward acceptance. As of April 1, 1996, NISO planned to approve it, notwithstanding negative votes by the American Society of Indexers and American Society for Information Science. This proposed standard addresses the indexing of electronic publications and the automated generation of indexes in addition to traditional aspects of indexing.

Chapter Content and Organization

Resources cited in this chapter reflect current trends in filing and indexing in North America. The chapter begins with general works in filing, indexing, directories, and online resources. Subsequent sections address standards and more specialized works on indexing and the library catalog, indexing of special formats and online materials, thesauri and controlled vocabularies, and automated indexing. The latter provides a representative sample of resources addressing indexing and the Internet. For filing rules and manuals, dictionaries, glossaries, bibliographies, periodicals, and organizations as sources of expertise, the relevant sections in the *Guide to Technical Services Resources* remain useful resources.

EA
General Works

EAA
General Works: Filing

EAA1. Buckland, Michael K., Barbara A. Norgard, and Christian Plaunt. "Filing, Filtering, and the First Few Found." *Information Technology and Libraries* 12, no. 3 (1993): 311–19.

Alphabetic filing, the norm in American card catalogs, has carried over to online catalogs, but online catalogs offer many other choices in ordering retrieval sets. The "Prototype for an Adaptive Library Catalog" project at the School of Library and Information Studies, University of California, Berkeley, aims to let users specify preferences in ordering the display of retrieval sets. References.

EAA2. Gill, Suzanne L. *File Management and Information Retrieval Systems: A Manual for Managers and Technicians.* 3rd ed. Englewood, Colo.: Libraries Unlimited, 1993.

This important work bridges the gap between library science and records management for businesses, showing how to apply indexing and classification methods to corporate records management. Like previous editions, it emphasizes creation of a filing procedures manual, with five case studies. Includes general overviews of new technologies: micrographics, personal computers and computer networks, and optical disks. Bibliography, glossary, and index.

EAB
General Works: Indexing

EAB1. Al-Kharashi, Ibrahim A., and Martha W. Evens. "Comparing Words, Stems, and Roots as Index Terms in an Arabic Information Retrieval System." *Journal of the American Society for Information Science* 45 (1994): 458–60.

Reports the results of using the experimental Micro-AIRS System (Microcomputer Arabic Information Retrieval System) to perform 29 searches by root, stem, and word of a database of 355 Arabic bibliographic records at King Abdulaziz University for Science and Technology, Saudi Arabia. Searching by root was judged most effective; searching by word, least effective. References.

EAB2. Blakeslee, Jan. "Indexing Encyclopedias and Multivolume Works." In *Managing Large Indexing Projects: Papers from the Twenty-fourth Annual Meeting of the American Society of Indexers, San Antonio, Texas, May 23, 1992*, 1–10. Port Aransas, Texas: American Society of Indexers, 1994.

Blakeslee discusses management of large indexing projects, based on personal experience. In an era of growing outsourcing, the discussion of bidding, scheduling, indexer-client relations, and selection of indexers should interest many librarians, probably more than will the coverage of indexing protocols and vocabulary control.

EAB3. Chu, Clara M., and Ann O'Brien. "Subject Analysis: The Critical First Stage in Indexing." *Journal of Information Science* 19 (1993): 439–54.

Novice indexers at University of California, Los Angeles, and Loughborough University were asked to perform subject analysis of three articles in science, social science, and the humanities and to identify the primary and secondary topics in natural language. The novices determined the primary topic much more easily for the articles in science and social science than for the humanities text; they performed poorly on all three texts in determining the secondary topic. Five factors influencing the ease of subject analysis were the discipline of the text, whether the text was factual or subjective, complexity of the subject, presence or absence of a bibliographic apparatus, and the clarity of the text. References.

EAB4. Fidel, Raya. "User-Centered Indexing." *Journal of the American Society for Information Science* 45 (1994): 572–76.

Contrasts document-oriented indexing, which attempts to summarize the content of a document, with user-oriented indexing, which anticipates the requests to which a document may be relevant. Calls for more research into searching behavior and automated indexing systems to support user-oriented indexing. References.

EAB5. Fugmann, Robert. "Representational Predictability: Key to the Resolution of Several Pending Issues in Indexing and Information Supply." In *Knowledge Organization and Quality Management: Proceedings of the Third International ISKO Conference, 20–24 June 1994, Copenhagen, Denmark*, edited by Hanne Albrechtsen and Susanne Oernager, 414–22. Frankfurt am Main, Germany: Indeks Verlag, 1994.

Representational predictability is presented as a basis for evaluating and improving the design of information systems, particularly the capabilities of controlled vocabulary vs. natural language text searching. References.

EAB6. Gerhard, Kristin H., Trudi E. Jacobson, and Susan G. Williamson. "Indexing Adequacy and Interdisciplinary Journals: The Case of Women's Studies." *College & Research Libraries* 54 (1993): 125–33.

When articles on women's studies were searched in online and print indexes, fifty-three of eighty-six journals were found to be inadequately indexed. Authors call for a comprehensive online index for women's studies. References.

EAB7. Giral, Angela, and Arlene G. Taylor. "Indexing Overlap and Consistency between the *Avery Index to Architectural Periodicals* and the *Architectural Periodicals Index*." *Library Resources & Technical Services* 37 (1993): 19–44.

A comparison of *Avery Index to Architectural Periodicals* and *Architectural Periodicals Index* showed that 71 percent of the sample articles were indexed by both. The two indexes differ in number and form of access points assigned. References.

EAB8. Hong Yi. "Indexing Languages, New Progress in China." *Knowledge Organization* 22, no. 1 (1995): 30–32.

Overview of indexing language research and practice in China, valuable to readers with no Chinese. Includes list of thesauri and classification schemes in use in China. References.

EAB9. Irving, Holly. "Cait—Computer-Assisted Indexing Tutor: Implemented for Training at NAL." *Agricultural Libraries Information Notes* 21, no. 4–6 (1995): 1–5.

Irving describes "cait," the Windows application developed for the computer-assisted training program introduced by the National Agricultural Library's Indexing Branch in 1995. "Cait" helps novice indexers progress to the journeyman level of expertise in less time and with less investment of training time by experienced indexers. The subject indexing module was developed first, reflecting NAL's immediate need to train more subject indexers for its AGRICOLA database. Mockups of "cait" screens are included.

EAB10. Kleinberg, Ira. "Making the Case for Professional Indexers: Where Is the Proof?" In *Indexing, Providing Access to Information: Looking Back, Looking Ahead, the Proceedings of the 25th Annual Meeting of the American Society of Indexers,* edited by Nancy C. Mulvany, 139–46. Port Aransas, Texas: American Society of Indexers, 1993.

In defense of human indexing, summarizes research on back-of-the-book indexing cited in *Library Literature* from 1988 through 1993, with secondary attention to research on indexing of online documents. Suggests further avenues for research. References.

EAB11. Liddy, Elizabeth D., and Corinne L. Jörgensen. "Reality Check! Book Index Characteristics that Facilitate Information Access." In *Indexing,*

Providing Access to Information: Looking Back, Looking Ahead, the Proceedings of the 25th Annual Meeting of the American Society of Indexers, edited by Nancy C. Mulvany, 125–38. Port Aransas, Texas: American Society of Indexers, 1993.

Describes a user study at the School of Information Studies, Syracuse University, that observed use of back-of-the-book indexes. Use is analyzed for three index variations: an index with no cross-references, a divided name-title and subject index, and an index with few concept words. Users had greater success with the first variation than with the last; their search strategies were unpredictable but showed a learning effect. References.

EAB12. Lunin, Lois F., and Raya Fidel, eds. "Perspectives on . . . Indexing." *Journal of the American Society for Information Science* 45 (1994).

Special section with nine articles presenting both research and opinions on indexing.

EAB13. Milstead, Jessica L. "Needs for Research in Indexing." *Journal of the American Society for Information Science* 45 (1994): 577–82.

Author describes the continuing need for research in indexing, including cognitive processes of indexers and users of indexes; vocabulary control; automated support for indexing; and print and electronic index display. References.

EAB14. Mulvany, Nancy C. *Indexing Books.* Chicago: University of Chicago Press, 1994.

A thorough, easy-to-use guide to creating back-of-the-book indexes, by a leader in the field. Emphasizes practical advice over theory; includes valuable overview of the book production process, indexing standards, and software for indexers. References and index.

EAB15. ———, ed. *Indexing, Providing Access to Information: Looking Back, Looking Ahead, the Proceedings of the 25th Annual Meeting of the American Society of Indexers.* Port Aransas, Texas: American Society of Indexers, 1993.

Four of the fifteen papers in this volume from the 1993 meeting in Alexandria, Va., are distinctly British or Australian in viewpoint; several others are very brief, present a single product or service, or are misleadingly titled.

EAB16. Quinn, Brian. "Recent Theoretical Approaches in Classification and Indexing." *Knowledge Organization* 21, no. 3 (1994): 140–47.

Selective review of recent work in classification and indexing theory, including theoretical work in automated indexing. References. For annotation, see also **DA8**.

EAB17. Robertson, Michael. "Foreign Concepts: Indexing and Indexes on the Continent." *The Indexer* 19 (1995): 160–72.

Robertson compares recent continental European back-of-the-book indexes with the draft International Standard, which reflects British influence. References.

EAB18. Rowley, Jennifer. *Organizing Knowledge.* 2nd ed. Aldershot, Hampshire, England; Brookfield, Vt.: Ashgate, 1992.

A practical guide to all aspects of cataloging and indexing. British orientation is noticeable but should not reduce its value to North American readers. Part III, "Subjects," covers subject analysis, classification, and manual and computer-aided indexing systems. References and index.

EAB19. Salminen, Airi, Jean Tague-Sutcliffe, and Charles McClellan. "From Text to Hypertext by Indexing." *ACM Transactions on Information Systems* 13, no. 1 (1995): 69–99.

Proposes a system in which document indexing is the basis for hypertext markup of the document. References.

EAB20. Savic, Dusko. "Designing an Expert System for Classifying Office Documents." *Records Management Quarterly* 28, no. 3 (1994): 20–29.

Savic describes use of an expert system to classify office documents and also provides a useful general discussion of expert systems technology. References.

EAB21. Soergel, Dagobert. "Indexing and Retrieval Performance: The Logical Evidence." *Journal of the American Society for Information Science* 45 (1994): 589–99.

An analysis of indexing characteristics with a scheme to evaluate the performance of retrieval systems. Recall and discrimination are the basic performance measures. Considers viewpoint-based and importance-based indexing exhaustivity, specificity, correctness, and consistency. References.

EAB22. Sukiasyan, Eduard. "Information-Retrieval Systems: Systems Analysis of Problems of Quality Management." In *Knowledge Organization and Quality Management: Proceedings of the Third International ISKO Conference, 20–24 June 1994, Copenhagen, Denmark,* edited by Hanne Albrechtsen and Susanne Oernager, 27–33. Frankfurt am Main, Germany: Indeks Verlag, 1994.

Applies a systems approach to traditional aspects of indexing quality, such as depth, specificity, and consistency.

EAB23. Tibbo, Helen R. "Indexing for Information." *Journal of the American Society for Information Science* 45 (1994): 607–19.

Tibbo argues that indexing schemes developed to support scientific research are not adequate for indexing research materials used in the humanities. She lists requirements for indexes to humanities material, including nonprint, archival and manuscript collections, and graphic materials. Indexing for humanities research lags behind indexing in the sciences because less funding is available in the humanities and because the materials are less structured and more varied. Tibbo calls for further development of humanities indexing that is not dependent on the scientific indexing model. References are particularly valuable.

EAB24. Weinberg, Bella Hass. "Why Postcoordination Fails the Searcher." *The Indexer* 19 (1995): 155–59.

Examines deficiencies of postcoordinated searching of databases: relationships between terms are not clear; Boolean searching is difficult; too many hits result in lower precision. References.

EAB25. Wellisch, Hans W. "Book and Periodical Indexing." *Journal of the American Society for Information Science* 45 (1994): 620–27.

Wellisch presents the basic functions of indexing and considers classic issues of policy, indexing languages, depth, naming, and format. He argues that fully automatic indexing systems cannot perform all the basic indexing functions.

EAB26. Wormell, Irene. "Subject Access Redefined: How New Technology Changes the Conception of Subject Representation." In *Knowledge Organization and Quality Management: Proceedings of the Third International ISKO Conference, 20–24 June 1994, Copenhagen, Denmark,* edited by Hanne Albrechtsen and Susanne Oernager, 431–39. Frankfurt am Main, Germany: Indeks Verlag, 1994.

Reviews the progress of the last ten years in subject access. Contrasts the consequences of using "physical document" or "semantic entity" as the basic construct. Considers the impact of trends in electronic publishing, including Standard Generalized Markup Language (SGML). References.

EAC
Directories

EAC1. *The Index and Abstract Directory: An International Guide to Services and Serials Coverage.* Birmingham, Ala.: EBSCO, 1989– . Irregular. ISSN 1041–1321.

This directory is now a serial, issued irregularly. The third edition, issued in 1994, lists 56,000 serial titles and about 1,300 abstracting and indexing services, about 900 of them "active." Listings include price, publisher, format, coverage, etc.

EAC2. *Register of Indexers Available.* Toronto: Indexing and Abstracting Society of Canada/Société canadienne pour l'analyse de documents, 1991– . Annual. ISSN 1199-004X.

Lists freelance indexers; text in English and French.

EAD
Sources of Expertise

ELECTRONIC DISCUSSION GROUPS

EAD1. *AAT-L: Art & Architecture Thesaurus Discussion List.* Electronic discussion group.

A forum established to facilitate communication between the *Art & Architecture Thesaurus* office and thesaurus users. Several postings weekly. To subscribe, send a message to LISTSERV@LISTSERV.UIC.EDU with the command SUBSCRIBE AAT-L [your name]. Send messages to AAT-L@LISTSERV.UIC.EDU

EAD2. *ASIS-L.* Electronic discussion group. List owners: Merri Beth Lavagnino, Debbie Lords, David Carlson, Jennifer Bates, and Cassandra Brush.

Unmoderated electronic discussion group for members of the American Society for Information Science. Heavy traffic. To subscribe, send a message to LISTSERV@ASIS.ORG with the command SUBSCRIBE ASIS-L [your name]. Send messages to ASIS-L@ASIS.ORG

EAD3. *AUTOCAT: Library Cataloging and Authorities Discussion Group.* Electronic discussion group. List owner: Judith Hopkins, State University of New York at Buffalo.

Best known as an electronic discussion group for descriptive catalogers, AUTOCAT also occasionally carries discussions of particular interest to the indexing community. A recent discussion thread dealt with librarians' reputation and status as indexers and classifiers. Frequently carries cross-postings from more specialized lists that may touch on thesaurus construction, automated indexing, etc. Very heavy traffic. To subscribe, send a message to LISTSERV@UBVM.CC.BUFFALO.EDU with the command SUBSCRIBE AUTOCAT [your name]. Send messages to AUTOCAT@UBVM.CC.BUFFALO.EDU

EAD4. *INDEX-L.* Electronic discussion group. List owner and moderator: Charlotte Skuster, Binghamton University Science Library.

Moderated discussion group for indexers at all levels of expertise; welcomes discussion of indexes of all kinds. To subscribe, send a message to LISTSERV@BINGVMB.CC.BINGHAMTON.EDU with the command SUBSCRIBE INDEX-L. Send messages to INDEX-L@BINGVMB.CC.BINGHAMTON.EDU. Archives available at URL: gopher://gopher.gasou.edu; follow the path: Georgia Southern University/Henderson Library/Other Organizations/INDEX-L.

EAD5. *INTERCAT.* Electronic discussion group.

Unmoderated electronic discussion group for communications about OCLC's project "Building a Catalog of Internet-Accessible Resources," or

INTERCAT, which was started with support from the U.S. Department of Education and now includes 225 participating libraries. Notable recurring threads include the comparative value of bibliographic cataloging for World Wide Web documents versus full-text retrieval engines featuring keyword searching. Unmoderated; archives. Heavy traffic (six or more messages each day). To subscribe, send a message to LISTSERV@OCLC.ORG with the command SUBSCRIBE INTERCAT [your name]. messages to INTERCAT@OCLC.ORG. Archives available at URL http://ftplaw.wuacc.edu/listproc/intercat/archive.html

ELECTRONIC PERIODICALS

EAD6. *LCCN: LC Cataloging Newsline.* Serial online. Washington, D.C.: Library of Congress, Cataloging Directorate, v. 1, Jan. 15, 1993– . Quarterly (irregular). ISSN 1066-8820.

For annotation, *see also* **CAD3**. Covers all aspects of cataloging activities at the Library of Congress. *LCCN* generally includes brief notices of changes in subject cataloging policy, projects to change groups of headings in Library of Congress Subject Headings, developments in automation at LC, etc.

WORLD WIDE WEB SITES AND HOME PAGES

EAD7. **American Society of Indexers (ASI).** URL: http://www.well.com/users/asi

This home page provides information about the society, indexing, meetings and educational opportunities, other organizations, and links to online resources of interest to indexers. The list of Internet indexes and search engines is particularly interesting.

EAD8. **Indexing and Abstracting Society of Canada (IASC). Société canadienne pour l'analyse de documents (SCAD).** URL: http://tornade.ere.umontreal.ca/~turner/iasc/home.html

Information about IASC, related organizations, and hot links to sites of interest to indexers.

EB
Standards

EB1. **American National Standards Institute.** *Guidelines for the Construction, Format, and Management of Monolingual Thesauri: An American National Standard.* Bethesda, Md.: NISO Press, 1994. ANSI/NISO Z39.19-1993.

This revision of *American National Standard Guidelines for Thesaurus Structure, Construction, and Use, ANSI Z39.19-1980* was developed by the National Information Standards Organization and approved by ANSI, August 30, 1993. References and index. For annotation, *see also* **DCE3**.

EB2. Anderson, James D. "Standards for Indexing: Revising the American National Standard Guidelines Z39.4." *Journal of the American Society for Information Science* 45 (1994): 628–36.

Discusses the 1993 draft of the *Proposed American National Standard Guidelines for Indexes and Related Information Retrieval Devices*, ANSI/NISO Z39.4 **(EB4)**, and presents the seven codrafters' professional backgrounds. Syntax, vocabulary management, and comprehensive planning and design are the three fundamental requirements for a good index. Looks at changes in the indexing environment, including the growth of automated indexing, and contrasts the roles of expert opinion, as expressed in advisory standards like Z39.4, and empirical research in moving a profession forward. References.

EB3. International Organization for Standardization. *Information and Documentation—Guidelines for the Content, Organization, and Presentation of Indexes: Draft International Standard, ISO/TC46/SC9*. Geneva, Switzerland: International Organization for Standardization, 1993. ISO/DIS 999.

Largely oriented to British indexing practice, this standard is of interest for comparative purposes. It addresses only indexes to books, periodicals, and other materials in print media.

EB4. National Information Standards Organization. *Proposed American National Standard Guidelines for Indexes and Related Information Retrieval Devices*. Bethesda, Md.: NISO, 1993. ANSI/NISO Z39.4-199X.

Drafted in 1993, this proposed standard would replace *American National Standard for Library and Information Sciences and Related Publishing Practices: Basic Criteria for Indexes, ANSI Z39.4-1984* for indexes to all kinds of documents in all formats and media. Major sections cover the scope of the standard; definitions; functions, types, and design of indexes; vocabulary; headings, entries, and search statements; display of index arrays; rules for alphanumeric arrangement; and relations between indexers, authors, and publishers. There are also four guides to the standard for print indexes to single documents, database and other continuing indexes, automatic indexing, and nondisplayed indexes designed for electronic searching. NISO intended to approve the revised standard during 1996; as of May 1, the draft was going through a second ballot. For information about its status, telephone NISO at 301-654-2512 or access the NISO World Wide Web page (URL: http://www.niso.org).

EC
Indexing and the Library Catalog

EC1. **Allen, Bryce.** "Improved Browsable Displays: An Experimental Test." *Information Technology and Libraries* 12 (1993): 203–8.

Allen reports a test of an online browser interface that substantially reduced the amount of scanning required to find subject headings. Subdivisions are suppressed in the initial display, but are displayed on demand. The browser interface did not affect search effectiveness, but made browsing more efficient. References.

EC2. **Brooks, Terence A.** "People, Words, and Perception: A Phenomenological Investigation of Textuality." *Journal of the American Society for Information Science* 46 (1995): 103–15.

This useful contribution to the literature on catalog user behavior reports two experiments that studied how textual factors influenced users' perception of bibliographic records. In one experiment, subjects were asked to match ERIC subject descriptors to ERIC database records; in the second, users were asked to rank the relevance of subject descriptors to bibliographic records. References.

EC3. **Cherry, Joan M., et al.** "OPACs in Twelve Canadian Academic Libraries: An Evaluation of Functional Capabilities and Interface Features." *Information Technology and Libraries* 13 (1994): 174–95.

The twelve Canadian online public access catalogs evaluated in this study performed best in the area of online screen display and worst in the area of help in subject searching. Authors propose a list of features of an ideal OPAC, which is used to assess the twelve existing ones. References.

EC4. **Kilgour, Frederick G.** "Effectiveness of Surname-Title-Words Searches by Scholars." *Journal of the American Society for Information Science* 46, no. 2 (1995): 146–51.

For annotation, *see* **CBF7**.

EC5. **Klatt, Mary J.** "An Aid for Total Quality Searching: Developing a Hedge Book." *Bulletin of the Medical Library Association* 82 (1994): 438–41.

Describes the subject hedge, a list of terms that supplement controlled vocabulary terms. A list of subject hedges, or hedge book, is a useful aid for online searching in medicine. References.

EC6. Matthews, Joseph R. "The Distribution of Information: The Role for Online Public Access Catalogs." *Information Services & Use* 14 (1994): 73–78.

Author argues that the commercial abstracting and indexing community could profit from distributing its products through OPACs. Presented at the 1994 NFAIS Annual Meeting in Philadelphia. References.

EC7. McGarry, Dorothy. "International Cooperation in Subject Analysis." *International Cataloguing and Bibliographic Control* 23, no. 4 (1994): 77–81.

Survey of IFLA conferences and projects on classification and indexing, including Library of Congress Subject Headings. Originally presented at the IFLA UBC/UNIMARC Seminar, Vilnius, Lithuania, June 2–4, 1994.

EC8. O'Brien, Ann. "Online Catalogs: Enhancements and Developments." *Annual Review of Information Science and Technology* 29 (1994): 219–42. Medford, N.J.: Learned Information, Inc. 1994.

Wide-ranging review of progress in the performance of online catalogs. Considers the role of keyword searching, multiple vocabularies, and Library of Congress Subject Headings. Sees poor subject access and lack of understanding of user behavior as major problems. Eighty-three references.

EC9. Studwell, William E. "Who Killed the Subject Code?" *Technical Services Quarterly* 12 (1994): 35–41.

Studwell expresses his regrets that a national subject cataloging code has not materialized in the U.S. He sees inadequate standardization in *Library of Congress Subject Headings* and its application. References.

EC10. Tibbo, Helen R. "The Epic Struggle: Subject Retrieval from Large Bibliographic Databases." *American Archivist* 57 (1994): 310–26.

Describes a retrieval study comparing the OCLC OLUC (Online Union Catalog) and the OPAC at the University of North Carolina at Chapel Hill, using a sample of USMARC AMC (archival and manuscripts control) records created at UNC Chapel Hill. Finds major differences in the two retrieval environments and notes problems posed by large retrieval sets. References.

EC11. Tillotson, Joy. "Is Keyword Searching the Answer?" *College & Research Libraries* 56 (1995): 199–206.

Experiments with keyword searching at Memorial University of Newfoundland and the University of Toronto confirmed the hypothesis that both keyword and controlled vocabulary searches should be supported in catalogs. Seventeen OPAC interfaces are evaluated. References.

EC12. Yee, Martha M., ed. *Headings for Tomorrow: Public Access Display of Subject Headings.* Chicago: American Library Association, 1992.

For annotation, *see* **DCA3**.

ED
Indexing Online Documents and Other Special Format Materials

ED1. Layne, Sara Shatford. "Some Issues in the Indexing of Images." *Journal of the American Society for Information Science* 45 (1994): 583–88.

A perceptive look at indexing of visual and pictorial images. Layne argues that indexing should provide access to useful groupings of images as well as to individual images, based on the four types of attributes of images: biographical, subject, exemplified, and relationship attributes. References.

ED2. McFadden, Thomas G. "Indexing the Internet." *Learned Publishing* 7, no. 1 (1994): 17–27.

McFadden shows how indexing of electronic publications presents many of the same issues as indexing of printed documents. He urges the indexing community to prepare for the challenges posed by electronic publishing. References.

ED3. Oernager, Susanne. "The Image Database: A Need for Innovative Indexing and Retrieval." In *Knowledge Organization and Quality Management: Proceedings of the Third International ISKO Conference, 20–24 June 1994, Copenhagen, Denmark,* edited by Hanne Albrechtsen and Susanne Oernager, 208–16. Frankfurt am Main, Germany: Indeks Verlag, 1994.

Argues that databases of digitized images will require interfaces for browsing and searching. Proposes an indexing and retrieval strategy that combines traditional subject indexing with term association clustering. References.

ED4. Peterson, Candace. "Newspaper Indexing: The San Antonio Express-News." In *Managing Large Indexing Projects: Papers from the Twenty-fourth Annual Meeting of the American Society of Indexers, San Antonio, Texas, May 23, 1992,* 11–14. Port Aransas, Texas: American Society of Indexers, 1994.

Report of a newspaper indexing project carried out by library staff at San Antonio College. The thesaurus used in the project is the *Subject Guide to IAC Databases,* issued by the Information Access Company and based on *Library of Congress Subject Headings.*

ED5. Svenonius, Elaine. "Access to Nonbook Materials: The Limits of Subject Indexing for Visual and Aural Languages." *Journal of the American Society for Information Science* 45 (1994): 600–606.

Svenonius argues that some nonbook materials defy subject indexing, either because the nonbook medium is being used for "nondocumentary" purposes or because the subject referenced is nonlexical. References.

ED6. Turner, J. "Indexing Film and Video Images for Storage and Retrieval." *Information Services & Use* 14 (1994): 225–36.

Discusses challenges in subject indexing of individual images from film and video. References.

EE
Thesauri and Controlled Vocabularies

EE1. Alvarado, Rubén Urbizagástegui. "Cataloging Pierre Bourdieu's Books." *Cataloging & Classification Quarterly* 19, no. 1 (1994): 89–105.

Terms from *Library of Congress Subject Headings* assigned to twenty-two books by Pierre Bourdieu, represented on eighty-eight records in the OCLC Online Union Catalog, were found to be inadequate for representing the content of Bourdieu's varied oeuvre. The LCSH headings assigned did not reflect Bourdieu's conceptual categories, showed ideological bias, and indicated a gap between librarians' vocabularies and those used by social scientists. References.

EE2. Bovey, J. D. "Building a Thesaurus for a Collection of Cartoon Drawings." *Journal of Information Science* 21 (1995): 115–22.

Describes construction of a thesaurus for the collection of 80,000 newspaper cartoons at the Center for the Study of Cartoon and Caricature, Kent University, England. Considers differences between indexing for a card catalog and an OPAC. References.

EE3. Chan, Lois Mai. *Library of Congress Subject Headings: Principles and Application.* 3rd ed. Englewood, Colo.: Libraries Unlimited, 1995.

This new edition of a standard work presents the *Library of Congress Subject Headings* as the world's foremost subject retrieval tool. Part 1, "Principles, Form, and Structure," covers the history of LCSH, its basis in literary warrant, and the construction of headings and use of subdivisions and cross references. Chapter 7, "Subject Authority Control and Maintenance," explains the process used by the Library of Congress and its cooperative subject cataloging partners to establish and revise headings for LCSH. Part 2, "Application," considers subject cataloging in general, special types of material such as serials and biography, and subject areas that receive special treatment in LCSH. Part 3, "Future Prospects," discusses LCSH as an online retrieval tool. Glossary, references, and index. For annotation, *see also* **DCB2**.

EE4. Chaplan, Margaret A. "Mapping LaborLine Thesaurus Terms to Library of Congress Subject Headings: Implications for Vocabulary Switching." *Library Quarterly*, 65 (1995): 39–61.

Terms from Labour Canada Library Services' LaborLine Thesaurus were mapped by human effort to LCSH in order to study the feasibility of automatic vocabulary switching to determine the correct LCSH terms for use in searching a catalog. Searching the database showed that only 40 percent to 60 percent of LaborLine terms realistically could be switched in this way. The author calls for richer cross-reference structure in LCSH and concludes that manual mapping may be more effective than automatic switching. References.

EE5. Chen, Hsinchun, et al. "Automatic Thesaurus Generation for an Electronic Community System." *Journal of the American Society for Information Science* 46, no. 3 (1995): 175–93.

The first half of the article is a very clear overview of issues in automatic thesaurus generation. The second half reports the automatic generation of a thesaurus for the Worm Community System, an electronic community system containing data used by molecular biologists studying the nematode worm *C. elegans*. The Worm thesaurus contains 2,709 personal names, 798 gene names, twenty experimental methods, and 4,302 subject descriptors, each of which has an average of ninety weighted neighboring terms. The thesaurus was tested successfully by six researchers. References. For annotation, *see also* **DCE6**.

EE6. Collantes, Lourdes Y. "Degrees of Agreement in Naming Objects and Concepts for Information Retrieval." *Journal of the American Society for Information Science* 46 (1995): 116–32.

Three groups of subjects were asked to name forty separate stimuli. The responses were compared with both the responses of the other subjects and the established terms for the stimuli in *Library of Congress Subject Headings*. The subjects had a low rate of agreement with each other and with *LCSH*. References.

EE7. Gascon, Pierre. "Le Répertoire de vedettes-matière de la Bibliothèque de l'Université Laval: sa génèse et son évolution." *Documentation et Bibliothèques* 39 (1993): 124–39 (1ère partie) and 40 (1994): 25–32 (2ème partie).

The subject heading vocabulary *Répertoire de vedettes-matière (RVM)* was developed at the Université Laval and is now accepted by the National Library of Canada as the Canadian national standard for French subject headings. *RVM* has also been adopted by the Bibliothèque nationale and other libraries in France. References.

EE8. Hemmasi, Harriette. "The Music Thesaurus Project at Rutgers University." *New Jersey Libraries* 28, no. 2 (1995): 21–23.

Hemmasi offers a description of the Music Thesaurus Project begun in 1991 at Rutgers and used to supplement the Library of Congress Subject Headings for music and musical terms.

EE9. Jones, Susan, and Micheline Hancock-Beaulieu. "Support Strategies for Interactive Thesaurus Navigation." In *Knowledge Organization and Quality Management: Proceedings of the Third International ISKO Conference, 20–24 June 1994, Copenhagen, Denmark,* edited by Hanne Albrechtsen and Susanne Oernager, 366–73. Frankfurt am Main, Germany: Indeks Verlag, 1994.

Jones and Hancock-Beaulieu suggest that support for direct, interactive use of thesauri can improve user satisfaction. They describe a prototype designed for such support in the City Interactive Knowledge Structure project, Department of Information Science, City University, London. References.

EE10. Jones, Susan, et al. "Interactive Thesaurus Navigation: Intelligence Rules OK?" *Journal of the American Society for Information Science* 46 (1995): 52–59.

Considers approaches to developing automated, interactive thesaurus navigation. Reports a study of how staff and students in the School of Informatics, City University, London, navigated the INSPEC thesaurus. Thesaurus navigation to refine or expand searches did not definitely increase retrieval effectiveness. References.

EE11. Kent, Robert E. "Implications and Rules in Thesauri." In *Knowledge Organization and Quality Management: Proceedings of the Third International ISKO Conference, 20–24 June 1994, Copenhagen, Denmark,* edited by Hanne Albrechtsen and Susanne Oernager, 154–60. Frankfurt am Main, Germany: Indeks Verlag, 1994.

Kent suggests how Formal Concept Analysis, as introduced by Rudolf Wille, can yield information about the word-sense comparability required to construct thesauri. References.

EE12. Library of Congress, Prints and Photographs Division. *Thesaurus for Graphic Materials.* Washington, D.C.: Cataloging Distribution Service, Library of Congress, 1995.

This tool contains the subject, genre, and physical characteristic terms compiled and used by the Prints and Photographs Division of the Library of Congress and by pictorial libraries and archives throughout the U.S. Includes references. The two parts, "TGM I: Subject Terms" and "TGM II: Genre and Physical Characteristic Terms," were previously published separately. "TGM I: Subject Terms" is also available online at URL: http://lcweb.loc.gov/rr/print/tgm1. LC plans to make "TGM Part II: Genre and Physical Characteristic Terms" available also.

EE13.   Massimiliano, Giurelli, and Giliola Negrini.   "A Tool to Guide the Logical Process of Conceptual Structuring." In *Knowledge Organization and Quality Management: Proceedings of the Third International ISKO Conference, 20–24 June 1994, Copenhagen, Denmark,* edited by Hanne Albrechtsen and Susanne Oernager, 342–49. Frankfurt am Main, Germany: Indeks Verlag, 1994.

Describes CLASTHES, a Macintosh program for faceted thesaurus generation. The program's underlying conceptual principles are based on Ingetraut Dahlberg's Systematifier principles for the organization of a conceptual system. References.

EE14.   Petersen, Toni, and Patricia J. Barnett, eds.   *Guide to Indexing and Cataloging with the Art & Architecture Thesaurus.* New York: Oxford University Press for the Getty Art History Information Program, 1994.

For annotation, *see* **DCE5**.

EE15.   Studwell, William E.   "A Tale of Two Decades, or, The Decline of the Fortunes of LC Subject Headings." *Western Association of Map Libraries Information Bulletin* 25 (1994): 73–74.

Studwell is highly critical of *Library of Congress Subject Headings;* he presents the perceived advantages of PRECIS and keyword searching.

EE16.   Weller, Carolyn R., and J. E. Houston, eds.   *ERIC Identifier Authority List (IAL) 1992.* Phoenix: Oryx Press, 1992.

Supplements the *Thesaurus of ERIC Descriptors,* 12th ed. (1990). Lists more than 44,000 identifiers alphabetically and in twenty subject category lists. Identifiers are more specific than descriptors, although an identifier may eventually be established as a descriptor.

EF
Automated Indexing

EF1.   Alexander, Michael.   "Digital Data Retrieval: Testing Excalibur." *Initiatives for Access News* no. 2 (1994). Also available online at URL: gopher://portico.bl.uk/11/portico/notice/blnews

Alexander describes the experience of the British Library's Initiatives for Access program with PixTex/EFS, a search engine developed by Excalibur Technologies, Inc., which uses Adaptive Pattern Recognition Technology to improve retrieval of indexed text converted from scanned images by optical character recognition (OCR) techniques.

EF2.   Brusch, Joseph A., and Toni Petersen.   "Automated Mapping of Topical Subject Headings into Faceted Index Strings Using the *Art & Architecture Thesaurus* as a Machine Readable Dictionary." In *Knowledge Organization and Quality Management: Proceedings of the Third International ISKO*

Conference, 20–24 June 1994, Copenhagen, Denmark, edited by Hanne Albrechtsen and Susanne Oernager, 390–97. Frankfurt am Main, Germany: Indeks Verlag, 1994.

A Getty Art History Information Program project at the Victoria & Albert Museum National Art Library, Great Britain, mapped topical subject headings into faceted index strings, using the *Art & Architecture Thesaurus*. Reviews project design and results. References.

EF3. Cohen, Jonathan D. "Highlights: Language- and Domain-Independent Automatic Indexing Terms for Abstracting." *Journal of the American Society for Information Science* 46 (1995): 162–74.

Cohen discusses the generation of index terms, or "highlights," from text. He claims that the method is effective in any language and shows experimental results for English, Russian, Japanese, German, Spanish, and Georgian. References.

EF4. Courtois, Martin P., William M. Baer, and Marcella Stark. "Cool Tools for Searching the Web: A Performance Evaluation." *Online* 19, no. 6 (1995): 14–32.

Authors discuss common characteristics of World Wide Web "databases," more commonly known as search engines, in contrast to bibliographic databases. They evaluate CUI W3 Catalog, Harvest, Lycos, Open Text Web Index, WebCrawler, World Wide Web Worm, and Yahoo as they performed in three sample searches. Good discussion of desirable characteristics in Web search services. One reference.

EF5. Crystal, David. "Some Indexing Decisions in the Cambridge Encyclopedia Family." *The Indexer* 19 (1995): 177–83.

Describes the automated indexing of the *Cambridge Encyclopedia* database; includes an overview of the *Encyclopedia's* publishing history.

EF6. Hodge, Gail M. *Automated Support to Indexing*. Philadelphia: National Federation of Abstracting and Indexing Services, 1992.

Wide-ranging survey of indexes to bibliographic databases. Chapters on current research projects and future technologies offer students ideas for their own research, while the chapter evaluating commercial indexing software (written by Sarah Seyn) holds more hands-on interest. Case studies are an important feature. References.

EF7. ———. "Computer-Assisted Database Indexing: The State-of-the-Art." In *Indexing, Providing Access to Information: Looking Back, Looking Ahead, the Proceedings of the 25th Annual Meeting of the American Society of Indexers*, edited by Nancy C. Mulvany, 33–44. Port Aransas, Texas: American Society of Indexers, 1993.

Overview of computer-assisted subject indexing. Hodge distinguishes between computer support for clerical activities, for quality control, and for intellectual activities in the indexing process. References.

EF8. Humphrey, Susanne. "The MedIndEx® Prototype for Computer Assisted MEDLINE® Database Indexing." In *Indexing, Providing Access to Information: Looking Back, Looking Ahead, the Proceedings of the 25th Annual Meeting of the American Society of Indexers,* edited by Nancy C. Mulvany, 45–54. Port Aransas, Texas: American Society of Indexers, 1993.

The MedIndEx (Medical Indexing Expert) project at the National Library of Medicine developed an interactive prototype for computer-assisted indexing of the MEDLINE database, using *Medical Subject Headings (MeSH)* as the thesaurus. References.

EF9. Moffatt, Alistair, and Timothy A. H. Bell. "In Situ Generation of Compressed Inverted Files." *Journal of the American Society for Information Science* 46 (1995): 537–50.

Moffatt and Bell offer an indexing algorithm for automatic generation of large compressed inverted indexes (concordances) in situ, avoiding common problems of other methods that require too much random access storage or too much temporary file space. A large text collection can be indexed using as little as 1 megabyte of memory. References.

EF10. Notess, Greg R. "Comparing Web Browsers: Mosaic, Cello, Netscape, WinWeb and InternetWorks Lite." *Online* 19, no. 2 (1995): 36–40.

Reference librarian's perspective on five well-known World Wide Web browsers.

EF11. ———. "Searching the World-Wide Web: Lycos, WebCrawler and More." *Online* 19, no. 4 (1995): 48–53.

Notess describes in detail five World Wide Web searching and indexing tools: Lycos, WebCrawler, World Wide Web Worm, Harvest Broker, and CUI, as well as Yahoo and Veronica for gopher, noting features that make each superior in various reference situations.

EF12. Woodruff, Allison Gyle, and Christian Plaunt. "GIPSY: Automated Geographic Indexing of Text Documents." *Journal of the American Society for Information Science* 45 (1994): 645–55.

GIPSY is an algorithm used to extract words and phrases containing geographic names or characteristics from text, to support geo-referenced indexing and retrieval. A prototype thesaurus has been constructed for the geographic area California. References.

F

Serials Management

Deborah E. Burke

Electronic technology has created the most visible changes in serials management in the past few years. Proliferating forms of electronic communication, from e-mail to Web pages, are changing the ways that everyone involved with serials is doing business. Relations with vendors are, in many cases, becoming more like partnerships. These changes are having an effect on how many libraries are configuring the restructuring efforts they are undertaking. Tight budgets emphasize the continuing problem with spiraling serials costs, especially in the sciences.

Electronic publishing raises many issues. Serials librarians have to cope with new arrangements for ordering and processing these materials. More electronic journal titles are becoming available on CD-ROM. Librarians must decide how best to make these titles accessible to their users. With rapidly changing hardware and software options, there is almost an embarrassment of riches from which to choose. Unfortunately, most libraries do not have an embarrassment of funds to accompany these choices.

The compiler especially wants to thank Marcia Tuttle for her kind assistance in the preparation of this chapter. Her guidance and expertise have been invaluable.

Serials management units have added e-mail and fax to conventional paper purchase orders. More vendors are now working with libraries in using electronic standards such as EDI (Electronic Data Interchange) and X12 (American National Standards Institute, Accredited Standards Committee's X12 formats for EDI) to boost efficiency for placing claims and paying invoices. Vendors' databases are accessible over the Internet for information verification and communication with representatives. Invoices on magnetic tape are offered as alternatives to manual payment of invoices. Major publishers and vendors now have sites accessible on the World Wide Web (WWW) for browsing and general information. Many libraries' technical services units now have WWW home pages with connections to standard "tools" such as currency conversion charts.

Aside from acquisition issues, electronic periodicals also raise questions about the scholarly publishing scene. The traditional roles of author, publisher, reviewer, vendor, and user change if the author can publish his or her work without intervention. Copyright issues also become more complicated when the product can be changed and revised almost at will, and perhaps not by the original author. Publishers are offering alternatives to print. Some journal publishers have tested buyers' reception to the inclusion of a CD-ROM along with the print version (sometimes at extra cost, which usually is not well received), others are offering magnetic tapes of their journals for libraries to mount on local mainframe computers, while others are experimenting with making journals accessible on the Internet. Many purely electronic journals (that is, those that exist only electronically) are currently free.

Relationships with vendors also are evolving. Libraries face reduced financial resources and must squeeze the most benefit out of every dollar. Vendors are finding that they must offer more services to meet libraries' needs and to outpace their competition. As libraries automate more processes, their expectations of vendors have risen. Vendors not only offer standard services of placing subscriptions, claiming issues that do not arrive, and paying publishers on behalf of their subscribers, they also may provide a variety of new services, including on-site special projects, tape invoice products, marking, labeling, mailing, and producing routing lists. Aside from these services, larger vendors also may offer article delivery services, table of contents notification services, and other value-added options. Vendors are also expected to participate in development of industry standards. No longer simply salespeople, serials vendors are forming partnerships with libraries as more libraries turn to outsourcing as a way to solve financial problems.

Many libraries across the U.S. are in the midst of restructuring or reorganizing efforts. These efforts may be driven by the parent organization's financial crunches or may be an internal effort to cope with personnel changes or automation needs. Numerous libraries have been using their automated system long enough to have outgrown it; they are contemplating migrating to new systems, which is just as much of a wrench as their first change from manual procedures to an automated process. As yet, not many libraries have identified serials management as an attractive function to be outsourced.

One aspect of serials that remains unchanged is their continuing rising cost, although purchasers hope that the innovation of electronic publishing will have a positive, price-lowering effect. Despite libraries' materials budgets being increasingly skewed toward serials and away from monographs, the budget realignments are not covering serials cost increases due to inflation or other reasons. Currency exchange rates and worldwide economic conditions also may have an effect on journal costs. Many libraries react to these changes with cancellation projects, which involve painful choices for all involved. Dan Tonkery (**FBB5**) notes that "everyone is canceling the same high priced journals." In these cases, librarians only can hope that union lists and article delivery services will join with interlibrary loan services to fill the gap or, less likely, that publishers will bring down their prices.

What is the future for serial specialists? Challenges will include continuing to accommodate changes brought by advancing technology. Gillian M. McCombs (**FDF6**) suggests a "point break" approach, a surfing term to describe the very best, most concentrated waves. Technology, including but not limited to the Internet, will provide a long, forceful and continuing wild ride. Most importantly, it is still true, as Marcia Tuttle stated in the Serials Management chapter of the *Guide to Technical Services Resources*, that "serials management is a library specialty in which some feel that change is the only constant" (p. 121).

Chapter Content and Organization

Chapter headings are the same as those in the *Guide to Technical Services Resources*. Both the novice and the expert should find articles of interest and usefulness. When several articles addressing the same topic have been published, such as libraries' experiences with converting to automated serials systems, the compiler has attempted to provide a diverse sampling. Most titles focus on North American issues, but a few works of international scope are included.

Titles cited in *GTSR* are not repeated here unless they have new editions, have undergone major changes, or, in the case of electronic resources, have different access information. These older references are still important; newer resources supplement but do not replace them. Some sections contain an introductory paragraph further explaining their scope. Finally, because of the dynamic nature of electronic resources, the reader is urged to check electronic addresses, since they quickly become dated.

FA
General Works

FAA
Textbooks, Guides, and Manuals

FAA1. Chen, Chiou-Sen Dora. *Serials Management: A Practical Guide.* Chicago: American Library Association, 1995.

An excellent and comprehensive overview of serials management from a practical perspective, concentrating on explanations and instructions. Chen includes especially useful sections addressing electronic journals' differences. A collection development chapter includes a section with suggestions for cooperation with selectors. Includes a "Selected Bibliography on Serials Management."

FAA2. *Serials Acquisitions Glossary.* Chicago: American Library Association, 1993.

Prepared by the Serials Section, Acquisitions Committee of the American Library Association's Association for Library Collections & Technical Services. The committee's purpose was to "gather and define the current, practical, and colloquial terms that underlie the communication between those librarians, vendors, and publishers who participate in the serials acquisitions chain."

FAB
Directories

FAB1. American Library Association, Association for Library Collections & Technical Services, Acquisitions Section, Publications Committee, Foreign Book Dealers Directories Series Subcommittee. *Book and Serial Vendors for Asia and the Pacific.* Foreign Book and Serials Vendors Directories, vol. 1. Chicago: American Library Association, 1995.

Information gathered from an Association of Research Libraries survey. Includes lists of libraries using various vendors in countries, with the vendors' addresses and services. Divided into three sections: Vendors by Country, Master Vendor List, and Libraries Responding to the Questionnaire. Available in hard copy and online. Gopher address: gopher.uic.edu/The Library/ALA/ ALA Divisions/ALCTS/The Book and Serials Vendors for Asia and the Pacific. URL: gopher://ala1.ala.org:70/11/alagophxiii/alagophxiiialcts/ alagophxiiialctsasia

FAB2. Whiffin, Jean I., ed. *International Directory of Serials Specialists.* Binghamton, N.Y.: Haworth Press, 1995.

Lists 144 serials specialists from forty-six countries. Arranged alphabetically by country, with additional indexes by personal name and by expertise. Information derived from questionnaires includes current position, business address, areas of expertise, and past accomplishments, such as conferences, papers, publications. Unfortunately, many notable serialists are not included. The editor states in her Introduction that additional country coverage is still required; she hopes that these data can be captured for future editions.

FAB3. Wilkas, Lenore Rae. *International Subscription Agents.* 6th ed. Chicago: American Library Association, 1994.

Agents are listed alphabetically by country of origin. This edition includes phone numbers, fax numbers, cable addresses, and toll-free phone numbers. Several agents have "prospered and grown" since the 5th edition, but many smaller and older companies are no longer in existence. Political upheavals also have removed some companies. Includes geographical and subject indexes. Useful for all libraries buying non-U.S. serials or books.

FAC
Bibliographies

FAC1. Bonario, Steve, and Ann Thornton. "Library-Oriented Lists and Electronic Serials." *Microcomputers for Information Management* 11, no. 3 (1994): 209–26.

Selected electronic discussion groups and electronic serials of interest to librarians. Provides names and addresses of the lists, access instructions, and a subject index.

FAC2. *CurrentCites*. Serial online. Berkeley: University of California Regents, Teri Andrews Rinne, ed. no.1– , 1991– . Monthly. ISSN 1060-2356.

Contributors scan over thirty journals for superior or thought-provoking articles on electronic publishing, networks and networking, and general technology subjects. They provide signed informative annotations and on-the-mark critical abstracts. Electronic links are included to some articles. To subscribe, send a message to LISTSERV@LIBRARY.BERKELEY.EDU with the command: SUB CITES [your name]. Archived at URL http://sunsite.berkeley.edu/CurrentCites/

FAC3. Hitchcock, Steve, Leslie Carr, and Wendy Hall. *A Survey of STM Online Journals 1990–1995: The Calm before the Storm.* Online. Multimedia Research Group, University of Southampton, United Kingdom, 1996. URL: http://journals.ecs.soton.ac.uk/survey/survey.html

A report surveying over 100 full-text, peer-reviewed journals in science, technology, and medicine. Analyzes the publication data, formats used, and some historical perspective of the journals. Highlights the fact that many new online journals are electronic versions of their paper counterparts. Includes hypertext links to the journals and to listings of source directories.

FAC4. Morrison, David F. W. "Bibliography of Articles Related to Electronic Journal Publications and Publishing." In *Directory of Electronic Journals, Newsletters and Academic Discussion Lists,* by Lisabeth A. King and Diane Kovacs, 35–43. Washington, D.C.: Association of Research Libraries, 1994.

Bibliography based upon citations published in the electronic journal *CurrentCites* **(FAC2)**. Comprehensive annotations signed by their contributors.

FAC5. *NewJour: Electronic Journals & Newsletters.* Serial online. San Diego: University of California, San Diego, 1995– . Co-owners: Ann Shumelda Okerson, Yale University, and James J. O'Connell, University of Pennsylvania. URL: http://gort.ucsd.edu/newjour

An Internet list for reporting and announcing new online electronic journals. Archive available on the World Wide Web goes back to August 1993, searchable and with a reverse chronological index. To subscribe, send a message to MJD@CCAT.SAS.UPENN.EDU with the command: SUBSCRIBE NEWJOUR [your name]. Archive maintained at *NewJour's* WWW site.

FAC6. Riddick, John F. "An Electrifying Year: A Year's Work in Serials, 1992." *Library Resources & Technical Services* 37, no. 3 (1993): 335–42.

A selective review and bibliography of serials literature published in 1992. Cataloging, electronic publishing, CD-ROM serials, collection development, serials pricing, and management are the major themes addressed in the accompanying essay.

FAD
Periodicals

Periodicals listed in *Guide to Technical Services Resources* are not included here unless they have new access information or have changed significantly in content or scope.

FAD1. *ACQNET.* Serial online. Boone, N.C.: Appalachian State University Library, Eleanor Cook, ed. no.1– , 1990– . Irregular. ISSN 1057-5308. URL: http://www.library.vanderbilt.edu/law/acqs/acqnet.html

For annotation, *see also* **BAC1.** *ACQNET* provides a "medium for acquisitions librarians and others interested in acquisitions work to exchange information, ideas, and to find solutions to common problems" and addresses many topics of interest to serials acquisitions. Archived at Web site.

FAD2. *Citations for Serial Literature.* Serial online. New York: Readmore, Inc., Marilyn Geller, ed. no. 1– , 1992– . Irregular. ISSN 1061-7434. URL: http://www.readmore.com/info/csl.html

SERCITES lists contents and abstracts, if available, for articles related to the serials information chain. Occasionally includes a section entitled "Serendipitous Citings" with relevant citations from journals not generally listed. A useful current awareness service for serials specialists. To subscribe, send a message to LISTSERV@SUN.READMORE.COM with the command: SUBSCRIBE SERCITES [your name]. Archived at Web site.

FAD3. *Newsletter on Serials Pricing Issues.* Serial online. Chapel Hill, N.C.: Marcia Tuttle, ed. no. 1–, Feb. 1989– . Irregular. ISSN 1046-3410. URL: http://sunsite.unc.edu/reference/prices/prices.html

Newsletter contains brief articles, reports, and news about serials pricing. This newsletter is highly regarded and widely cited in library literature. Contributors are primarily librarians, publishers, subscription agents, and scholars. Journal is available on the Internet, Blackwell's CONNECT, Readmore's ROSS, or in paper format for EBSCO customers. To subscribe, send a message to LISTPROC@UNC.EDU with the command: SUBSCRIBE PRICES [your name]. Archived at Web site.

FAE
Sources of Expertise

ELECTRONIC DISCUSSION GROUPS

FAE1. *SERIALST.* Electronic discussion group. Burlington: University of Vermont Library. Moderator: Birdie MacLennan. URL: http://www.uvm.edu/~bmaclenn/serialst.html

Moderated discussion group with over 2,400 members in thirty-four countries, covers all aspects of serials management, including selection, acqusition, cataloging, and binding; provides announcements of conferences and workshops. An excellent current awareness resource. To subscribe, send message to LISTSERV@UVMVM.UVM.EDU, with the command: SUBSCRIBE SERIALST [your name]. Send messages for posting to SERIALST@UVMVM.UVM.EDU. Archived at Web site.

PROFESSIONAL ASSOCIATIONS

One significant change over the last few years has been the development of gophers and World Wide Web home pages by professional associations. The following sites are of particular interest to serials specialists. More complete descriptions of the associations can be found in *GTSR*.

FAE2. Association of College and Research Libraries (ACRL). URL: http://www.ala.org/acrl.html or URL: gopher://gopher/ala.org

FAE3. Association for Library Collections and Technical Services (ALCTS). URL: http://www.ala.org/alcts.html or URL: gopher://gopher.ala.org:70/11/alagophxiii/alagophxiiialcst

FAE4. North American Serials Interest Group (NASIG). URL: http://nasig.ils.unc.edu or URL: gopher://nasig.ils.unc.edu:6050

NASIG's gopher contains postings from NASIG-L (the associations electronic discussion group), committee listservs, and an ftp site. The Web site was added in early 1996; some links (including the *NASIG Newsletter* and conference proceedings) are limited to NASIG members. Provides links to other Web resources of interest to serials specialists.

FAF
Conference Proceedings

FAF1. Flowers, Janet L. "Systems Thinking about Acquisitions and Serials Issues and Trends: A Report on the 1993 Charleston Conference." *Library Acquisitions: Practice & Theory* 18, no. 2 (1994): 227–38.

Flowers summarizes topics covered in a "thematic approach," looking at conference presentation topics from four angles: perspective, context (operating environments of editors, publishers, vendors, libraries), evaluation of performance, and communication (focusing on electronic data interchange).

FAF2. Holley, Beth, and Mary Ann Sheble. *A Kaleidoscope of Choices: Reshaping Roles and Opportunities for Serialists: Proceedings of the North American Serials Interest Group, Inc., Ninth Annual Conference, June 2–5, 1994, University of British Columbia, Vancouver, B.C.* Binghamton, N.Y.: Haworth Press, 1995. Also published as *The Serials Librarian* 25, nos. 3/4 (1995).

Reports on plenary sessions addressing the impact of technology on segments of the industry and the changes in skills needed to manage that impact. Includes reports of concurrent sessions, workshops, and preconference sessions.

FAF3. McMahon, Suzanne, Miriam Palm, and Pam Dunn, eds. *If We Build It: Scholarly Communications and Networking Technologies: Proceedings of*

the North American Serials Interest Group, Inc., Seventh Annual Conference, June 18–21, 1992, the University of Illinois at Chicago. Binghamton, N.Y.: Haworth Press, 1993. Also published as *The Serials Librarian* 23, nos. 3/4 (1993).

Reports on plenary, breakout, and workshop sessions. This conference included a joint day with the Society for Scholarly Publishing. Conference focused on the "phenomenal growth of electronic technology" and the changes in services, economic models, and cooperation that this growth requires of librarians, publishers, and vendors.

FAF4. McMillan, Gail, and Marilyn L. Norstedt, eds. *New Scholarship: New Serials: Proceedings of the North American Serials Interest Group, Inc., Eighth Annual Conference, June 10–13, 1993, Brown University, Providence, R.I.* Binghamton, N.Y.: Haworth Press, 1994. Also published as *The Serials Librarian* 24, nos. 3/4 (1994).

Papers from plenary and concurrent sessions, addressing current topics, with an "emphasis on research and its relationship to publishing," especially in electronic publishing. Summary reports of preconference and conference workshops.

FAF5. Sheble, Mary Ann, and Beth Holley, eds. *Serials to the Tenth Power: Tradition, Technology, and Transition: Proceedings of the North American Serials Interest Group, Inc., Tenth Anniversary Conference, June 1–4, 1995, Duke University, Durham, N.C.* Binghamton, N.Y.: Haworth Press, 1996. Also published as *The Serials Librarian* 28, nos. 1/2 and 3/4 (1996).

Tenth anniversary conference of NASIG. Plenary sessions presented the impact of electronic publishing on the dissemination of knowledge, new copyright issues, and overviews of future issues, including Internet security and continuing access to information. Summary reports of preconference, concurrent, and workshop sessions. Index.

FB
Management
of Serials Units

FBA
Organization

FBA1. Boardman, Edna M. "How to Run a Tight Ship in the Magazine Stacks." *The Book Report* 13 (Jan./Feb. 1995): 17–18.

Very brief, but informative article lists job duties of all members of the school library staff. Provides a useful overview.

FBA2. Burrows, Toby, and Philip G. Kent, eds. *Serials Management in Australia and New Zealand: Profile of Excellence.* Binghamton, N.Y.: Haworth Press, 1993. Also published as *Australian & New Zealand Journal of Serials Librarianship* 3, nos. 3/4 (1993).

This collection of fifteen papers focuses on serials management in Australasia, but should interest any librarian experiencing change. Unusual factors include the "tyranny of distance" causing slow delivery and claims responses despite a sophisticated electronic infrastructure. Contributions are from all library types, with great range in their users, services, and placement on the automation continuum.

FBA3. Cook, Eleanor I., and Pat Farthing. "A Technical Services Perspective of Implementing an Organizational Review While Simultaneously Installing an Integrated Library System." *Library Acquisitions: Practice & Theory* 19, no. 4 (1995): 445–61.

For annotation, *see* **AHC3**.

FBA4. Grochmal, Helen M. "Achieving Success as a Serials Librarian: Some Advice." *The Serials Librarian* 24, no. 2 (1993): 85–89.

Practical, specific guidelines for evaluating a serials department or for beginning a new position. A particularly useful, brief introduction for novice serials specialists.

FBA5. Jones, Wayne, ed. *Serials Canada: Aspects of Serials Work in Canadian Libraries.* Binghamton, N.Y.: Haworth Press, 1995. Also published as *The Serials Librarian* 26, nos. 3/4 (1995).

This volume includes ten articles addressing "diverse aspects of serials in Canada." Includes economic aspects of electronic journals, rising serial prices, the development of the Canadian Periodicals Price Index, and an overview of the Canadian Serials Industry Systems Advisory Committee (CSISAC).

FBA6. Ten Have, Elizabeth Davis. "Serials in Strategic Planning and Reorganization." *Serials Review* 19, no. 2 (1993): 7–12, 48.

Michigan State University Libraries' Technical Services moved from a department organized by both form and function to a department with acquisitions teams and cataloging teams. Author describes the factors leading to the reorganization and the planning process involved in designing and implementing the changes.

FBB
Relationships with External Organizations

FBB1. "Between a Rock and a Hard Place: The Future of the Subscription Agent." *The Serials Librarian* 24, nos. 3/4 (1994): 113–27.

Includes four papers presented as a concurrent session at the 8th annual North American Serials Interest Group conference. Representatives from four companies describe the current state and future roles of subscription agents. Especially helpful for perspective on the range of agents' viewpoints. Includes: "The Role of the Specialized Vendor in a Changing Market," by Jane Maddox (Otto Harrassowitz); "Mega Vendor: Threat or Promise?" by John E. Cox (B. H. Blackwell, Ltd.); "Future Value-Added Services: Remaining Competitive in a New Market," by J. T. Stephens (EBSCO Industries, Inc.); and "Access vs. Ownership: Strategic Implications for Agents," by Adrian W. Alexander (Faxon Company).

FBB2. Duranceau, Ellen. "Vendors and Librarians Speak on Outsourcing, Cataloging, and Acquisitions." *Serials Review* 20, no. 3 (Fall 1994): 69–83.

Two sections of this "Balance Point" article are of special interest to serials librarians. John Baker's "Blackwell's in a New Marketplace" describes the new PLUS journal consolidation service, which provides many additional services and is targeted for markets in "areas of the world where postal services can be somewhat unreliable . . ." Nancy Gershenfeld's contribution gives an account of Microsoft Corporation's arrangement with Readmore to provide on-site serials management and staff.

FBB3. Kennedy, Kit. "Access Blues: A Song We Are All Singing." *Journal of Library Administration* 20, no. 1 (1994): 49–58.

Presented at the 1994 University of Oklahoma conference. Kennedy addresses the who, what, where, when, and why of changes in the vendors' and libraries' relationships. A thought-provoking perspective. The refrain of the song "Access Blues" is "And how much will it cost?"

FBB4. Reich, Vicky. "Marketing to Libraries: What Works and What Doesn't." *The Serials Librarian* 23, nos. 3/4 (1993): 197–206.

Written from the viewpoint of academic libraries but also applicable to other types of libraries. Explores how acquisitions and collection development are changing, suggests what libraries need from publishers, and gives tips about possible successful marketing strategies.

FBB5. Tonkery, Dan. "Reshaping the Serials Vendor Industry." *The Serials Librarian* 25, nos. 3/4 (1995): 65–72.

Recent investments in digital technology are causing management and administrators to seek a "technology payoff," savings realized from these investments. As virtual libraries of the future emerge, they will need "virtual agents [to] serve as the link between the resources of many traditional publishers." Presented at the 1994 NASIG Conference.

FC
Serials Publishing

FCA
Printed Serials

FCA1. Gans, Alfred. *Serials Publishing and Acquisitions in Australia.* Binghamton, N.Y.: Haworth Press, 1993. Also published as *Australian & New Zealand Journal of Serials Librarianship* 4, no. 2 (1993).

Australian publishers and agents address current practices, future trends, possible future problems. Australian serials librarians describe various aspects of serials processing.

FCA2. Stankus, Tony. "Looking Trojan Gift Horses in the Mouth: Are Special Issues Special Enough to Pay Extra Money for and Bind?" *Technicalities* 15, no. 9 (Sept. 1995): 8–10.

This "Making Sense of Serials" column looks at "some typical special issue specimens": annual convention issues, subspecialty symposium issues, anniversary issues, Festschriften, buyer's guide issues, year-in-review issues.

FCA3. Woodward, Hazel and Stella Pilling, eds. *The International Serials Industry.* Aldershot, Hampshire, England; Brookfield, Vt.: Gower, 1993.

An excellent resource, this collection provides a "detailed examination and overview of the scholarly information chain and of the international serials industry." Contributors from all sectors of the serials industry address issues of common concern from differing perspectives. Topics include serials pricing, electronic publishing media, quality versus quantity, serials and developing countries, journal versus article delivery, and national and international cooperation.

FCB
Electronic Serials

FCB1. Cramer, Michael D. "Licensing Agreements: Think before You Act." *College & Research Libraries News* 8 (Sept. 1994): 496–97.

Most electronic publications include some sort of license agreement defining the purchaser's responsibilities. Cramer provides a list of suggestions to follow when acquiring products with license agreements. This brief article is obligatory reading for anyone contemplating such an agreement.

FCB2. Dworaczek, Marian, and Victor G. Wiebe. "E-Journals: Acquisition and Access." *The Acquisitions Librarian* 12 (1994): 105–21.

The authors identify differences in acquisitions procedures for electronic journals, questions of access including publicity and "user-friendliness," and different types of technical access.

FCB3. *Electronic Journals in ARL Libraries: Issues and Trends.* Elizabeth Parang and Laverna Saunders, comps. SPEC Kit, no. 202. Washington, D.C.: Association of Research Libraries, Office of Management Studies, 1994.

Task force reports from six ARL member libraries plus "Report to the CIC Library Directors" from the Committee on Institutional Cooperation's Task Force on the CIC Electronic Collection. Includes a list of selected readings. A companion publication to SPEC Kit no. 201 **(FCB4)**; both include SPEC survey results.

FCB4. *Electronic Journals in ARL Libraries: Policies and Procedures.* Elizabeth Parang and Laverna Saunders, comps. SPEC Kit, no. 201. Washington, D.C.: Association of Research Libraries, Office of Management Studies, 1994.

Survey of seventy-five ARL libraries' practices for collection development, acquisition, cataloging, and preservation of electronic journals. Public access to the journals, personnel needs, and training methods also were included in the survey. Includes sample collection policies, sample gopher and OPAC screens, and a list of selected readings. A companion publication to SPEC Kit no. 202 **(FCB3)**.

FCB5. Germain, J. Charles. "Publishing in the International Marketplace." *Library Acquisitions: Practice & Theory* 19, no. 2 (1995): 225–29. For annotation, *see* **BH3**.

FCB6. Keating, Lawrence R. II, Christa Easton Reinke, and Judi A. Goodman. "Electronic Journal Subscriptions." *Library Acquisitions: Practice & Theory* 17 (Winter 1993): 455–63.

Tutorial which focuses on the "practical aspects of processing electronic subscriptions," using experiences from the University of Houston Libraries. Figures include a sample serial order request form and work-flow diagrams.

151

FD
Serials Processing

FDA
World Wide Web Sites and Home Pages

FDA1. *Cornell University Library Technical Services Manual.* URL: http://www.library.cornell.edu/tsmanual
For annotation, see **CAE3**.

Many libraries' technical services units have initiated World Wide Web (WWW) home pages or gopher sites. Quality and readability of these pages, which are frequently "under construction," vary enormously. Most are primarily for local staff use and include internal documents, local procedures, restructuring plans, and links to online tools. Few are devoted solely to serials; many include items of serials-related interest along with cataloging and acquisitions tools. Two sites that are good examples of the different types of pages and arrangements are listed.

FDA2. MacLennan, Birdie. *Serials in Cyberspace: Collections, Resources, and Services.* Online. Burlington: University of Vermont. URL: http://uvm.edu:80/~bmaclenn or URL: http://www.uvm.edu/~bmaclenn
"An assortment of places to go, things to see . . ." Electronic links to academic/research sites in the U.S. and other English-speaking countries, selected e-journal titles, and miscellaneous collections and resources. Most links include brief annotations or remarks.

FDA3. Technical Processing Online Tools (TPOT). University of California at San Diego Libraries. Online. URL: http://tpot.ucsd.edu/
For annotation, see **CAE9**.

FDB
Management of Serials Processing

FDB1. Bevis, Mary D., and Sonja L. McAbee. "NOTIS as an Impetus for Change in Technical Services Departmental Staffing." *Technical Services Quarterly* 12, no. 2 (1994): 29–43.
A look at shifting staff responsibilities, especially for nonprofessional staff, following the installation of NOTIS Acquisitions/Serials module.

Provides a useful perspective on staffing issues for any library using support staff in serials management. For annotation, see also **AHC2**.

FDB2. Intner, Sheila. "Outsourcing: What Does It Mean for Technical Services?" *Technicalities* 14, no. 3 (March 1994): 3–5.

Articles about outsourcing usually concentrate solely on cataloging issues. Intner also addresses serials issues, identifying sources of outsourcing already in place (but not named as outsourcing), and pointing out potential benefits of combining suppliers of serials articles and suppliers of serial issues. For annotation, see also **AHB5**.

FDC
Subscription Agent Selection and Evaluation

FDC1. Kent, Philip G. "How to Evaluate Serials Suppliers." *Library Acquisitions: Practice & Theory* 18, no.1 (1994): 83–87.

Kent examines price and service as two criteria for evaluating vendors. He provides a checklist of service criteria, suggestions for getting the best prices, and offers examples from the Australian CSIRO Library Network's experiences.

FDC2. Merriman, John. "The Work of Subscription Agents." *The Serials Librarian* 24, no. 1 (1993): 1–24.

Merriman emphasizes services offered by members of the Association of Subscription Agents. The author discusses factors in play between the agent and the publisher, the agent and the librarian, and lists problem areas in each of their offices that might keep the journal from getting to the reader. He also provides aims, activities, and membership information of serials interest groups in the United Kingdom, United States, Australia, South Africa, and the Netherlands.

FDC3. Presley, Roger L. "Firing an Old Friend, Painful Decisions: The Ethics between Librarians and Vendors." *Library Acquisitions: Practice & Theory* 17 (1993): 53–59.

For annotation, see **BI5**.

FDD
Subscription Agents' Databases

Subscription agents' databases listed in *Guide to Technical Services Resources* (B. H. Blackwell's *CONNECT*, EBSCO Subscription Services' *EBSCONET*, and Readmore Academic Subscription Services' *ROSS*) remain available on the Internet to their customers. Faxon has replaced their *Datalinx* with a new database, *Faxon Source Online* **(FDD4)**. *Harrassowitz Online* **(FDD5)** is newly available. In addition, many subscription agents are keeping current by providing World Wide Web home pages.

FDD1. Blackwell Group. Online. Oxford, England: Blackwell Group. URL: http://www.blackwell.co.uk

Blackwell's home page has links to all of its companies describing their individual features, customer contacts, and service specialties. The B. H. Blackwell Ltd. Periodicals Service page also includes information about *CONNECT*. The Blackwell North America, Inc. section of the document also includes information about *Bridges to Blackwell,* their collection of online databases, and a "Reference Shelf," providing links to Web sites of library interest.

FDD2. EBSCO Information Services. Online. Cary, Ill.: EBSCO Information Services. URL: http://www.ebsco.com/

EBSCO's home page gives detailed information about EBSCO Subscription Services, including *EBSCONET*. The *EBSCOdoc* section describes document delivery services and allows the visitor to sample the resources available. Registration is required for actual use of the document delivery service through this medium. Within *EBSCOdoc,* a "Reference Desk" provides some library tools and electronic links to other library sites.

FDD3. Faxon Company Home Page. Online. Westwood, Mass.: Faxon Company. URL: http://www.faxon.com/

Provides organizational information about Faxon Company, Inc., Dawson Holdings PLC, Faxon Canada, Ltd., DataTrek, Inc. Price projections are available. Brief tutorials on Archie, gopher, Veronica, WAIS, World Wide Web, CATALIST, Hytelnet, LIBS, and LIBTEL. Also supplies numerous and well-organized electronic links to electronic journals and publishers' home pages. Especially useful links to standards organizations' pages, such as SISAC (Serials Industry Systems Advisory Committee), NISO (National Information Standards Organization), ISO (International Organization for Standardization), and ISSN International Center.

FDD4. *Faxon Source Online.* Online database. Westwood, Mass.: The Faxon Company.

Faxon's new subscription management system, with "full searching, ordering and claiming capability." Available to Faxon customers.

FDD5. *Harrassowitz Online.* Online. Weisbaden, Germany: Harrassowitz.

Harrassowitz Online is a password-protected new service available over the Internet providing bibliographic details for serials titles, account information, online ordering and claiming, publishers' dispatch information. Several levels of access can be assigned and screens are "user friendly." Customers also may request management reports and copies of statements or invoices, or send e-mail messages to Harrassowitz using this service.

FDD6. *ReadiCat.* Online database. New York: Readmore Inc.

Readmore's database of over 100,000 journal, periodical, and CD-ROM titles with price and frequency information. Restricted to Readmore customers and guest users.

FDD7. Readmore Webserver. Online. New York: Readmore, Inc. URL: http://www.readmore.com/

Readmore's home page is divided into three major sections. "Readmore Electronic Services" gives information about the *ROSS* database, *BACKSERV* service (back issues), Knight-Ridder's *UnCover* document delivery service, and *REMO*, the Readmore serials management system. "Readmore Publisher Services" provides publisher links and a "New Titles Review Service," which includes links to online journals. The "Readmore Information Library" supplies extensive links to WWW sites of interest to librarians, including electronic journals, library systems vendors, worldwide library catalog listings, library-related organizations and companies, professional organizations, and standards organizations.

FDE
Claiming and Other Acquisition Processes

FDE1. *BACKSERV.* Electronic discussion group. Readmore, Inc. List administrators: Marilyn Geller and Amira Aaron.

An unmoderated listserv designed to facilitate exchange of serials back issues. Open to subscribers and non-subscribers; no requirement to be a Readmore customer. To subscribe, send a message to LISTSERV@SUN.READMORE.COM, with the command: SUBSCRIBE BACKSERVE [your name]. Messages may be posted to BACKSERV@SUN.READMORE.COM. Postings maintained for three months on an archive available at URL gopher://gopher.readmore.com:70/11/backserv

FDE2. Kascus, Marie, and Faith Merriman. "Using the Internet in Serials Management." *College & Research Libraries News* 56 (March 1995): 148–50, 176.

The authors give examples of ways use of the Internet has helped them quickly resolve serials problems, including communicating with peers in making policy or procedure decisions, checking vendor databases for claim information, and resolving bindery preparation questions.

FDE3. **Meiseles, Linda, and Emerita M. Cuestra.** "Multiformat Periodicals: A New Challenge for the Periodicals Manager." *The Serials Librarian* 24, no. 2 (1993): 19–29.

Increasing numbers of serial titles are arriving with nonprint supplementary materials. The authors conducted a survey on *SERIALST* (**FAE1**), a serials discussion group on the Internet. This article includes a copy of the survey and the results. Two of the recommendations the authors make are to develop local procedures for handling these materials in order to ensure consistency, and to label physical issues as containing software.

FDF
Automation of Serials Processing

FDF1. **Banach, Patricia.** "Migration from an In-House Serials System to INNOPAC at the University of Massachusetts at Amherst." *Library Software Review* 12, no.1 (Spring 1993): 35–37.

This brief overview describes planning, record loading processes, and final cleanup of records. Predictive check-in data was not tape loaded but subsequently keyed in manually.

FDF2. **Carter, Ruth C., and Paul B. Kohberger, Jr.** "Using SPSS/PC+ and NOTIS Downloaded Files of Current Subscription Records at the University of Pittsburgh." *Library Resources & Technical Services* 37, no. 2 (April 1993): 227–35.

Current serial subscription information was downloaded from the NOTIS system to SPSS/PC+ on an IBM-compatible personal computer. Various lists and data files were created to help with serials decision making, including holdings by geographical area and active subscriptions from a particular publisher.

FDF3. **Chiang, Belinda.** "Migration from Microlinx to NOTIS: Expediting Serials Holdings Conversion through Programmed Function Keys." *The Serials Librarian* 25, nos. 1/2 (1994): 115–31.

This article details the design and implementation of a project to add holdings and order information to Queens College Library of the City University of New York's NOTIS database, called CUNY+. Librarians using other software with recording capability may be able to adapt these techniques. Many helpful screen examples are included.

FDF4. **Davis, Trisha, and James Huesmann.** *Serials Control Systems for Libraries.* Essential Guide to the Library IBM PC, vol. 12. Westport, Conn.: Mecklermedia, 1994.

The authors provide an overview of basic functions and features of microcomputer-based serials control systems. They then apply this overview to five systems available at the time of publication, offering a wide range of features. The authors acknowledge the speed of change in PC-based systems and recommend that readers use their methodology to evaluate currently available systems. The methodology will provide a continuing practical template for any library examining a serial control system. Includes glossary.

FDF5. Harri, Wilbert. "Implementing Electronic Data Interchange in the Library Acquisitions Environment." *Library Acquisitions: Practice & Theory* 18, no. 1 (1994): 115–17.

Harri defines EDI, offers potential advantages for its use, and lists elements essential for its implementation: a fully integrated library system, participation of vendors, EDI software, ability to communicate electronically, and, especially, completion of the X12 standards.

FDF6. McCombs, Gillian M. "The Internet and Technical Services: A Point Break Approach." *Library Resources & Technical Services* 38, no. 2 (April 1994): 169–77.

McCombs emphasizes the need for flexibility and creativity in the future technical services unit. Offers an especially interesting list of action steps to take and pitfalls to avoid. "Point break" is a surfing term. "Point breaks frequently produce long rides that start with a fast takeoff. Once into a point break there is no time to relax . . ." For annotation, *see also* **AA9**.

FDF7. Paul, Sandra K. "EDI/EDIFACT." *Library Resources & Technical Services* 39, no. 2 (April 1995): 180–83.

Paul defines EDI, EDIFACT, X12 and gives a brief history of their development. The effect of the U.S. government facing a mandate from President Clinton that "all U.S. government procurement [must] be accomplished through electronic commerce" by the year 1997 will prompt other businesses and the library community to implement electronic means to accomplish more and more routine operations.

FDF8. Richter, Linda, and Joan Roca. "An X12 Implementation in Serials: MSUS/PALS and Faxon." *Serials Review* 20, no. 1 (Spring 1994): 13–24, 42.

For this pilot project, MSUS/PALS, central processing site for a fifty-five member library consortium in Minnesota, agreed to exchange data with Faxon on behalf of North Dakota State University (NDSU), a consortium member and Faxon client. This article comprehensively details plans and processes for the project. Includes sample files and reports as figures. Sidebars clearly define and explain EDI and X12, plus a bibliography is included.

FDF9. Saffady, William. "Automated Acquisitions and Serials Control." In *Introduction to Automation for Librarians,* 347–73. 3rd ed. Chicago: American Library Association, 1994.

Gives an overview of acquisitions and serials processes, with mention of some specific products. Discusses automation implementation alternatives:

custom services such as the PHILSOM system; and commercial services such as Faxon's LINX, EBSCO's EBSCONET, INNOVACQ, and NOTIS. Includes list of Additional Readings.

FDF10. "SISAC Item Contribution Identifier: New SISAC Code." *Computers in Libraries* 13, no. 1 (1993): 23–24.

Brief article prepared by the Serials Industry Systems Advisory Committee (SISAC) explaining the SISAC Item Contribution Identifier (SICI) which is a "means of unique identification of a serial publication [or a portion thereof] when referring to it in an Electronic Data Interchange." Publishers, vendors, and library automation systems are adding SISAC barcodes to their products.

FE
Serials Pricing

FE1. Alexander, Adrian W. "Periodical Prices, 1990–1992." *Library Acquisitions: Practice & Theory* 17 (Spring 1993): 3–19.

FE2. ———. "Periodical Prices, 1991–1993." *Library Acquisitions: Practice & Theory* 18 (Spring 1994): 23–41.

FE3. ———. "Periodical Prices, 1992–1994." *Library Acquisitions: Practice & Theory* 19 (Spring 1995): 63–82.

An annual comparative price study compiled and published by the Faxon Company. Each article compares "price changes, patterns and trends in a variety of categories" over a three-year period. Analyzes and measures average prices based on active titles in Faxon's database of over 200,000 titles worldwide as well as weighted prices reflecting order activity.

FE4. Alexander, Adrian W., and Kathryn Hammell Carpenter. "U.S. Periodical Price Index for 1993." *American Libraries* 24 (May 1993): 390–93.

FE5. ———. "U.S. Periodical Price Index for 1994." *American Libraries* 25 (May 1994): 450–54.

FE6. ———. "U.S. Periodical Price Index for 1995." *American Libraries* 26 (May 1995): 446–54.

Annual studies by Library Materials Price Index Committee (LMPIC) of ALA's Association of Library Collections and Technical Services (ALCTS). Studies use a selected sample of periodical titles published in the U.S., with price information supplied by the Faxon Company. Includes price index comparison table.

FE7. Brooke, F. Dixon Jr., and Allen Powell. "EBSCO 1995 Serials Price Projections." *Serials Review* 20, no. 3 (Fall 1994): 85–94.

Projections for the coming year's price increases. Projections for libraries invoiced in U.S. dollars and other currencies, European libraries, Australian libraries, and Canadian Libraries. Charts and graphs of cost histories, cost of a typical subscription list over the past five years and charts comparing journal costs and currency fluctuation for various types of libraries.

FE8. Fisher, Janet H., and John Tagler, eds. "Perspectives on Firm Serials Prices." *Serials Review* 19, no. 4 (Winter 1993): 63–72.

This article in the "Balance Point" series looks at the importance of firm serials prices early in the budget planning cycle, a particularly important, but often overlooked, aspect. An ARL librarian, a subscription agent, and representatives from a university press and a commercial publisher give their perspectives. Includes March 1993 letter from ninety-five ARL directors to STM publishers and subscription agencies requesting firm price quotes by September 1. Interesting charts of typical time lines that compare library cycles with publishers' cycles.

FE9. Ketcham, Lee, and Kathleen Born. "The Art of Projecting: The Cost of Keeping Periodicals." *Library Journal* 118 (April 15, 1993): 42–48.

FE10. ———. "Projecting Serials Costs: Banking on the Past to Buy for the Future." *Library Journal* 119 (April 15, 1994): 44–50.

FE11. ———. "Serials vs. the Dollar Dilemma: Currency Swings and Rising Costs Play Havoc with Prices." *Library Journal* 120 (April 15, 1995): 43–49.

Library Journal annual Periodical Price Surveys use five years of cost history information supplied by EBSCO Subscription Services. Three Institute for Scientific Information (ISI) databases *(Arts and Humanities Citation Index, Social Sciences Citation Index,* and *Science Citation Index)* comprise the core list of journals surveyed. The authors categorize the information by subject and country of origin. Includes projections of cost trends for the coming year.

FF
Resource Sharing, Union Listing, Serials Holdings

FF1. Mullis, Albert A. "Access to Serials: National and International Cooperation." In *The International Serials Industry,* edited by Hazel Woodward and Stella Pilling, 233–58. Aldershot, Hampshire, England; Brookfield, Vt.: Gower, 1993.

Summarizes interlibrary lending arrangements, union lists, and databases in the United Kingdom, France, Germany, the United States and Canada, and Scandinavia.

FF2. Schaffner, Ann C., ed. "Perspectives on the Future of Union Listing." *Serials Review* 19, no. 3 (1993): 71–78.

Five contributors to this "Balance Point" article examine changing factors and new developments affecting the need for union lists. Maintenance of holdings lists for efficient resource sharing is critical in times of massive cancellation projects, despite the expense involved. Preservation information is also an important incentive for maintaining union lists. Competition from databases offering article-level access, such as CARL UnCover, may affect future uses.

FF3. Tsui, Susan L. "Periodical Records Conversion: From Union List to Statewide Network." *The Serials Librarian* 26, no. 2 (1995): 95-112.

Step-by-step procedures used to convert the University of Dayton Libraries' brief periodical records to MARC records with the use of their OCLC Union List tape. This conversion also allowed the Libraries to participate in the Ohio Library Information Network (OhioLINK).

G

Collection Management

Genevieve S. Owens

As Peggy Johnson noted in her introduction to the Collection Management chapter of the *Guide to Technical Services Resources*, the literature in this field is voluminous. While information has continued to flow freely in the years since the *Guide's* publication, readers will note that this chapter does not cite a definitive compilation of experts' essays on the topic as a whole. No latter-day equivalent of *Collection Management: A New Treatise* (Greenwich, Conn.: JAI Press, 1991), the two-volume set edited by Charles B. Osburn and Ross Atkinson has been published. The rapid pace of change in libraries and the larger technological and cultural environments make writing definitive, comprehensive pieces on collection management difficult.

Does this state of affairs leave students, researchers, and practitioners wandering through a wilderness of disjointed professional literature? Thankfully, due to the labors of the authors cited in this

---

The compiler is grateful to her colleagues at Bucknell University's Bertrand Library for their assistance with this project. Special thanks go to the hardworking interlibrary loan staff, who cheerfully procured and personally delivered a host of material. The compiler also thanks her husband, Brent, for enduring her long hours with this chapter with his usual patience and sympathy.

chapter, the answer is a resounding "no." While recent literature has often focused on very specific aspects of collection management, it also has concerned itself with identifiable issues and trends. The most obvious theme in the work, as witnessed by the size of the Nonprint Media section of this chapter, is the advent of electronic information resources. Collection managers are struggling to master the nuts and bolts of new technologies like the Internet and electronic texts. They are exploring ways to apply and adapt well-established collecting policies and principles to those technologies. In the midst of these very practical activities, collection managers also are questioning what the technology means for the short- and long-term future of collections and working to move beyond the myths and rhetoric that surround the electronic revolution. This topic's significance is reinforced by the theme of the first ALCTS Advanced Collection Management and Development Institute, held in 1993. Indeed, the Institute's proceedings, *Collection Management and Development: Issues in an Electronic Era* **(GAE3),** while general enough to earn a place in the Overview of Collection Management section, also serve as a valuable introduction to many important facets of this issue.

While modern technologies figure prominently in the recent collection management literature, postmodern themes run through much of it, as well. In some instances, postmodernity manifests itself in pieces on collecting in more inclusive ways, in selecting materials that support diversity and multiculturalism. In women's studies, one of the first areas in which the need for inclusive collecting came to the fore, it is intriguing to see three articles devoted to the assessment of indexing sources. As this coverage demonstrates, the field has clearly moved from the margins in which it began to a more central place in the realm of scholarly inquiry. The notion of margins and the mainstream also shapes two thoughtful articles on canons and collection managers' roles in both assigning materials to and keeping them from these two extremes.

In other instances, postmodern themes emerge in articles on what might be termed the *culture wars*, pieces on the struggle between proponents of all-inclusive collecting and those who would collect more conservatively. For librarians, many if not most of whom embrace the American Library Association's intellectual freedom principles, the debate often centers more on what constitutes inclusive, unbiased collecting rather than the need for such programs, *per se*. Despite this chapter compiler's efforts to confine the Censorship and Intellectual Freedom section to pieces on collections matters rather than broader discussions, one is struck by its size and this topic's continued, even heightened, importance at the century's end. Its connection with

ture of New York State Agriculture and Rural Life." *Library Resources & Technical Services* 37 (Oct. 1993): 434–43.

The approach described in this article is of critical importance in the area of selection for preservation, subsequently dubbed the "whole-discipline approach" or the "core literature approach." The methodology involves "a partnership of librarians and scholars seeking to identify and preserve the most significant agricultural literature of a state," in this example, New York State. Has been linked subsequently with Gwinn's report for USAIN, *A National Preservation Program for Agricultural Literature* **(HD2)**.

STATEWIDE PRESERVATION PLANNING

The following is a partial list of the statewide preservation programs that have been published from 1992 to 1995, funded in large part by the National Endowment for the Humanities Division of Preservation and Access or by the National Historical Publications and Records Commission:

HD9. California State Library. *The California Preservation Program.* Sacramento: California State Library, 1995.

HD10. Colorado Preservation Alliance. *On the Road to Preservation: A State-Wide Preservation Action Plan for Colorado.* Denver: Colorado Preservation Alliance, 1993.

HD11. Florida State Historical Records Advisory Board. *Historical Records Advisory Board Strategic Plan.*Tallahassee: Florida State Historical Records Advisory Board, 1994.

HD12. Massachusetts Task Force on Preservation and Access to the Library, Archives, Public Records, and Governmental Communities and the Citizens of Massachusetts. *Preserved to Serve: The Massachusetts Preservation Agenda.* Boston: Massachusetts Board of Library Commissioners, 1992.

HD13. Michigan. State Historical Records Advisory Board. *Strategies to Preserve Michigan's Historical Records.* Lansing: Michigan History Magazine, 1994.

HD14. Ohio Historical Records Advisory Board; Statewide Preservation Planning Committee. *The Ohio 2003 Plan: A Statement of Priorities and Preferred Approaches to Historical Records Programs in Ohio;* and *To Outwit Time: Preserving Materials in Ohio's Libraries and Archives.* Columbus: Ohio Historical Society, the State Library of Ohio, 1995.

HD15. Oklahoma Historical Records Advisory Board. *To Save Our Past: A Strategic Plan for Preserving Oklahoma's Documentary Heritage.* Oklahoma City: Oklahoma Department of Libraries, 1994.

HD16. Rhode Island Council for the Preservation of Research Resources. *Bricks and Mortar for the Mind: Statewide Preservation Program for Rhode Island.* Providence: Rhode Island Department of State Library Services, 1992.

HD17. South Carolina. State Historical Records Advisory Board. *Palmetto Reflections: A Plan for South Carolina's Documentary Heritage.* Columbia: South Carolina Department of Archives and History, 1994.

HD18. Walter, Katherine L. *Saving the Past to Enrich the Future: A Plan for Preserving Information Resources in Kansas.* Topeka: Kansas Library Network, 1993.

HE
Copyright

HE1. Bruwelheide, Janis H. *The Copyright Primer for Librarians and Educators.* 2nd ed. Chicago: American Library Association; Washington, D.C.: National Education Association, 1995.

A good first source for basic copyright information. Covers general information; fair use; library copying under the law; classroom copying; college and university copying and the library reserve room; musical, dramatic, and nondramatic performance; off-air taping; videotape; the electronic environment; software; multimedia; Internet; distance education concerns. Appendixes and bibliographical references. Index. *See also* **HE2.**

HE2. ———. *Copyright Issues for the Electronic Age.* Syracuse, N.Y.: ERIC Clearinghouse, April 1995. EDO-IR-95-3.

A two-page set of questions and answers about existing copyright law. Defines multimedia and addresses right of transmission and library exceptions. References.

HE3. *Intellectual Property and the National Information Infrastructure: The Report of the Working Group on Intellectual Property Rights.* Bruce Lehman, Commissioner of Patents and Trademarks. Washington, D.C.: Information Infrastructure Task Force, 1995. Available online at URL: http://www.upsto.gov/web/ipnii/

Commonly known as "The White Paper," this important document relates to the National Information Infrastructure (NII), intellectual property, copyright, patents, and information technology. The report provides explanations of legal issues of copyright in cyberspace. Makes recommendations for changes in laws that affect intellectual property, including clarification of existing rights; application of fair use privileges; technological protection systems; and support of [then] pending legislation.

HE4. *Reproduction of Copyrighted Works by Educators and Librarians.* Circular 21. Washington, D.C.: Copyright Office, Library of Congress, 1995. Available online at URL: ftp://ftp.loc.gov/pub/copyright/circs/circ21
   A retitling of the 1977 "Circular R21" entitled *Copyright and the Librarian,* which covers the text of the Copyright Act of 1976, sections 106 and 504.

HF
Disaster Control, Recovery, Insurance, and Security

HF1. George, Susan C., comp. *Emergency Planning and Management in College Libraries.* CLIP Note, no. 17. Chicago: Association of College & Research Libraries, American Library Association, 1994.
   A small volume (142 pages) covering the issues of disaster control and contingency planning in college libraries. Includes several emergency plans and safety documentation examples from U.S. college libraries. Bibliographic references.

HF2. Kahn, Miriam B. *Disaster Response and Prevention for Computers and Data.* Columbus, Ohio: MBK Consulting, 1995.
   This resource and the following piece by Kahn **(HF3)** are companion volumes in loose-leaf binder format. This title covers "basic information for responding to a disaster involving water damaged computers, hardware and software." Bibliography.

HF3. ———. *First Steps for Handling & Drying Water Damaged Materials.* Columbus, Ohio: MBK Consulting, 1994.
   For annotation, *see also* **HF2**. A "how-to" for handling wet books, papers and photographs; uses clear and basic line drawings with accompanying text. Bibliography. Readers also may wish to consult the author's *Disaster Prevention and Response for Special Libraries: An Information Kit* (Washington, D.C.: Special Libraries Association, 1995).

HF4. **Trinkaus-Randall, Gregor.** *Protecting Your Collections: A Manual of Archival Security.* Chicago: Society of American Archivists, 1995.

A good addition to the literature addressing a perennial problem. This eighty-four-page volume updates Timothy Walch's *Archives & Manuscripts: Security* (Chicago: Society of American Archivists, 1977) and includes security issues, such as a theft and destruction of materials, and disasters, such as fire and flood. Illustrated, with references and a bibliography; appendixes to supplement text.

HG
Environmental Control and Pest Management

HG1. *Bugs, Mold & Rot. [Proceedings of] A Workshop on Residential Moisture Problems, Health Effects, Building Damage, and Moisture Control.* Washington, D.C.: National Institute of Building Sciences, 1991.

HG2. *Bugs, Mold & Rot II. [Proceedings of] A Workshop on Control of Humidity for Health, Artifacts, and Buildings.* Washington, D.C.: National Institute of Building Sciences, 1993.

Not library or archival literature, *per se,* but both sources (**HG1** and **HG2**) are nevertheless important sources of information on controlling the indoor environment, especially relative humidity and its relationship to fungal growth and insect pests. Bibliographic references.

HG3. **Harmon, James D.** *Integrated Pest Management in Museum, Library, and Archival Facilities: A Step by Step Approach for the Design, Development, Implementation, & Maintenance of an Integrated Pest Management Program.* 1st ed. [Indianapolis]: Harmon Preservation Pest Management, 1993.

A loose-leaf format that, in this edition, is 140 pages. A useful addition to the literature. Bibliographic references.

HG4. *IPI Storage Guide for Acetate Film.* Rochester, N.Y.: Image Permanence Institute, 1993.

A valuable set of management tools that enables the collection manager to approximate the life span of acetate films related to the onset of "vinegar syndrome," the chemical deterioration leading acetate films to exude a vinegar-like odor. The tools included in this set are a booklet, wheel, graphs, and table which provide for a practical and an ingenious method for translating IPI's research into management action. Bibliography.

HG5. McCrady, Ellen. "Indoor Environment Standards: A Report on the NYU Symposium." *Abbey Newsletter* 19 (Dec. 1995): 93–98.

A good and succinct summary, with editorial comment, of the various scientific discussions regarding standards for temperature, relative humidity, and allowable fluctuations of temperature and relative humidity in repositories. This is an area with considerable controversy and, at this writing, not resolved. *Abbey Newsletter* (18 [Aug./Sept. 1994]: 45) published the Smithsonian Institution Conservation Analytical Laboratory (CAL) press release, an early volley in the debate. Other publications have also covered the controversy. Bibliographic references.

HG6. Price, Lois Olcott. *Managing a Mold Invasion: Guidelines for Disaster Response.* Technical Series, no. 1. Philadelphia: Conservation Center for Art and Historic Artifacts, 1994.

There is little practical advice on controlling mold in collections accessible to laypersons. This important contribution covers causes, response techniques, and treatment recommendations. Bibliography.

HG7. Puglia, Steven. "Cost-Benefit Analysis for B/W Acetate: Cool/Cold Storage vs. Duplication." *Abbey Newsletter* 19 (Sept. 1995): 71–72.

A valuable and succinct consideration of the costs and benefits of duplicating acetate-based black-and-white photographic materials versus storing in cold conditions. Refers to the *IPI Storage Guide for Acetate Film* **(HG4)** as a basis for prediction of acetate life expectancy and recommended action.

HG8. Reilly, James M., Douglas W. Nishimura, and Edward Zinn. *New Tools for Preservation: Assessing Long-Term Environmental Effects on Library and Archives Collections.* Washington, D.C.: Commission on Preservation and Access, 1995.

A brief (thirty-five pages) but very significant report by Image Permanent Institute researchers that explains a new approach to considering environmental storage conditions for library and archival materials. Using the concepts and modeling of PI (preservation index) and TWPI (time weighted preservation index) and special monitoring equipment, the authors explain how shelf life expectancies of stored library or archival materials can be estimated at those ambient conditions. Appendixes. Bibliographic references. *See also* Sebera on isoperms **(HG9)**.

HG9. Sebera, Donald K. *Isoperms: An Environmental Management Tool.* Washington, D.C.: Commission on Preservation and Access, 1994.

A brief (sixteen pages) but critically important explanation, for the preservation manager and library administrator, of the scientific principles of passive preservation using the storage environment. Discusses the effects of relative humidity and temperature and their combined effects on the longevity of collections. Stresses how the isoperm theories can be used as quantitative management tools by the administrator. Bibliographic references. Also available online at URL: http://www-cpa.stanford.edu/cpa/reports/isoperm/isoperm.html

HG10. Trinkley, Michael. *Preservation Concerns in Construction and Remodeling of Libraries: Planning for Preservation.* Columbia: South Carolina State Library, 1992.

An informative volume discussing many basic building construction and design issues that affect the completed building's friendliness to the library collections. Discusses essential architectural details such as site selection, floor coverings, electrical systems, roofing, and other related factors. Bibliographic references.

HH
Imaging and Preservation Reproduction

HH1. Bagnall, Roger. *Digital Imaging of Papyri.* Washington, D.C.: Commission on Preservation and Access, 1995.

An eight-page report summarizing the collaboration of the Commission and the Advanced Papyrological Information System (APIS, a project of a group formed through the American Society of Papyrologists) to use digital technologies to improve access to collections of papyri, a very fragile medium. This report is the result of a conference at the University of Michigan in March of 1994. Bibliographic references.

HH2. Conway, Paul. "Digitizing Preservation." *Library Journal* 119 (Feb. 1, 1994): 42–45.

A general introduction to imaging and optical storage. Conway suggests the potential of digital imaging systems for libraries and also identifies problem areas that are not yet resolved. He asks the question, "The technology is here—but why convert paper and microfilm into bits and bytes?" Sidebar lists pilot imaging projects.

HH3. ———. *Preservation in the Digital World.* Washington, D.C.: Commission on Preservation and Access, 1996.

This provocative report provides an intellectual rationale for maintaining the centrality of preservation concepts and ethics in an increasingly digital environment. Conway argues that digital imaging technology is more than another reformatting option. Imaging involves transforming the very concept of format; thus, a digital world transforms traditional preservation concepts from protecting the physical integrity of the object to specifying the creation and maintenance of the object. Extensive bibliography.

HH4. ———. "Selecting Microfilm for Digital Preservation: A Case Study from Project Open Book." *Library Resources & Technical Services* 40 (Jan. 1996): 67–77.

This piece focuses on selection of materials for conversion to digital images in Project Open Book. *See also* **HH5.**

HH5. Conway, Paul, and Shari Weaver. *The Setup Phase of Project Open Book: A Report to the Commission on Preservation and Access on the Status of an Effort to Convert Microfilm to Digital Imagery.* Washington, D.C.: Commission on Preservation and Access, 1994.

A twenty-four page report on the progress made at Yale University on a system to convert analog images on preservation microfilm to digital images. This is a critically important project, results of which will inform the profession about increasing the accessibility of materials on microfilm. Bibliographic references. *See also* **HH4.** Also available online at URL: http://www-cpa.stanford.edu/cpa/reports/conway.html

HH6. *Digital Imaging Technology for Preservation: Proceedings from an RLG Symposium Held March 17 and 18, 1994, Cornell University, Ithaca, New York.* Nancy E. Elkington, ed. Mountain View, Calif.: Research Libraries Group, Inc., 1994.

A series of papers and "tutorials" that serve to define the key issues related to the use of digital imaging technology for preservation, in the context of long-term access to preserved information. This is a very important set of papers from key personnel at a very important conference. Also of interest are the bibliographic references from each paper and a list of "vendor contacts." The publication of *RLG Digital Image Access Project* **(HH22)** continues discussion of the changes in access to digital images, especially scanned photographs.

HH7. Elkington, Nancy E., ed. *RLG Archives Microfilming Manual.* Mountain View, Calif.: Research Libraries Group, 1994.

Essential source of information for preservation microfilming of archival materials, covering every aspect of the process including program management, vendor selection to materials preparation, archival control, the filming itself, and quality assurance. Nine appendixes cover important issues that supplement the text. Well-organized and illustrated. Bibliography and index.

HH8. Ester, Michael. "Image Quality and Viewer Perception." *Visual Resources* 7 (1991): 51–63.

Often-cited article related to digital resolution levels necessary for adequate reader acceptability. *See also* the author's *Draft White Paper on Digital Imaging in the Arts and the Humanities.* (San Marino, Calif.: Getty Conservation Institute, Getty Art History Information Program. Initiative on Electronic Imaging and Information Standards, March 3–4, 1994; available from the Getty Conservation Institute.)

HH9. Fox, Lisa L., ed. *Preservation Microfilming: A Guide for Librarians and Archivists.* 2nd ed. Chicago: American Library Association, 1996.

A welcome update to the 1987 version edited by Nancy E. Gwinn. In this larger edition (394 pages), each chapter has been updated and the number and scope of the appendixes expanded. Illustrated, bibliography, and index.

HH10. Gertz, Janet. *Oversize Color Images Project, 1994–95: Final Report of Phase I.* Washington, D.C.: Commission on Preservation and Access, 1995.

Results of the first phase of a project "to identify acceptable preservation and digital access techniques for dealing with oversize, color images associated with text." The report summarizes the project's goals, describes the five large (up to 23-inch × 25.5-inch) maps selected for the project, reviews project methodology and vendor selection, and summarizes results, including user evaluation, bibliographic control issues, and costs. Appendixes; bibliographic references. *See also* **HH11.**

HH11. ———. "Selection for Preservation: A Digital Solution for Illustrated Texts." *Library Resources & Technical Services* 40 (Jan. 1996): 78–83.

Another report on the Oversize Color Images Project **(HH10)** with an emphasis on the importance of the marriage of technical and selection-for-preservation issues.

HH12. Graham, Peter S. *Intellectual Preservation: Electronic Preservation of the Third Kind.* Washington, D.C.: Commission on Preservation and Access, 1994.

A brief (seven-page) but widely cited paper on the importance of security, encryption, and document identification in the context of electronic text mutability. Outlines a system of time-stamping documents. Bibliographic references. Also available online at URL http://www-cpa.stanford.edu/cpa/reports/graham/intpres.html

HH13. *Guidelines for Electronic Preservation of Visual Materials.* Part I. Submitted to the Library of Congress by Picture Elements, Inc. Washington, D.C. and Berkeley, Calif.: Picture Elements, Inc., 1995.

A paper prepared for the Library of Congress to address several practical and theoretical aspects in the conversion of texts and images to digital images. (The title's term "visual materials" is more often referred to as "images.") The report can be read as a basic primer of digital imaging basics and assumptions. The report points to the superiority of gray-scale over binary digital images for the best results in the reproduction of text and illustrations. The report is twenty-three pages of text, with appendixes including examples of images, glossary and brief bibliography.

HH14. Jones, C. Lee. *Preservation Film: Platform for Digital Access Systems.* Washington, D.C.: Commission on Preservation and Access, 1993.

A three-page position paper encouraging those creating preservation microfilm to produce film of the highest quality (resolution and density) possible to enable the conversion of analog filmed images to digital formats. The writer advocates that libraries and archives move beyond films created by the Kodak MRD cameras (no longer being manufactured, but still very much in use) to cameras that render images with higher resolution than the MRDs.

HH15. Kenney, Anne R. "Digital-to-Microfilm Conversion: An Interim Preservation Solution." *Library Resources & Technical Services* 37 (Oct. 1993): 380–401. Erratum in *Library Resources & Technical Services* 38 (Jan. 1994): 87–95.

Summarizes the results of a research project at Cornell University to demonstrate the technical feasibility of producing preservation-quality computer output microform (COM) from digital imagery. The resultant film is capable of meeting ANSI/AIIM standards for image quality. Other nontechnical issues are discussed. Bibliographic references. *See also* Kenney and Personius (**HH17**).

HH16. Kenney, Anne R., and Stephen Chapman. *Tutorial: Digital Resolution Requirements for Replacing Text-Based Material: Methods for Benchmarking Image Quality.* Washington, D.C.: Commission on Preservation and Access, 1995.

An important work introducing resolution requirements for paper-based documents converted to digital imagery. Format is truly a "tutorial" approach with chapter headings reading "Where to Begin," "How to Determine Digital Image Quality," "Scanning Methodologies and Compression Techniques," and sections covering benchmarking of resolution requirements for bitonal and grayscale materials. Discusses the digital quality index, visual inspection, and recommended guidelines for scanning monographs and serials; agency records; manuscript materials; and halftone and continuous tone images. Twenty-two pages. Bibliographic references.

HH17. Kenney, Anne R., and Lynn K. Personius. *A Testbed for Advancing the Role of Digital Technologies for Library Preservation and Access: Final Report by Cornell University to the Commission on Preservation and Access.* Washington, D.C.: Commission on Preservation and Access, 1993.

Final report of the Cornell University research project that explored the technical feasibility of producing preservation-quality computer output microform from digital imagery. Identifies technical, practical, and policy issues. *See also* **HH15**.

HH18. McClung, Patricia A. *Digital Collections Inventory Report.* Washington, D.C.: Council on Library Resources, Commission on Preservation and Access, 1996.

This report (prepared in 1995) provides "a snapshot of scanning projects at a particular point in time." McClung determined how much digitizing of library collections was planned, under way, or completed. Provides project descriptions and contact information. Includes brief analysis of questions and needs related to the development of inventory tool. It is intended to stimulate discussion.

HH19. *Preservation of Electronic Formats & Electronic Formats for Preservation.* Janice Mohlhenrich, ed. Fort Atkinson, Wis.: Highsmith, 1993.

Published papers from six nationally recognized experts in digital imaging from the 1992 Wisconsin Preservation Program (WISPPR). The six papers, with the question-and-answer responses from the conference, make for a good introduction to the issues. Illustrated. Annotated bibliography and index.

HH20. *Preserving Scientific Data on Our Physical Universe: A New Strategy for Archiving the Nation's Scientific Information Resources.* Washington, D.C.: National Academy Press, 1995.

A study commissioned by the National Archives, the National Oceanic and Atmospheric Administration, the National Aeronautics and Space Administration, and the National Research Council (U.S.) to study the long-term preservation of masses of scientific data and information from the federal government and stored in electronic formats. Bibliographic references. A related report issued concurrently is **HH25.**

HH21. Rhyne, Charles S. *Computer Images for Research, Teaching, and Publication in Art History and Related Disciplines.* Washington, D.C.: Commission on Preservation and Access, 1996.

Twelve-page paper advocating participation by scholars from key disciplines, such as art history, in the process of computer imaging in those disciplines. The paper discusses the effects of cost and rapidly changing technology, and the potential value of computer images in scholarship. Stresses value of quality and accuracy of the images converted to digital imagery. Bibliographic references.

HH22. *RLG Digital Image Access Project: Proceedings from an RLG Symposium held March 31 and April 1, 1995, Palo Alto, California.* Patricia A. McClung, ed. Mountain View, Calif.: Research Libraries Group, 1995.

Reflections of the ongoing work by this project's participants to understand the complexities of converting photograph collections to digital images for preservation and/or enhanced access. The primary issues are intellectual control and retrieval, rather than technical reformatting. Bibliographic references.

HH23. Robinson, Peter. *The Digitization of Primary Textual Sources.* Office for Humanities Communication Publications, no. 4. Oxford: Oxford University Office for Humanities Communication, 1993.

Often cited in the literature related to the conversion of analog texts to digital formats. Bibliographic references.

HH24. Rothenberg, Jeff. "Ensuring the Longevity of Digital Documents." *Scientific American* 272 (Jan. 1995): 42–47.

An interesting discussion of the inherent difficulties of preserving documents in digital form, written for the well-informed layperson. Bibliographic references.

HH25. *Study on the Long-term Retention of Selected Scientific and Technical Records of the Federal Government: Working Papers.* Washington, D.C.: National Academy Press, 1995.

For annotation, *see* **HH20**.

HH26. Task Force on Archiving of Digital Information. *Preserving Digital Information.* Version 1.0. Washington, D.C.: Commissioned by the Commission on Preservation and Access, and the Research Libraries Group, 1995.

First version of very important work being done to outline the problems of preserving (archiving) digital information. Sets out roles and responsibilities and makes recommendations. Appendixes. Bibliographic references.

HH27. Trader, Margaret P. "Preservation Technologies: Photocopies, Microforms, and Digital Imaging—Pros and Cons." *Microform Review* 22 (Summer 1993): 127–34.

An overview of the competing reformatting options and technologies facing libraries. Literature review, with a conclusion about the probable coexistence of mixed formats well into the future. Bibliographic references.

HH28. Van Houweling, Douglas E., and Michael J. McGill. *The Evolving National Information Network: Background and Challenges. A Report of the Technology Assessment Advisory Committee to the Commission on Preservation and Access.* Washington, D.C.: Commission on Preservation and Access, 1993.

Concise explanatory survey of the structure of the national information network, bringing together in one source a great deal of information about the components of this infrastructure. Explains, in context with some history, many terms that are increasingly in common usage. Written before the "World Wide Web" was a household term, but of real value for background knowledge. Appendix. Also available online at URL http://www-cpa.stanford.edu/cpa/reports/evolv/index.html

HH29. Waters, Donald J., and Anne Kenney. *The Digital Preservation Consortium: Mission and Goals.* Washington, D.C.: Commission on Preservation and Access, 1994.

A six-page description of the need, mission, and goals of the eleven-member consortium of research libraries (at time of its writing) within the development of the National Information Infrastructure and the Internet. In brief, the mission of the DPC is "to advance the use and utility of digital technology for preserving and improving access to intellectual works of national and international importance." Also available online at URL http://www-cpa.stanford.edu/cpa/reports/dpcmiss.html

HI
Management and Organization of Preservation Programs

HI1. *Automating Preservation Management in ARL Libraries.* Patricia Brennan and Jutta Reed-Scott, comps. SPEC Kit, no. 198. Washington, D.C.: Association of Research Libraries, Office of Management Services, 1993.

The results of a mid-1993 survey among ARL libraries to ascertain applications in preservation units. Summarizes responses and provides examples of MIS and fiscal management systems; production statistics; library binding; replacement and reformatting; condition surveys and needs assessment; and environmental monitoring.

HI2. Baird, Brian J. "Motivating Student Employees: Examples from Collections Conservation." *Library Resources & Technical Services* 39 (Oct. 1995): 410–16.

Collects and summarizes principles from the personnel management literature to make a useful addition to the literature of library materials conservation. Includes tips and practical approaches for motivating student employees and increasing their productivity. Bibliographic references. For a fuller approach to the broader subject, one might consult Black **(HI4)** and Kathman and Kathman **(HI8)**.

HI3. Banks, Jennifer. *Options for Replacing and Reformatting Deteriorated Materials.* Washington, D.C.: Association of Research Libraries, Office of Management Services, 1993.

One of a series of brief monographs ("resource guides") written for ARL and other libraries that are undertaking reviews of the preservation programs. Provides program overview, defines terms, outlines current practices in libraries, and outlines strategies in replacement and reformatting. Appends significant documents from other institutions. Bibliographic references.

HI4. Black, William K., ed. *Libraries and Student Assistants: Critical Links.* Binghamton, N.Y.: Haworth Press, 1995. Also published as *Journal of Library Administration* 21, nos. 3/4 (1995).

A far-reaching collection of twelve papers covering many aspects of student employees in libraries, including their changing roles, recruiting, training, motivating, rewarding, and more. One paper is an extensive annotated bibliography on student employees in academic libraries.

HI5. DeStefano, Paula. "Use-Based Selection for Preservation Microfilming." *College & Research Libraries* 56 (Sept. 1995): 409–18.

The author reviews the literature related to the practices of the national preservation microfilming strategy that rely heavily on subject-based, collection-based, "clean sweep" approaches in selecting materials for preservation microfilming in research libraries. Circulation (use) is extolled and defended as a pragmatic and sensible alternative approach to selection for preservation. Bibliographic references.

HI6. Gertz, Janet, et al. "Preservation Analysis and the Brittle Book Problem in College Libraries: The Identification of Research-Level Collections and Their Implications." *College & Research Libraries* 54 (May 1993): 227–38.
For annotation, *see* **GE32**.

HI7. Haynes, Douglas. "Pro-Cite for Library and Archival Condition Surveys." *Library Resources & Technical Services* 39 (Oct. 1995): 427–33.
Haynes explains how the software Pro-Cite was used at the Classics Library at the University of Texas at Austin in a condition survey of 24,000 monographs. Advantages of this software include the ability to perform Boolean searching; disadvantages include inability to perform mathematical and statistical compilations and to generate charts or graphs.

HI8. **Kathman, Michael, and Jane McGurn Kathman, comps.** *Managing Student Employees in College Libraries.* CLIP Note, no. 20. Chicago: Association of College & Research Libraries, American Library Association, 1994.
Well-trained student workers, who know how to handle library materials, are essential. For annotation, *see* **IC8**.

HI9. Kellerman, Lydia Suzanne. "Moving Fragile Materials: Shrink-Wrapping at Penn State." *Collection Management* 18, nos. 1/2 (1993): 117–28.
A description of a practical application of shrink-packaging for library preservation purposes, an application increasingly used for phased treatments as well as for protection during collections moving. *See also* Stagnitto **(HI15)** for a similar project.

HI10. *Managing Preservation: A Guidebook.* A Cooperative Publication of the State Library of Ohio and the Ohio Preservation Council. [Columbus]: State Library of Ohio, Ohio Preservation Council, 1995.
For annotation, *see* **HA6**.

HI11. *Managing the Preservation of Serial Literature: An International Symposium. Conference held at the Library of Congress, Washington, D.C., May 22–24, 1989.* Merrily A. Smith, ed. Sponsored by the International Federation of Library Associations and Institutions and the Library of Congress. Munich, New York: K. G. Saur, 1992.
A significant publication about a daunting problem. In six parts covering preservation decision making in serials management; preservation in original

format; preservation in secondary format; information needs for managing serials preservation; regional, national, and international serials bibliographic programs; and cooperative international serials preservation. Bibliographies and index.

HI12. "MLA's Statement on the Significance of Primary Records." *Abbey Newsletter* 19 (Dec. 1995): 101–2. Originally published in MLA's *Profession 95* (1995): 27–28.

A reprinting of the Modern Language Association's advocacy statement for the retention of original materials in original formats. The issuance of this document in a previous draft format resulted in lively discussion within preservation and collection development ranks. Recommended articles (from two scholars who served on MLA's ad hoc committee) include Ruth Perry's "Embodied Knowledge" (originally published in *Harvard Library Bulletin New Series* 3 [Summer 1993]: 57–62) and G. Thomas Tanselle's "The Latest Forms of Book-Burning" (originally published in *Common Knowledge* 2 [Winter 1993]: 172–77).

HI13. Ogden, Barkley W., and Maralyn Jones. *CALIPR: An Automated Tool to Assess Preservation Needs of Books and Document Collections for Institutional or Statewide Planning.* Computer software. Sacramento: California State Library, 1991.

Widely cited and used tool for collections condition surveys. Includes disks and a manual. For other recent research on use of condition survey methodology, see Haynes **(HI7)**. For an example of a project that used the CALIPR methodology, see *Preservation Needs Assessment: A Management Tool: Final Report of the 1992–1993 Central New York Preservation Needs Assessment Project* (Syracuse: Central New York Library Resources Council, 1993).

HI14. O'Neill, Edward T., and Wesley L. Boomgaarden. "Book Deterioration and Loss: Magnitude and Characteristics in Ohio Libraries." *Library Resources & Technical Services* 39 (Oct. 1995): 394–408.

The results of a large-scale sampling of monographs from a ninety-year publishing period (1851–1939) held in Ohio libraries, to ascertain both physical condition and availability. The study's interest lies in its approach using a geographical area (an entire state), its findings on comparable conditions of exact editions in that geographical area among varied types of libraries, and the documented impact of the nationwide preservation microfilming effort on sample titles. Bibliographic references.

HI15. Stagnitto, Janice. "The Shrink-Wrap Project at Rutgers University Special Collections and Archives." *Book and Paper Group Annual* 12 (1993): 56–60.

For annotation, *see* **HI9**.

HJ
Media (Nonbook and/or Nonpaper) Preservation and Reproduction

HJ1. Boyle, Deirdre. *Video Preservation: Securing the Future of the Past.* New York: Media Alliance, 1993.

Summarizes the results of a 1991 Media Alliance survey and symposium which included major repositories of videotape in 1991. A good summary of the issues in preservation and storage of videotape. Includes a glossary of video terms, lists of video preservation facilities, and bibliographic references.

HJ2. Child, Margaret S. *Directory of Information Sources on Scientific Research Related to the Preservation of Sound Recordings, Still and Moving Images and Magnetic Tape.* Washington, D.C.: Commission on Preservation and Access, 1993.

A fourteen-page directory "compiled to provide information about the current work of laboratories and researchers in non-print media in order to help explore and define possible areas requiring preservation work." Divided into Part I: "Laboratories and Organizations" (such as AIC, AMIA, ARSC, CCI, IPI, etc.) and Part II: "Sources of Information" (databases, serials, monographs, articles, conference proceedings). Also available online at URL http://www-cpa.stanford.edu/cpa/reports/child.html

HJ3. *Encyclopedia of Recorded Sound in the United States.* Guy A. Marco, ed. New York: Garland, 1993.

A significant publication (910 pages) in which preservation is covered with an entry by Susan G. Swartzburg entitled "Preservation of Sound Recordings" (pp. 542–46).

HJ4. *Film Preservation 1993: A Study of the Current State of American Film Preservation. A Report of the Librarians of Congress.* Annette Melville and Scott Simmon, comps. Washington, D.C.: Library of Congress, National Film Preservation Board, 1993. Available online at URL: http://lcweb.loc.gov/film/study.html

A mandated report (four volumes in three) on the state of American film preservation. Extensive research. Includes hearings held in Los Angeles and Washington, D.C. *See also* the plan, *Redefining Film Preservation* **(HJ10)**, which followed this study.

227

HJ5. *Knowing the Score: Preserving Collections of Music* Mark Roosa and Jane Gottlieb, comps. Chicago: Association for Library Collections & Technical Services, 1994.

A volume with chapters by knowledgeable professionals about the unique preservation and conservation requirements of music scores. Bibliographic references. Index.

HJ6. Lindner, Jim. "Confessions of a Videotape Restorer; Or, How Come These Tapes All Need to Be Cleaned Differently?" *AMIA Newsletter* no. 24 (April 1994):9–10.

A description of creating preservation copies of important videotape, by an experienced professional in the business.

HJ7. Palmer, Joseph W. "Saving Those Historic Videotapes: It May Already Be Too Late." *Public Libraries* 33 (March/April 1994): 99–101.

A nontechnical, yet informative, article on creating preservation copies of videotapes.

HJ8. *Permanence, Care, and Handling of CDs.* Online. Kodak Home Page. URL: http://www.kodak.com/daiHome/techInfo/permanence.shtml

A practical guide to CD longevity and care. Contents include discussion of how long CDs (audio, CD-ROM, data discs, and photo CDs) can last, safe handling, and storage conditions, as well as CD permanence in perspective. Looks at writable CDs as well as commercial products, long-term aging as well as catastrophic failure.

HJ9. *Preservation of Archival Materials: A Report of the Task Force on Archival Selection to the Commission on Preservation and Access.* Washington, D.C.: Commission on Preservation and Access, 1993.

A seven-page summary, with recommendations, of the primary preservation concerns in archives repositories. Complements the Commission's extensive literature related to the preservation of published books and serials. Also available online at URL: http://www-cpa.stanford.edu/cpa/reports/arcrept.html

HJ10. *Redefining Film Preservation: A National Plan: Recommendations of the Librarian of Congress in Consultation with the National Film Preservation Board.* Annette Melville and Scott Simmon, comps. Washington, D.C.: Library of Congress, 1994. Available on online at URL: http://lcweb.loc.gov/film/plan.html

The published "action plan to save America's motion picture heritage," this thirty-page report, with recommendations (with nearly fifty additional pages of supporting documentation), concludes wide-ranging discussions mandated by the National Film Preservation Act of 1992. *See also* the study, *Film Preservation 1993* **(HJ4)**, which preceded the plan.

HJ11. Ritzenthaler, Mary Lynn. *Preservation of Archival Records: Holdings Maintenance at the National Archives.* Technical Information Paper, no. 6. Washington, D.C.: National Archives and Records Administration, 1990.

Looks at preservation policies and practices at the U.S. National Archives.

HJ12. ———. *Preserving Archives and Manuscripts.* Chicago:The Society of American Archivists, 1993.

A thorough revision of Ritzenthaler's important *Archives and Manuscripts: Conservation* (SAA, 1984). An essential reference and background source for this area. Covers deterioration and its causes; storage and housing; conservation; and related topics. Bibliography. *See also* the author's earlier *Preservation of Archival Records* **(HJ11)**.

HJ13. *Safe Handling, Storage, and Destruction of Nitrate-Based Motion Picture Films.* Kodak Publication H-182. Rochester, N.Y.: Eastman Kodak, [n.d.].

Brief (eight-page) guide for those who must handle potentially dangerous and impermanent nitrate films. The title is descriptive. The volume also includes information about reformatting films, evaluating the content of the films, and where to find additional information and advice.

HJ14. Schrock, Nancy Carlson, and Mary Campbell Cooper. *Records in Architectural Offices: Suggestions for the Organization, Storage and Conservation of Architectural Office Archives.* 3rd rev. ed. Cambridge: Massachusetts Committee for the Preservation of Architectural Records, 1992.

An important work in its third edition, covering these primary areas: current practice, organizing office records, preserving office records, summary recommendations, and a section on available resources for further information and assistance. Appendix and bibliography.

HJ15. Smolian, Steven. *Preservation Appraisal of Sound Recordings Collections.* Frederick, Md.: Smolian Sound Studios, 1995.

A pamphlet designed to assist the sound-collection archivist in examining the collection from a preservation perspective. Covers the range of sound archives preservation, including analysis of format and condition; preservation transfer; and quality control issues.

HJ16. Tuttle, Craig A. *An Ounce of Preservation: A Guide to the Care of Papers and Photographs.* Highland City, Fla.: Rainbow Books, 1995.

A small volume (111 pages) for general audiences. Bibliographic references. Index.

HJ17. Van Bogart, John W. C. *Magnetic Tape Storage and Handling: A Guide for Libraries and Archives.* Washington, D.C.: Commission on Preservation and Access; St. Paul, Minn.: National Media Laboratory, 1995.

An excellent and concise (thirty-four pages) introduction to the basics of magnetic tape components, management of tape collections, life expectancy projections, and preservation principles. Includes the Ampex Recording Media Corporation's "Guide to the Care and Handling of Magnetic Tape," a brief bibliography, and a good glossary. See also **HJ18**.

HJ18. ———. *Media Stability Studies: Final Report.* NML Technical Report RE-0017. St. Paul, Minn.: National Media Laboratory, 1994.

Provides additional and more technical information on magnetic tape, going beyond the information provided in **HJ17**.

HJ19. Watt, Marcia, and Lisa Biblo. "CD-ROM Longevity: A Select Bibliography." *Conservation Administration News* 60 (Jan. 1995): 11–13.

An unannotated list of about thirty-five sources on the topic. Despite the dates on the publications (the most recent of which is from 1993), the list can be of value in doing research on life expectancy issues in this medium.

HJ20. Wilhelm, Henry, with contributing author Carol Brower. *The Permanence and Care of Color Photographs: Traditional and Digital Color Prints, Color Negatives, Slides, and Motion Pictures.* Grinnell, Iowa: Preservation Publishing Company, 1993.

Landmark and long-awaited work by the authority on stability, permanence, and storage of color photographs. This volume is the result of years of research and testing by Wilhelm and others. This edition is nearly 750 pages in length, profusely illustrated, supplemented by many graphs and tables, and studded with practical recommendations. Bibliographies. Index.

HK
Microform Publications

HK1. *Bibliographic Guide to Microform Publications: 1992.* Boston, Mass.: G. K. Hall, 1986– . ISSN 0891-3749.

Latest volume of a publication that lists the non-serial microforms created and cataloged by the New York Public Library and the Library of Congress. Includes information on the microforms' polarity, dimensions, color, emulsion, generation, and base.

HK2. Schwartz, Werner. *The European Register of Microform Masters—Supporting International Cooperation.* Washington, D.C.: Commission on Preservation and Access, 1995.

An explanation and update of the effort to create and maintain a dynamic master list of master microforms created by European repositories, as part of the Commission on Preservation and Access's International Program. Bibliographic references.

HK3. Subcommittee on Contract Negotiations for Commercial Reproduction of Library & Archival Materials. "Contract Negotiations for the Commercial Microforms Publishing of Library and Archival Materials: Guidelines for Librarians and Archivists." *Library Resources & Technical Services* 38 (Jan. 1994): 72–85.

Comprehensive set of guidelines, from librarians' or archivists' perspective, for consideration when working with a microform publisher for commercial publication of library and archival materials. Bibliographic references.

HL
Photocopiers and Photocopying

HL1. Association for Library Collections & Technical Services, Reproduction of Library Materials Section, Subcommittee on Preservation Photocopying Guidelines, Copy Committee. "Guidelines for Preservation Photocopying." *Library Resources & Technical Services* 38 (July 1994): 288–92.

Documented guidelines for the creation of "full-size, paper-based reformatting operations for complete volumes using electrostatic copying technologies." These guidelines cover all the essential points in describing the characteristics of "preservation photocopies" (paper, binding, image permanence, image adhesion, image placement, margins, image quality, and notice of copy) as well as the procedures (preparation/collation, disbinding/cutting, illustrations, bibliographic description, inspection, and disposition of the original). Bibliography and appendix, in which an example of the "Notice of Copy" is reproduced.

HL2. Buyers Laboratory, Inc. "Test Reports on 15 Photocopiers." *Library Technology Reports* 30 (Sept./Oct. 1994): 541–673.

Library Technology Reports' periodic review of photocopiers for the library market.

HL3. Wright, Dorothy W. "Selecting a Preservation Photocopy Machine." *College & Research Libraries News* 55 (Jan. 1994): 14–18.

A straightforward methodology for selecting a photocopy machine to be used for preservation replacement photocopying in the library. In addition to other important points of a good copier, this paper stresses the need for a well-tuned machine and provides a succinct explanation for the "peel test" method of checking the fused image. Bibliographic references.

HL4. *Xerographic Toner Adhesion Method.* Version 9. Chester, S.C.: Sequa Chemicals, Inc. 1995. Loose-leaf.

A brief (six-page) explanation of how this company tests toner adhesion in photocopies. A sensible method for testing the relative permanence of photocopied documents, in light of the dearth of other available techniques. See, however, the "peel test" methodology **(HL3)**. This methodology is also advocated by Norvell M. M. Jones in *Archival Copies of Thermofax, Verifax, and Other Unstable Records,* National Archives Technical Information Paper, no. 5 (Washington, D.C.: National Archives and Records Administration, 1990).

HM
Staff Training and User Awareness

HM1. Cloonan, Michèle Valerie. *Global Perspectives on Preservation Education.* Munich: K. G. Saur, 1994.

The result of information obtained from fourteen countries in regard to preservation education in formal library school curricula; regional centers and networks; large or national libraries; and other methods of reaching students at all levels. Published under the auspices of the IFLA PAC Core Programme and with that group's perspective. Bibliography. Index.

HM2. Coleman, Christopher D. G., comp. *Preservation Education Directory.* 7th ed. Chicago: Association for Library Collections & Technical Services, American Library Association, 1995.

Lists preservation programs, courses, internships, and workshops in library schools and elsewhere. Reflects an increased activity and opportunity for preservation education since the previous (1990) edition.

HM3. Ford, Helen. *The Education of Staff and Users for the Proper Handling of Archival Materials.* A RAMP Study with Guidelines. Paris: UNESCO, 1991. PGI-91\WS\17.

Presents effective suggestions for educating handlers on the safe care and handling of materials. RAMP studies are available from the West Virginia Library Commission, Cultural Center, Charleston, WV 25305.

HM4. George, Gerald. *"Difficult Choices": How Can Scholars Help Save Endangered Research Resources?* Washington, D.C.: Commission on Preservation and Access, 1995.

A report that summarizes eight years of work with the Commission's scholarly advisory committees on how best to proceed in the preservation of research resources. The twenty-eight-page work indicates the "understandable reluctance of scholars to make choices because of the unpredictability of research needs" and the recommendation to focus on cooperative approaches to preservation of research materials. Bibliography. Membership lists of the Commission's scholarly advisory committees. For progress in a specific discipline, one could compare this report to the work *Preserving the Anthropological Record,* Sydel Silverman and Nancy J. Parezo, eds. 2nd ed. (New York: Wenner-Gren for Anthropological Research, Inc., 1995).

HM5. Intner, Sheila S. "Preservation Training for Library Users." *Technicalities* 14 (Sept. 1994): 7–10.

This article and and the following resource **(HM6)** are companion articles with titles descriptive of their content. Both discuss the issues with common sense and practicality. This piece discusses the daunting task of reaching library users with a positive message about the cumulative effects of usage and handling. Bibliographic references.

HM6. ———. "Training Staff for Preservation." *Technicalities* 14 (July 1994): 4–6.

A companion article to **HM5**. This article on library personnel's role in preservation groups strategies into the three areas of processing, handling, and storing materials for best effect.

HM7. Walters, Tyler O. "Breaking New Ground in Fostering Preservation: The Society of American Archivists' Preservation Management Training Program." *Library Resources & Technical Services* 39 (Oct. 1995): 417–26.

A description of the SAA Preservation Management Training Program, a significant three-year effort in continuing education for practicing archivists. Bibliographic references.

HN
Standards, Specifications, and Guidelines

Most resources cited below are office standards intended to assure consistency and quality in products and practices. These published standards are under almost constant review by organizations affected by them. Some resources, for example, the *AIIM Buying Guide* **(HN24)**, suggest specific specifications or offer guidelines to apply when selecting equipment and developing practices and procedures.

LIBRARY BINDING

HN1. **National Information Standards Organization.** *Data Elements for Binding of Library Materials.* National Information Standards Series. Bethesda, Md.: NISO Press, 1995. ANSI/NISO Z39.76-199x.

PAPER

HN2. ———. *American National Standard for Permanence of Paper for Publications and Documents in Libraries and Archives.* National Information Standards Series. New Brunswick, N.J.: Transaction Publishers, 1993. ANSI/NISO Z39.48-1992.

HN3. ———. *Environmental Guidelines for the Storage of Paper Records.* Bethesda, Md.: NISO Press, 1995. TR-01-1995.

MICROGRAPHICS

HN4. **American National Standards Institute.** *American National Standard for Imaging Media (Film): Thermally Processed Silver Microfilm: Specifications for Stability.* New York: American National Standards Institute, 1994. ANSI/NAPM IT9.19-1994.

HN5. ———. *American National Standard for Imaging Media: Photographic Activity Test.* New York: American National Standards Institute, 1994. ANSI/NAPM IT9.16-1993.

HN6. ———. *American National Standard for Imaging Media: Photographic Films, Papers, and Plates: Glossary of Terms Pertaining to Stability.* New York: American National Standards Institute, 1992. ANSI/NAPM IT9.13-1992.
Revision of the 1991 document, which was withdrawn prior to publication due to technical inconsistencies.

HN7. ———. *American National Standard for Imaging Media (Photography): The Effectiveness of Chemical Conversion of Silver Images against Oxidation: Methods for Measuring.* New York: American National Standards Institute, 1993. ANSI/NAPM IT9.15-1993. (National Association of Photographic Manufacturers.)

HN8. ———. *American National Standard for Imaging Media: Processed Safety Photographic Film: Storage.* New York: American National Standards Institute, 1994. ANSI/NAPM IT9.11-1993. (Revision and redesignation of ANSI IT9.11-1991.)

HN9. ———. *American National Standard for Photography: Determination of Residual Thiosulfate and Other Related Chemicals in Processed Photographic*

Materials: Methods Using Iodine-amylose, Methylene Blue and Silver Sulfide. New York: American National Standards Institute, 1993. ANSI/ISO 417-1993; ANSI/NAPM IT9.17-1993. (Revision and redesignation of ANSI PH4.8-1985.)

HN10. ———. *American National Standard for Photography: Film Dimensions: Film for Documentary Reproduction.* New York: American National Standards Institute, 1993. ANSI/ISO 7247-1993; NAPM IT1.52-1994. (Revision and redesignation of ANSI PH1.52-1990.)

HN11. ———. *American National Standard for Photography: Photographic Films: Specifications for Safety Film.* New York: American National Standards Institute, 1992. ANSI/ISO 543-1990, ANSI IT9.6-1991. (Revision and redesignation of ANSI PH1.25-1984.)

HN12. ———. *American National Standard for Photography: Processed Vesicular Photographic Film—Specifications for Stability.* New York, N.Y.: American National Standards Institute, 1992. ANSI/ISO 9718-1991. ANSI IT9.12-1991. (Revision and redesignation of ANSI PH1.57-1985.)

HN13. Association for Information and Image Management (U.S.). *Recommended Practice for Microfilming Printed Newspapers on 35mm Roll Microfilm.* Silver Spring, Md.: Association for Information and Image Management, 1994. ANSI/AIIM MS111-1994. (Revision of MS111-1987.)

HN14. ———. *Standard for Information and Image Management: Microfiche.* Silver Spring, Md.: Association for Information and Image Management, 1992. ANSI/AIIM MS5-1992. (Revision of MS5-1991.)

HN15. ———. *Standard for Information and Image Management; Microfilm Package Labeling.* Silver Spring, Md.: Association for Information and Image Management, 1993. ANSI/AIIM MS6-1993.

HN16. ———. *Standard for Information and Image Management: Micrographics—Splices for Imaged Microfilm—Dimensions and Operational Constraints.* Silver Spring, Md.: Association for Information and Image Management, 1992. ANSI/AIIM MS18-1992.

HN17. ———. *Standard Recommended Practice—File Format for Storage and Exchange of Images—Bi-level Image File Format, Part 1.* Silver Spring, Md.: Association for Information and Image Management, 1993. ANSI/AIIM MS53-1993.

HN18. ———. *Standard Recommended Practice: Identification of Microforms.* Silver Spring, Md.: Association for Information and Image Management, 1993. ANSI/AIIM MS19-1993. (Revision of MS19-1987.)

HN19. National Information Standards Organization (U.S.). *Information on Microfiche Headers: A Draft American National Standard.* Bethesda, Md.: NISO Press, 1995. ANSI/NISO Z39.32-199x. (Revision of ANSI Z39.32-1981, in draft at this writing.)

OTHER MEDIA

HN20. Association for Information and Image Management (U.S.). *Glossary of Imaging Technology.* Silver Spring, Md.: Association for Information and Image Management, 1994. ANSI/AIIM TR2-1992.

HN21. ———. *Performance Guideline for the Legal Acceptance of Records Produced by Information Technology Systems.* Silver Spring, Md.: Association for Information and Image Management, 1994. ANSI/AIIM TR31-1994.

HN22. ———. *Resolution as It Relates to Photographic and Electronic Imaging.* Silver Spring, Md.: Association for Information and Image Management, 1993. Technical Report for Information and Image Management, AIIM TR26-1993.

HN23. **Standards for Electronic Imaging Technologies, Devices, and Systems. Proceedings of a Conference Held 1–2 February 1996, San Jose, California.** Bellingham, Wash.: SPIE Optical Engineering Press, 1996.

EQUIPMENT

HN24. *AIIM Buying Guide.* Silver Spring, Md.: Association for Information and Image Management. 1994– . Annual. Continues in part *AIIM Buying Guide and Membership Directory.*

HN25. Association for Information and Image Management (U.S.). *Electronic Imaging Output Printers.* Silver Spring, Md.: Association for Information and Image Management, 1993. Technical Report for Information and Image Management, AIIM TR29-1993.

HN26. ———. *Recommended Practice for Quality Control of Image Scanners.* Silver Spring, Md.: Association for Information and Image Management, 1988. ANSI/AIIM MS44-1988.

HN27. ———. *Standard Recommended Practice: Monitoring Image Quality of Roll Microfilm and Microfiche Scanners.* Silver Spring, Md.: Association for Information and Image Management, 1993. ANSI/AIIM MS49-1993.

I

Access Services

Julie Wessling

Access services mirrors the changing face of libraries. What is access services? What functions are handled by access services staff? How does access services relate to the lengthy debates about access and ownership issues in libraries? With the circulation function of checking out material to users remaining as the one common responsibility of departments labeled Access Services, it is becoming increasingly difficult to predict the structure of access services in any given library. Circulation services took on many of the characteristics of technical services with the implementation of automated integrated library systems during the late 1970s and into the 1980s. Although many heads of circulation, or access services, continued to report to a head of public services, access services sometimes moved organizationally into technical services. During the late 1980s and early 1990s, when materials were being gathered for the *Guide to Technical Services Resources*, access services was recognized as an expanded circulation department, incorporating services such as interlibrary loan and document delivery in order to offer the user a single access point for needed material. In general, "document delivery" refers to physical delivery of material, whether held on site or obtained from another library or commercial supplier, directly to the user.

Since 1993, the variety of access services models existing in libraries has expanded. This continuing transition reflects the migration of many libraries toward a virtual environment in which the end user has desktop access to the local catalog and also to catalogs in other libraries, document delivery from a range of commercial suppliers, links to a myriad of electronic databases, and access to full-text material. With the assistance of sophisticated integrated library systems and emerging technology, circulation services are expanding to include self-checkout machines, user-generated electronic requests for materials in a local collection, user-generated holds and renewals, electronic reserve operations with remote access, and delivery of locally held material directly to patrons. Interlibrary loan (ILL), with new expanded document delivery functions, is sometimes organized as a separate department within the library. However, ILL is just as likely to be coupled with collection development or acquisitions, rather than reference or circulation, in recognition of its relation to overall collection management. Advances in technology promise to streamline ILL processing and blend it seamlessly with commercial delivery options. Collections maintenance, or stacks maintenance operations, reflect an increasing awareness of preservation issues. Policies relating to library security issues reflect a broad-based concern with an increasingly violent society. Library security is no longer limited to problem patrons or theft of library materials.

Chapter Content and Organization

This chapter includes areas of responsibilitiy that traditionally have been included in an access services department—circulation, course reserve, collection maintenance, interlibrary loan, document delivery, and library security. Information sources included are presented under six section headings: General Works, including items of general interest to access services, resources on copyright, periodicals, and sources of expertise; Circulation, including course reserves, with special attention given to electronic reserve; Collection Maintenance; Interlibrary Loan; Document Delivery; and Library Security.

Resources dealing with interpretation of copyright law have been included in the general works section in recognition of its central importance to most units included in access services, especially reserve operations, interlibrary loan, and document delivery. Copyright is a complex issue. The compiler has sought to include resources that provide a framework for understanding copyright legislation and also resources that can assist with daily operations requiring compliance with copyright regulations.

IA
General Works

IA1. Bielefield, Arlene, and Lawrence Cheeseman. *Maintaining the Privacy of Library Records: A Handbook and Guide.* New York: Neal-Schuman, 1994.

Outlines the constitutional and statutory rights of library users. Chapters cover the First Amendment, the right to receive information and ideas, the right to privacy, copyright, the Americans with Disabilities Act, codes of ethics, censorship, and patron rules of conduct. Useful resource for developing and applying policies, including explanation to the public.

IA2. Sapp, Gregg, ed. *Access Services in Libraries: New Solutions for Collection Management.* New York: Haworth Press, 1992. Also published as *Collection Management* 17, no. 1/2 (1992).

The editor includes fifteen essays that capture the variety of organization and importance of access services departments in all types and sizes of libraries. Topics covered include management issues, storage facilities, circulation policies, impact of technology and electronic information, service issues, work flow analysis, and the future of access services.

IA3. Smith, Kitty. *Serving the Difficult Customer: A How-to-Do-It Manual for Library Staff.* New York: Neal-Schuman, 1993.

Many practical examples and advice for dealing with a variety of customers. The author covers patron and coworker behaviors and attitudes that can hinder communication and disrupt other patrons' use of the library. The bibliography is especially useful for identifying additional books, videos, and audio cassettes for staff training in customer service.

IA4. Walling, Linda Lucas, and Marilyn M. Irwin. *Information Services for People with Developmental Disabilities: The Library Manager's Handbook.* Westport, Conn.: Greenwood Press, 1995.

Access services staff is often responsible for providing assisted access to collections for patrons with disabilities. This useful book reviews ramifications of the Americans with Disabilities Act for libraries and provides practical advice for assisting individuals with disabilities.

IA5. Walters, Suzanne. *Customer Service: A How-to-Do-It Manual for Librarians.* New York: Neal-Schuman, 1994.

Clear text with practical suggestions for effective customer service. The circulation desk may be the only stop a user makes in the library; customer service skills are on display here all the hours a library is open.

IAA
Copyright Resources

IAA1. Association of Research Libraries (ARL). 21 Dupont Circle, Suite 800, Washington, DC 20036. Executive Director: Duane E. Webster. Phone: 202-296-2296; fax: 202-872-0884.

ARL maintains a selection of copyright resources and links to other servers through both a WWW home page and a gopher. URL: http://arl.cni.org/scomm/copyright/copyright.html and URL: gopher://arl.cni.org following the menu choices: /Scholarly communication/Copyright and Intellectual Property/

IAA2. Bennett, Scott, Kenneth Frazier, and Laura Gasaway. "Special Section: Fair Use and Copyright." *Computers in Libraries* 14, no. 5 (May 1994): 18–32.

Three separate articles reviewing the issues involving libraries and fair use, highlighting electronic reserve and interlibrary loan requirements.

IAA3. Bruwelheide, Janis H. *The Copyright Primer for Librarians and Educators.* 2nd ed. Chicago: American Library Association; Washington, D.C.: National Education Association, 1995.

For annotation, *see* **HE1**.

IAA4. Copyright Clearance Center (CCC). 222 Rosewood Drive, Danvers, MA 01923. Phone: 508-750-8400; fax: 508-750-4744; e-mail: info@copyright.com. URL: http://www.copyright.com/

Requests for permission to reproduce copyrighted material should be directed to the copyright holder. An alternative is to use the Copyright Clearance Center (CCC). The CCC has the right to grant permission and collect fees for certain publications. Libraries wishing to utilize the Copyright Clearance Center for payment of royalties may now access it through the CCC home page, where one may search through CCC catalogs, identify publications for which the CCC handles fees and permission, view royalty information, sign up as a customer, and report activity for ILL or course pack copying.

IAA5. Gasaway, Laura N., and Sarah K. Wiant. *Libraries and Copyright: A Guide to Copyright Law in the 1990s.* Washington, D.C.: Special Libraries Association, 1994.

Textbook in copyright law for librarians. Explains the why behind the law and provides guidelines a librarian needs to follow to be in compliance. Excellent index to help find the specific section that will answer a question at hand, although one may need to read a fair amount of legal analysis before locating the answer.

IAA6. Lutzker, Arnold P., Esq. *Commerce Department's White Paper on National and Global Information Infrastructure. Executive Summary for the Library and Educational Community.* Washington, D.C.: Fish & Richardson, P.C., Sept. 20, 1995. Available online at URL: http://sunsite.berkeley.edu/Copyright/analysis.html and URL: gopher://arl.cni.org/ under the menu choices: /Information Policy/Copyright.

Provides a clear review and analysis of *Intellectual Property and the National Information Infrastructure: The Report of the Working Group on Intellectual Property Rights* **(HE3)**, which sets forth recommendations on changes to intellectual property laws. Provides an extensive discussion of current laws and policies, most notably copyright, as they relate to digital information and new technology. This review was prepared at the request of the Association of Research Libraries, American Library Association, American Association of Law Libraries, Medical Library Association, and Special Libraries Association.

IAA7. Mathews, Michael, and Patricia Brennan, eds. *Copyright, Public Policy, and the Scholarly Community.* Washington, D.C.: Association of Research Libraries, 1995.

Prepared as a resource to stimulate discussions about the viability of the copyright law in an electronic, networked environment. It serves as a sampler of the range of viewpoints, containing papers by a university librarian, a political scientist speaking on behalf of scholarly societies, a director of a large scholarly society with an active publishing program, and two lawyers involved in working policies for managing university use of copyrighted works.

IAA8. Samuelson, Pamela. "Copyright and Digital Libraries." *Communications of the ACM* 38 (April 1995): 15–21, 110.

The author addresses the uneasy balance that exists between authors, publishers, libraries, and consumers of information in relation to existing law. She suggests that the fair-use concept should not be abandoned and it is not advantageous to any of the above groups to move to a strict pay-per-use scheme.

IAA9. U.S. Copyright Office, Library of Congress. Washington, DC 20559. URL: http://lcweb.loc.gov/copyright and URL: gopher://marvel.loc.gov:70/11/copyright

The Library of Congress electronic information system, LC MARVEL, includes a copyright section with announcements and information about the most recent regulations related to copyright. Frequently requested Copyright Office circulars may be viewed or downloaded using either LC's WWW home page or gopher.

IAB
Periodicals

Journals listed in *GTSR* are not included here, unless title or focus has changed.

IAB1. *Journal of Interlibrary Loan, Document Delivery & Information Supply.* New York: Haworth Press. v. 4– , 1993– . Quarterly. ISSN 1042-4458. Continues *Journal of Interlibrary Loan & Document Delivery Supply.*

Deals with practical issues related to ILL and expanding roles of interlibrary loan librarians. The title was expanded to include *Document Delivery* in response to broadening role of the interlibrary loan function.

IAB2. *OCLC Systems & Services.* Westport, Conn.: Meckler. v. 9– , 1993– . Quarterly. ISSN 1065-075X. Continues *OCLC Micro.*

Practical information for interlibrary loan staffs who use OCLC for searching or processing requests. For annotation, *see also* **AF3**.

IAB3. *Public & Access Services Quarterly.* New York: Haworth Press. v. 1– , 1995– . Quarterly. ISSN 1056-4942.

Includes articles on public and access services issues from both the practical and the philosophical viewpoint. Covers any topic that involves the direct interaction of librarians and patrons.

IAC
Sources of Expertise

PROFESSIONAL ORGANIZATIONS

IAC1. American Library Association. URL: http://www.ala.org

See also annotation **CAE17**. The ALA committees and discussion groups listed below are of particular interest to access services staff. Meetings of the committees and discussion groups at ALA provide an opportunity to discuss common issues and concerns with access service colleagues. In addition, these groups sponsor programs and pre-conferences of special interest.

Ad Hoc Subcommittee on Copyright works closely with the ALA Legislative Committee on various aspects of the copyright law and promotes forums for understanding copyright issues.

Association of College and Research Libraries, Copyright Committee serves as a clearinghouse for information about copyright issues that impinge upon academic libraries.

Association of Specialized and Cooperative Library Agencies, Interlibrary Cooperation and Networking Section, Interlibrary Cooperation Discussion Group considers interlibrary cooperation and the statewide development of library service, emphasizing the interdependence of all types of libraries.

Association of Specialized and Cooperative Library Agencies, Libraries Serving Special Populations Section includes forums and discussion groups to review issues involved with serving populations with special needs including blind, physically handicapped, deaf, developmentally disabled, etc.

Library Administration and Management Association, Building and Equipment Section, Safety and Security of Library Buildings Committee deals with issues related to the safety and security of persons and property in library buildings. A Library Safety/Security Discussion Group provides a forum for librarians interested in safety and security issues as they relate to the design, construction, renovation, and equipment of library facilities.

Library Administration and Management Association, Library Storage Discussion Group provides a forum for exchange of ideas on the planning, design, development, operation, management and/or dismantling of library collection storage.

Library Administration and Management Association, Systems and Services Section, Circulation/Access Services Committee studies and evaluates procedures and issues related to the management of circulation and access services in all types of libraries. A Circulation/Access Services Discussion Group provides a forum for issues relating to the management of circulation and access services operations in all types of libraries.

Reference and User Services Association, Management and Operation of Public Services Section, Interlibrary Loan Committee considers current aspects of interlibrary loan service and recommends solutions to interlibrary loan problems. An Interlibrary Loan Discussion Group provides a forum for issues pertaining to interlibrary loan and resource sharing among all types of libraries. (RUSA was formerly known as Reference and Adult Services Division.)

IAC2. International Federation of Library Associations and Institutions (IFLA), Section on Document Delivery and Interlending. URL: http://www.nlc-bnc.ca/ifla/VII/s15/sidd.htm

Based on recommendations from this committee, the IFLA Office for International Lending, located at the British Library Document Supply Center, issues international lending guidelines. In 1995, they issued "IFLA Fax Guidelines" (seven-page handout) also published in IFLA's *Guide to Centres of International Lending* **(IDB3).** International lending questions should be directed to IFLA Offices for UAP and International Lending, c/o The British Library, Document Supply Centre, Boston Spa, Wetherby, West Yorkshire LS23 7BQ, United Kingdom. (E-mail: ifla@bl.uk).

CONFERENCES

IAC3. International Conference on Interlending & Document Supply. Every two years.

This conference is usually held in conjunction with a national library convention. The fourth conference was held in 1995 in Calgary, Canada, in conjunction with the Canadian Library Association. The fifth conference is scheduled for August 1997 in Denmark in conjunction with the 1997 IFLA Conference. This comprehensive conference offers an international showcase for state-of-the-art interlibrary loan and document delivery options. Recent conferences have had heavy focus on technology.

ELECTRONIC DISCUSSION GROUPS

IAC4. *ARIEL-L* Electronic discussion group.

An *Ariel* **(IDE1)** users' electronic discussion forum offering practical tips and assistance for implementing *Ariel,* integrating it into work flow, and troubleshooting software or hardware problems with *Ariel.* Enhancements and software upgrades also are announced. To subscribe, send a message to LISTSERV@IDBSU.IDBSU.EDU with the command SUBSCRIBE ARIEL-L [your name].

IAC5. *AVISO-L Mailing List:* AVISO Inter-Library Loan Software Mailing List. Electronic discussion group.

An electronic discussion forum for *AVISO* **(IDE2)** users; to subscribe, send a message to LISTSERV@MORGAN.UCS.MUN.CA with the command SUBSCRIBE AVISO-L [your name]. Archives available at URL http://www.mun.ca/lists/aviso-l/

IAC6. *CircPlus.* Electronic discussion group. List owner: Daniel Lester, Boise State University.

Discussion forum for issues relating to circulation, reserve, stack maintenance, and similar activities in libraries. To subscribe, send a message to LISTSERV@IDBSU.IDBSU.EDU with the command SUBSCRIBE CIRCPLUS [your name].

IAC7. *CNI-Copyright.* Electronic discussion group. Moderator and list owner: Mary Brandt Jensen, University of South Dakota.

Discussions have included electronic reserve in academic libraries and electronic transmission of documents. To subscribe, send a message to LISTPROC@CNI.ORG with the command SUBSCRIBE CNI-COPYRIGHT [your name]. For annotation, *see also* **HAB5**.

IAC8. *DocDel-L.* Electronic discussion group. Hosted by EBSCOdoc, part of EBSCO Information Services.

DocDel-L serves as a broad forum for issues related to document delivery, including pricing, document delivery and the academic library, providers, and the future of document delivery. The list is mediated by EBSCO; the discussions to date do not reflect a bias—they have been rich and diverse. EBSCOdoc intends to invite "guest speakers" on a regular basis to add to general discussion. To subscribe, send a message to LISTPROC@WWW.EBSCODOC.COM with the command SUBSCRIBE DOCDEL-L [your name].

IAC9. *ILL-L.* Electronic discussion group. List owner: Mary Hollerich.

Interlibrary loan discussion forum covering all ILL issues and concerns along with practical application of policies and procedures related to document delivery and resource sharing. To subscribe send a message to LISTPROC@USC.EDU with a command SUBSCRIBE ILL-L [your name]. [Compiler's note: List owner, Mary Hollerich, has accepted a new position at Northwestern University and expects to transfer the list from University of Southern California to Northwestern sometime in the fall of 1996; list address will change at that time.]

IAC10. *SAFETY.* Electronic discussion group. List owner: Ralph Stuart, University of Vermont.

Unmoderated discussion forum covering environmental, health, and safety issues and problems on college and university campuses. A moderated version of the list with lower traffic levels, called *EnSafety,* is available for a $50 annual subscription fee. To subscribe, send a message to LISTSERV@UVMVM.UVM.EDU with the command SUBSCRIBE SAFETY [your name]. Send messages to SAFETY@UVMV.UVM.EDU. Archived at URL gopher://hazard.com and URL http://hazard.com

IAC11. *SAVEIT-L.* Electronic discussion group.

Unmoderated list for *SAVE-IT* users. To subscribe, send a message to LISTSERV@LEHIGH.EDU with the command SUBSCRIBE SAVEIT-L [your name]. Archives available at URL ftp://ftp.lehigh.edu/pub/listserv/save-it-l/Archives

IAD
Bibliographies

IAD1. *Copyright and Intellectual Property Resources.* Online. International Federation of Library Associations and Institutions. URL: http://www.nlc-bnc.ca/ifla/11/cpyright.htm

IFLA maintains an online bibliography on copyright and intellectual property; provides hot links to some resources.

IB
Collection Maintenance

The main objective of collection maintenance, also called stack maintenance or stack management, is to ensure that all library materials are maintained in good condition and are readily accessible to library users. Recent trends in collection maintenance revolve around broadening service goals of libraries and space limitations most libraries are experiencing for housing on-site collections. There is also increasing awareness of preservation responsibilities involved in routine shelving and stack maintenance functions.

The importance of returning material to the shelves in a timely manner cannot be overemphasized. Nothing is more frustrating to a user than to confirm that a book is owned, verify that it is not checked out, and yet not be able to locate it on the shelf. For patrons to be successful in finding a book, both accuracy of shelving and timeliness of shelving come into play. Users experiencing rapid delivery from electronic sources for some information needs are much less tolerant of delayed return of on-site collections to the shelves. Shelving continues to be the weak link in allowing onsite collections to maximize their role as the main course on a user's plate.

As collections continue to grow and libraries need to devote a greater amount of space to technology, more and more libraries are planning use of storage facilities. Access to materials in storage and rapid, dependable delivery for users remain an important part of the planning. Libraries are asking vendors of online catalogs to implement an automated request function for users to obtain materials from storage facilities. Some consortiums, for example academic libraries in California and Ohio, are planning common storage facilities for library materials. An essential ingredient to this planning is the user's ability to request electronically the delivery of stored materials to a convenient location. Most new library buildings are planned with the capability to accommodate movable compact shelving.

IB1. Baird, Brian J. "Motivating Student Employees: Examples from Collections Conservation." *Library Resources & Technical Services* 39, no. 4 (1995): 410–16.

For annotation, *see* **HI2**.

IB2. Bellanti, Claire Q. "Access to Library Materials in Remote Storage." *Collection Management* 17, no. 1/2 (1992): 93–104.

Reviews policies and procedures for providing users with quick and easy access to library materials that are housed in remote storage facilities.

IB3. **East, Dennis.** "User Views of Compact Shelving in an Open Access Library." *Collection Management* 18, no. 3/4 (1994): 71–85.

Looks at use of compact shelving in a music library and sound recordings archives. Provides useful background information for reaching a decision on whether to use compact shelving; also, includes a user survey of most liked and most disliked features. Gives overview of the installation of electrically and mechanically assisted compact shelving along with procedures for relocating materials during the process.

IB4. **O'Connor, Phyllis.** "Remote Storage Facilities: An Annotated Bibliography." *Serials Review* 20, no. 2 (1994): 17–26.

Useful for staff who are planning for and implementing a remote storage program. Covers planning the storage facility, selecting materials for storage, and implementing the storage program. Includes both theoretical and practical articles.

IB5. **Sam, Sherrie, and Jean A. Major.** "Compact Shelving of Circulating Collections." *College & Research Libraries News* 54, (Jan. 1993): 11–12.

Good review of issues involved with using compact (movable) shelving with high use collections.

IB6. **Sharp, S. Celine.** "A Library Shelver's Performance Evaluation as It Relates to Reshelving Accuracy." *Collection Management* 17, no. 1/2 (1992): 177–92.

As some libraries contemplate outsourcing the shelving of library materials, this study of factors that influence accuracy and quickness of student shelvers is a useful one to review. The data provide an indication of a link between job standards and evaluation to shelving results.

IB7. **Swartzburg, Susan G.** *Preserving Library Materials: A Manual.* 2nd ed. Metuchen, N.J.: Scarecrow Press, Inc., 1995.

This comprehensive manual of preservation issues includes practical information on shelving practice and protecting materials from damage. For annotation *see also* **HA10**.

IC
Circulation

Small libraries, or libraries with a small amount of lending activity, may continue to operate effectively with a well-designed manual circulation system. It is unusual, however, to find a library of any size that has not automated its circulation function. An automated circulation system is an essential piece of any integrated library system; the design of the circulation module and its transaction file parameters continue to provide valuable management data for developing collections. Now, as libraries add electronic resources, many networked for remote access, staff need to build a patron database able to support patron verification for multiple purposes, extending beyond eligibility for using an on-site collection and designed for automatic verification in a networked environment.

The design of the circulation module in any integrated library system is important for supporting an evolving number of user-initiated services. The transaction file not only provides status information in the online public catalog, indicating whether an item is checked out or available on the shelf, but also may link to a variety of user-initiated request functions. A patron, automatically verified by a user validation check in the patron database, is able to request an item, place a hold, or request a renewal without staff intervention. The patron database also should support authorization for access to remote databases and a variety of electronic resources. Some libraries have installed self-checkout machines, letting even in-house users eliminate a stop at the circulation desk.

The circulation activity receiving much attention in the current literature is electronic reserve. Technology exists to make reserve readings available to students in any location but the implementation of this service is complex, especially the need to interpret copyright legislation. Before expanding services into electronic reserve, the librarian should read further about fair-use guidelines and follow discussions concerning application of copyright law to electronic resources. This is a changing area with new interpretations surfacing frequently.

IC1. Bosseau, Don L. "Anatomy of a Small Step Forward: The Electronic Reserve Book Room at San Diego State University." *Journal of Academic Librarianship* 18, no. 5 (1993): 366–68.

Author reviews implementation of an electronic reserve book room at San Diego State University, touching on political aspects, copyright concerns, operational characteristics, and future plans.

IC2. Bosseau, Don L., Beth Shapiro, and Jerry Campbell. "Digitizing the Reserve Function: Steps toward Electronic Document Delivery." *Electronic Library* 13, no. 3 (1995): 217–23.

This is a useful composite of issues involved in electronic reserve; it is based on papers presented at EBSCO's Executive Seminar for research library directors at the 1995 ALA Annual Conference. Bosseau covers technology requirements, Shapiro addresses design and implementation of an electronic reserve service, and Campbell discusses copyright considerations.

IC3. **Delaney, Thomas.** "Electronic Reserve: The Library Goes to the People." In *The Internet Library: Case Studies of Library Internet Management and Use,* edited by Julie Still, 1–11. Westport, Conn.: Mecklermedia, 1994.

This paper describes the local development of a computer program that provides user-friendly, electronic access to noncopyrighted reserve readings using a campus network. The program also accommodates electronic submission of reserve lists from faculty.

IC4. **Enssle, Halcyon R.** "Reserve On-line: Bringing Reserve into the Electronic Age." *Information Technology and Libraries* 13, no. 3 (1994): 197–201.

Another look at a locally developed electronic reserve program for making reserve materials electronically accessible from outside the library. This project includes noncopyrighted material only.

IC5. **Fouty, Kathleen G.** *Implementing an Automated Circulation System: A How-to-Do-It Manual.* New York: Neal-Schuman, 1994.

This practical guide follows through all steps of planning for and implementing an automated circulation system. It includes evaluation of the process and strategies for helping both staff and the public adjust to a new system. The information is useful for staff in academic, public, special, and school libraries and covers basic steps fundamental to most implementation projects. Includes useful checklists and guidelines. References cited are of limited use; most date back to the 1980s.

IC6. **Hupp, Stephen L.** "The Reserve Department Revisited: A Study of Faculty and Student Use of the Capital University Library Reserve Department." *Public & Access Services Quarterly* 1, no. 1 (1995): 81–94.

This article offers a valuable introduction for understanding use of reserve materials in a typical academic library where a traditional, manual reserve operation is in place. It provides information for anticipating amount of use to assist in planning a new reserve service.

IC7. **Jensen, Mary Brandt.** "Electronic Reserve and Copyright." *Computers in Libraries* 13 (1993): 40–45.

Interesting review by an associate professor of law, who compares application of Copyright Law Section 106 (exclusive rights), Section 107 (fair use), Section 108 (reproduction by libraries), and Section 110 (performances and displays) and comments on application for electronic reserve and parallel activity in broadcasting and cable television.

IC8. **Kathman, Michael, and Jane McGurn Kathman, comps.** *Managing Student Employees in College Libraries.* CLIP Note, no. 20. Chicago, Association of College & Research Libraries, American Library Association, 1994.

Academic libraries frequently employ student workers in access services. This guide, a revision of an earlier publication with the same title, includes examples of policies and procedures from small university and college libraries

249

for student employment, covering such issues as dismissal, orientation, training, supervision, and performance review.

IC9. O'Connor, Phyllis, Susan Wehmeyer, and Susan Weldon. "The Future Using an Integrated Approach: The OhioLINK Experience." *Journal of Library Administration* 21, no. 1/2 (1995): 109–20.

The authors give an overview of the OhioLINK project with a detailed look at the online borrowing function, which is based on automated verification of patron eligibility for loan privileges from any of the participating libraries.

IC10. Seaman, Scott. "Impact of Basic Books v. Kinko's Graphics on Reserve Services at the University of Colorado, Boulder." *Journal of Interlibrary Loan, Document Delivery & Information Supply* 5, no. 3 (1995): 111–18.

This article reviews the impact of commercially produced course packets on the volume of reserve; statistics indicate a surprising increase in the number of reserve photocopies after the availability of course-packet services.

GUIDELINES AND STANDARDS

IC11. American Library Association. Intellectual Freedom Committee. *Guidelines for the Development of Policies and Procedures Regarding User Behavior and Library Usage.* Chicago: American Library Association, January 24, 1993.

Guidelines (three pages) for developing local policies and procedures to ensure access to information and services for all members of the community, along with procedures for dealing with user behavior problems.

IC12. *Effective Library Signage.* Kate W. Ragsdale and Donald J. Kenney, comps. SPEC Kit, no. 208. Washington, D.C.: Association of Research Libraries, Office of Management Services, 1995.

This kit discusses all aspects of library signage management including design, construction, location and installation, vandalism, temporary signs, and ADA compliance; documents the variety of sign management methods currently used in ARL libraries. Access services may or may not be responsible for signage; in all cases they are in a position to be aware of shortcomings in any existing sign program.

IC13. *Fair Use Guidelines for Electronic Reserve Systems. [Draft].* Memorandum from Electronic Reserves Drafting Sub-Group to Conference on Fair Use Participants, April 19, 1996.

This document is included here because of its significance for electronic reserves. It was prepared by a group of negotiators participating in the Conference on Fair Use (CONFU), which is sponsored by the National Information Infrastructure Task Force's Working Group on Intellectual Property Rights. Members include: Kenny Crews, Indiana University; Laura Gasaway,

University of North Carolina Law Library; Isabella Hinds, Copyright Clearance Center; Carol Risher, Association of American Publishers; and Mary Jackson, Association of Research Libraries. The draft recommends permissible fair uses of copyrighted works in electronic reserves systems and reflects compromises by all interested stakeholders. At the time of this writing, comments are being solicited. Once consensus is reached, participants likely will seek endorsement from their respective organizations.

IC14.   Mitchell, Eugene S., comp.   *Library Services for Non-affiliated Patrons.* CLIP Note, no. 21. Chicago: Association of College & Research Libraries, American Library Association, 1994.

Sample policies and procedures for dealing with service to the non-affiliated library user. Collected from small university and college libraries nationally. Includes examples of information sheets, application and registration forms, identification cards, recourse letters, and reciprocal agreements.

IC15.   National Information Standards Organization.   *Proposed American National Standard Format for Circulation Transactions Standards.* Bethesda, Md.: NISO, 1993. NISO Z39.70-199X.

The circulation transaction format defined by this standard identifies the data elements and their format that may be contained in any set of circulation transaction files and that are suitable for exchange between library systems. The standard provides a structured order chart, data dictionary, and transaction definitions; however, it does not define the communications format for the data elements.

IC16.   *Providing Public Service to Remote Users.* Craig Haynes, comp. SPEC Kit, no. 191. Washington, D.C.: Association of Research Libraries, Office of Management Services, 1993.

Results of a survey on the 1991 interlibrary lending and borrowing costs of seventy-six U.S. and Canadian research libraries. Examines cost differences among libraries in different geographic regions and between public and private institutions. Includes circulation issues and loan policies for distance learners and other remote users.

ID
Interlibrary Loan

The revised *National Interlibrary Loan Code* **(IDD3)** adopted in 1993 reflects the changing role of interlibrary loan in libraries. The 1980 code described interlibrary loan as "an adjunct to, not a substitute for, collection development." The 1993 code reveals a major shift in thinking in language that states, "interlibrary borrowing is an integral element of collection development for all libraries, not an ancillary option." With planned dependence on interlibrary loan, there is renewed interest in patron satisfaction with service and user surveys and questions about turnaround time, cost, and reliability. Technology advances in ILL reflect this new emphasis on user convenience and renewed concern for processing ILL requests quickly. Innovations address efficiencies in both requesting and lending aspects of ILL. Software packages for management of ILL transactions are proliferating. Most boast handling of copyright compliance, automatic updating of requests for status information, management of invoices and billing procedures, links with OCLC or other transmission options, and generation of statistics and management reports.

OCLC is the most heavily used service for transmission of ILL requests in the United States. Recent enhancements in OCLC ILL reflect libraries' interest in streamlining the ILL process and include many more links to commercial document suppliers, automatic upload of local patron electronic requests into PRISM ILL (IPT), custom holdings for automatic generation of preselected libraries into suggested lender strings, and introduction of the ILL Fee Management (IFM) service that promises elimination of paper invoices for participating libraries. Use of telefacsimile has become common for speeding delivery of photocopied items to requesting libraries. The Internet is increasingly used as a source of digitized articles. Ariel, the document transmission workstation developed by the Research Libraries Group, has become a common configuration for digital transmission of documents.

IDA
General Works

IDA1. **Baker, Shirley K., and Mary E. Jackson.** *The Future of Resource Sharing.* New York: Haworth Press, 1995. Also published as *Journal of Library Administration* 21, no.1/2 (1995).

Fifteen essays that explore resource-sharing questions, with particular emphasis on interlibrary loan. Topics include economic decision models, consortial agreements, copyright issues, possibilities of technology, and future directions, such as ARL's national project to revamp interlibrary loan and document delivery. This compilation provides an overview of resource-sharing issues and offers a framework for ILL practitioners to define their role in providing library services.

IDA2. ———. *Maximizing Access, Minimizing Cost: A First Step toward the Information Access Future.* Washington, D.C.: Association of Research

Libraries, 1993. Also available on the ARL gopher URL: arl.cni.org as a menu choice under Interlibrary Loan.

Prepared for the ARL Committee on Access to Information Resources November 1992 and revised February 1993, this seventeen-page "white paper" critiques the current interlibrary loan process and describes an ideal interlibrary loan and document delivery system. It includes eleven activities the authors identify as necessary to make progress toward the ideal model. The paper has been shared widely in the library community and with bibliographic utilities, local system providers, document delivery suppliers, and other related vendors; resulting discussions have heightened interest in resolving both technical and operational issues.

IDA3. *Benchmarking Interlibrary Loan: A Pilot Project.* OMS Occasional Paper, no. 18. Washington, D.C.: Association of Research Libraries, Office of Management Services, 1995.

Compilation of three reports on the ILL Benchmarking Projects, undertaken with support from the Council of Library Resources. Describes in detail the key elements to any successful benchmarking process while specifically examining interlibrary loan.

IDA4. Bustos, Roxann, comp. *Interlibrary Loan in College Libraries.* CLIP Note, no. 16. Chicago: Association of College and Research Libraries, American Library Association, 1993.

Compilation of policies, procedures, and forms primarily for ILL borrowing activities. Includes ILL and fax policies; special policies for faculty ILL, dissertations, and free periodical articles; consortia policies; sample state reimbursement agreement; reciprocal agreement request forms; and examples of patron request and notification forms.

IDA5. Cornish, Graham. "Training Opportunities for Interlibrary Loan and Document Supply Staff." *Journal of Education for Library and Information* 35, no. 2 (Spring 1994): 138–46.

This article provides an international look at training opportunities or, more accurately, lack of training opportunities for interlibrary loan practitioners. It delineates the aspects of interlibrary loan where training is needed and lists the known providers of available training and resources.

IDA6. Gould, S. "A Voucher Scheme to Simplify Payment for International Interlibrary Transactions." *Interlending & Document Supply* 23, no. 1 (1995): 15–19.

The author outlines the voucher scheme proposed by the IFLA Office for Universal Availability of Publications, which would facilitate paying for interlibrary loan transactions between libraries in different countries. The proposal would standardize the procedure and minimize the enormous range of payment methods now in operation among libraries.

IDA7. Hebert, Francoise. "Service Quality: An Unobtrusive Investigation of Interlibrary Loan in Large Public Libraries." *Library and Information Science Research* 16 (Winter 1994): 3–21.

Evaluates the quality of ILL in large public libraries, finding that the customer valued different factors than the librarian. Points out the need to balance all measures in evaluation of the quality of ILL; for example, reliability of the service was more important to users than actual turnaround time.

IDA8. Ison, Jan. "Patron-Initiated Interlibrary Loan: A Model for Service." *Illinois Libraries* 76, no. 3 (Summer 1994): 155–58.

An interesting review of the Illinois library system that provided the capability for patron-generated requests for materials held in other participating libraries, but—at the time the article was written—librarians continued to focus on mediated assistance and directing users to alternative materials in the home library. Those libraries allowing patrons to initiate their own requests experienced considerable success and a great deal of patron satisfaction.

IDA9. Jackson, Mary E. "Document Delivery over the Internet: Electronic Document Delivery Systems Used by Libraries." *Online* 17 (1993): 14–21.

The author reviews development of the Research Libraries Group's document transmission workstation, Ariel, contrasting it with other digital transmission efforts and use of fax for transmitting copies of articles.

IDA10. Kimmel, Janice L. "ILL Staffing: A Survey of Michigan Academic Libraries." *RQ* 35 (Winter 1995): 205–16.

Analyzes the results of a survey examining the relationship between ILL volume and staffing patterns. Useful data for identifying the most efficient levels of staff and their relationship to other areas of the library.

IDA11. Kinnucan, Mark T. "Demand for Document Delivery and Interlibrary Loan in Academic Settings." *Library & Information Science Research* 15 (Fall 1993): 355–74.

Report of a study by the author while he was a visiting scholar at OCLC. He interviewed faculty members and graduate students at three Ohio universities to determine their willingness to pay for fast delivery of an article from a commercial supplier. Findings indicate users are willing to pay for information they believe will be useful, but they are not willing to pay very much; usually not more than $6.00 for interlibrary loan or $4.00 to receive the same article from a commercial supplier. For most users, the time factor was less important than the cost.

IDA12. Lombardo, Nancy, and Peggy Jobe. "What's MIME is Yours." *Colorado Libraries* 19 (Winter 1994): 19–22.

The authors describe the use of Multipurpose Internet Mail Extension (MIME) for electronic transmission of documents over the Internet. This is offered as an alternative to using Ariel and does not require specialized software or hardware. MIME allows documents to be transmitted as attachments to e-mail messages.

IDA13. *OCLC's Resource Sharing Strategy.* Online. Dublin, Ohio: OCLC, Inc. 1994. URL: http://www.oclc.org/oclc/man/7959rx.htm

This six-page "white paper" outlines OCLC's vision of resource sharing between now and the year 2000, clearly articulating the role they see for OCLC. Because of the major role OCLC has traditionally played in interlibrary loan traffic and more recently in links to commercial suppliers, this is an important document for understanding the evolution of ILL services.

IDA14. Perrault, Anna H., and Marjo Arseneau. "User Satisfaction and Interlibrary Loan Service: A Study at Louisiana State University." *RQ* 35 (Fall 1995): 90–100.

Authors provide findings from a survey of user expectations and overall satisfaction with interlibrary loan service. Includes information for replicating the survey and measuring service objectives in any interlibrary loan operation.

IDA15. Sessions, Judith, et al. "OhioLINK Inter-Institutional Lending Online: The Miami University Experience." *Library Hi Tech* (Issue 51) 13, no. 3 (1995): 11–24, 38.

The authors describe the addition of patron-initiated interlibrary loan to the OhioLINK project, offering users the ability to place online requests for material held at other libraries in the consortium. The article promotes patron-initiated, inter-institutional lending as an essential piece for providing users in Ohio with a virtual library. The process automates both the ILL and circulation functions, blurring the two roles and facilitating rapid access to materials for the user.

IDA16. Smith, Jane. "Electronic ILL at Colorado State University." *OCLC Systems & Services* 9, no. 3 (1993): 31–37.

The author describes development and implementation of a user-initiated ILL electronic request system at Colorado State and comments on the development of an ILL link in OCLC's FirstSearch databases. The development of local patron electronic request systems and the ability to capture requests in electronic databases for relay to an ILL office for handling are important steps in moving toward a seamless interface for the user to identify and order needed documents simultaneously.

IDA17. ———. "Points of View: ILL PRISM Transfer." *OCLC Systems & Services* 10, no. 2/3 (1994): 34–36.

The author describes the linking of a local electronic patron request system with PRISM OCLC, using OCLC's new IPT (ILL PRISM transfer) service. This eliminates the need for ILL staff to rekey electronic requests from users into the OCLC system for transmission to other libraries.

IDA18. Toyofuku, Anthony, and Colby Riggs. "The Ant*ill: Using Perl to Automate the ILL Lending Process." *OCLC Systems & Services* 11, no. 2 (1995): 37–41.

The authors describe the development of software to automatically search incoming OCLC lending requests in the University of California–Irvine local OPAC (Innovative Interfaces). This allows a lending library to attach call numbers and circulation status to incoming requests, greatly speeding the handling process. The program code is included.

IDB
Textbooks, Guides, and Manuals

In addition to the following materials, ILL practitioners need up-to-date procedure manuals for online interlibrary loan systems, lending policies of all libraries to which requests are sent, and all consortium, state, or regional lending codes that apply.

IDB1. Boucher, Virginia. *Interlibrary Loan Practices Handbook.* 2nd ed. Chicago: American Library Association, 1996.

An extensive revision of the 1984 publication, updating technological options and information on resources. This essential how-to-do-it manual for ILL practitioners will likely maintain its prime spot on the desk of ILL staff in libraries of all sizes. Especially well-designed for training new ILL staff.

IDB2. Gilmer, Lois C. *Interlibrary Loan: Theory and Management.* Englewood, Colo.: Libraries Unlimited, 1994.

This comprehensive text includes basic information for running an ILL operation and also provides a clear, concise review of issues and trends. Topics include a historical overview of ILL, networking issues, copyright applications, organization and administration of ILL, policies and procedures for operation, integration of document delivery, and speculation on future trends. Includes useful bibliography.

IDB3. International Federation of Library Associations and Institutions. *Guide to Centres of International Lending.* 5th ed. Boston Spa: IFLA Offices for UAP and International Lending, 1995.

Includes lending policies of libraries in over 190 countries. Entries are arranged alphabetically according to the English name of the country. Information for most countries includes a summary of national guidelines, full address including telephone, fax, and e-mail; types of forms accepted, charges and methods of payment, loan periods, and any restrictions. The guide includes "International Lending Principles and Guidelines for Procedure," and "The IFLA Fax Guidelines."

IDB4. International Federation of Library Associations and Institutions. *Guide to Document Delivery Centres.* Boston Spa: IFLA Offices for UAP and International Lending, forthcoming in late 1996.

This companion volume to the *Guide to Centres of International Lending* **(IDB3)** will provide corresponding access and supply information for commercial document suppliers.

IDB5. Massis, Bruce, and Winnie Vitzansky, eds. "Interlibrary Loan of Alternative Materials: A Balanced Sourcebook." *Journal of Interlibrary Loan & Information Supply* 3, no.1/2 (1992): 1–196.

This sourcebook, compiled by the IFLA Section of Libraries for the Blind, gives information for borrowing Braille books and other resources for the visually impaired from more than fifty libraries worldwide. Provides specific interlending information for these specialized resources held in twenty-nine countries.

IDC
Directories

IDC1. *Ariel Address List and Directory.* Online. URL: http://www.rlg.org/aridir.html

An electronic directory of Internet addresses for registered Ariel sites. Provides IP (Internet Protocol) address, machine names, and site name.

IDC2. *Gabriel: Gateway to Europe's National Libraries.* Online. URL: http://portico.bl.uk/gabriel/en/welcome.html

Web server for the Europe's national libraries represented in the Conference of European National Libraries (CENL), hosted by the British Library World Wide Web Service. Provides entries in English, French, and German.

IDC3. Jones, D. Lee, ed. *The Library Fax/Ariel Directory: More than 10,560 Fax Sites and 555 Ariel Sites.* 9th ed. Kansas City, Mo.: CBR Consulting Services, Inc., 1995. (10th ed. expected in 1996). Previously published under the title *Directory of Telefacsimile Sites in North American Libraries.*

With the eighth edition, 1994, this directory of library fax sites began to include Ariel sites with their Internet addresses for electronic transmission of documents. The directory covers over 5,500 libraries in the United States, Canada, and Mexico.

IDC4. Morris, Leslie H. *Interlibrary Loan Policies Directory.* 5th ed. New York: Neal-Schuman, 1995.

Includes lending policies of over 1,425 academic, public, and special libraries in the U.S., Puerto Rico, and Canada. Useful tool for reviewing and comparing interlibrary loan policies. Each entry includes ILL address, acceptable methods of transmission, average turnaround time, materials loaned, and billing procedures.

IDC5. Schuyler, Michael. *OPAC Directory: An Annual Guide to OnLine Public Access Catalogs and Databases.* Westport, Conn.: Meckler, 1993– .

Especially useful to libraries without access to one of the major bibliographic utilities for identifying libraries holding needed materials.

IDD
Guidelines and Standards

IDD1. American Library Association, Association of College & Research Libraries, Rare Books & Manuscripts Section, Ad Hoc Committee on the Loan of Rare and Unique Materials. "Guidelines for the Loan of Rare and Unique Materials." *College & Research Libraries News* 54 (May 1993): 267–69.

These guidelines were approved by the ACRL Standards & Accreditation Committee, ACRL Board of Directors, and the ALA Standards Committee in February 1994.

IDD2. American Library Association, Reference and Adult Services Division, Management of Public Services Section, Interlibrary Loan Committee. "Guidelines and Procedures for Telefacsimile and Electronic Delivery of Interlibrary Loan Requests." *RQ* 34, 1 (Fall 1994): 32–33. Also available on the ALA gopher at URL: gopher://gopher.ala.org:70/11/alagopherxiii/alagopherxiiirasd

These guidelines address the needs of libraries that use fax and electronic document delivery (EDD) for ILL borrowing and lending processes and cover equipment, uniform operation of equipment, request formatting, and guidelines for responses to requests.

IDD3. ———. "National Interlibrary Loan Code for the United States 1993." *RQ* 33 (Summer 1994): 477–79. Also available on the ALA gopher at URL: gopher://gopher.ala.org:70/11/alagopherxiii/alagopherxiiirasd

Provides general guidelines for the requesting and supplying of materials between libraries. A complete rewrite of the 1980 "National Interlibrary Loan Code," it recognizes ILL as an integral element of collection development for all libraries, not an ancillary option. The revised code is much more access-oriented than previous codes, omitting the list of items not to be requested, and explicitly stating that any materials, regardless of format, may be requested from another library. This code may be used as a model for development of state, regional, or local interlibrary loan codes.

IDD4. Association of Research Libraries. *Transborder Interlibrary Loan: Shipping Interlibrary Loan Materials from the U.S. to Canada.* Online. Washington, D.C.: Association of Research Libraries, 1995. URL: gopher:// arl.cni.org under the menu choices: /Access to Research Resources/Interlibrary Loan and Document Delivery/Shipping Interlibrary Loan Materials from the U.S. to Canada.

Includes effective methods of shipping and returning materials to Canadian libraries. Print copies sent only to those without gopher access; contact Mary Jackson at mary@cni.org or 202-296-2296.

IDD5. National Information Standards Organization. *Interlibrary Loan Data Elements.* Bethesda, Md.: NISO Press, 1995. ANSI/NISO Z39.63-1989.

The Z39.63 standard is an important element in the evolution of a virtual library environment where the user has seamless access to documents regardless of their location. The refinement and application of this protocol should be adhered to in any local system development as well as become a piece of library online vendor development. Approved by the American National Standards Institute on February 2, 1989.

IDD6. Turner, Fay. "Document Ordering Standards: The ILL Protocol and Z39.50 Item Order." *Library Hi Tech* 13, no. 3 (1995): 25–38.

As multiple options emerge for patron-initiated requests and the need for users to have seamless access to multiple databases and suppliers is apparent, there is a great deal of interest in clarifying the relationship of ILL protocol to Z39.50 item ordering. Initiatives such as the ARL North American Interlibrary Loan and Document Delivery (NAILDD) project have raised interest not only among librarians but also among bibliographic utilities and library software vendors to restructure and integrate ILL. Turner contrasts the ILL protocol standard and Z39.50 item ordering. Even though there is an overlap of functions, the author argues that the two standards can be used to complement rather than compete with each other. This article presents the technical issues in understandable language, providing excellent background for librarians to see clearly what is at stake and to engage them in the discussion.

IDE
Application Software Resources for ILL Management and Communication

IDE1. *Ariel.*

Ariel, the Research Libraries Group document transmission workstation, is used to transmit copies of articles digitally over the Internet. The proprietary software for *Ariel* and information on equipment specifications is available from RLG. *Ariel* is not restricted to RLG libraries; many academic libraries, special libraries, and a variety of multi-type library consortiums use *Ariel* for rapid transmission of documents. Information on *Ariel* is available on the Research Libraries Group Web server at URL: http://lyra.stanford.edu/ariel.html. *ARIEL-L* **(IAC4)** is an online discussion group for *Ariel* users.

IDE2. *AVISO.*

AVISO, originally developed for use by Canadian libraries, has recently been modified for use by American libraries, including the ability to track U.S.

copyright compliance. *AVISO* is a full management software for ILL records and also supports communication of requests. *AVISO-L* **(IAC5)** is an electronic discussion forum for *AVISO* users. For more information, contact ISM Library Information Services, 3300 Bloor St. W., 16th Floor, West Tower, Etobicoke, Ontario, Canada M8X 2X2.

IDE3. OCLC Online Computer Library Center, Inc.

OCLC plays a significant role in interlibrary loan. Currently, most major libraries in the United States transmit requests on OCLC; more than 500,000 interlibrary loan transactions are handled monthly. OCLC offers a monthly basic statistics package on subscription basis for participating libraries. OCLC is testing additional statistical options, for example PILLAR, which support customized management reports. Information on OCLC products is available by contacting OCLC directly or electronically at URL http://www.oclc.org/ or URL gopher://oclc.org. These offer extensive resources for news, reports, and documentation relating to OCLC. Includes, for example, a walk-through program to explain how the ILL Fee Management (IFM) program works. In addition, *PRISM News* is found online and can also be searched from the PRISM OCLC ILL module; it has updated information, for example, a comprehensive list of libraries using IFM for handling charges.

IDE4. *Patron Request System (PRS).*

PRS is an ILL management software, developed at Brigham Young University and designed to track and provide statistics for ILL borrowing and ILL lending activity. Special features include automatic generation of patron notices and full copyright handling. The program is designed to interface with OCLC via OCLC microenhancer software. There is provision to accommodate RLIN ILL activity. For additional information, contact Kathy Hansen, Head, Interlibrary Loan, Brigham Young University Library, Provo, UT 84602.

IDE5. Research Libraries Group, Inc.

RLG utilizes the Research Libraries Information Network (RLIN) to transmit ILL requests among members. For information on the RLIN ILL system, contact the Research Libraries Group or visit its WEB page at URL http://www.rlg.org

IDE6. *SAVE-IT.*

SAVE-IT is an ILL management software that works with OCLC, using the savescreen function of OCLC Passport software to save data from ILL workforms. It interfaces with the OCLC microenhancer software for updating requests. *SAVE-IT* supports copyright compliance and generates statistics that can be customized for management reports. *SAVEIT-L* **(IAC11)** is an electronic discussion list for *SAVE-IT* users. For more information, contact Patrick Brumbauch, Interlibrary Software and Services, Inc., P.O. Box 1958, Cleveland, OH 44106.

IE
Document Delivery

ILL costs are the focus of much professional attention. The benchmark ARL/RLG ILL cost study **(IE19)** made it clear that commercial document delivery could play an important, cost-effective role in providing information to users. This study changed the way both librarians and users view ILL and opened the door for aggressive use of document suppliers as an alternative to traditional ILL.

The North American Interlibrary Loan and Document Delivery (NAILDD) project emphasizes the blurring of interlibrary loan and document delivery and focuses on the need to encourage technical improvements in all aspects of online ILL and document delivery operations, providing the user with a dependable source of documents beyond onsite collections. It emphasizes the need for national standards and the importance of system design capabilities to improve interlibrary loan and document delivery services for users. The NAILDD project has played an important role in encouraging dialogue among library vendors, software developers, document delivery suppliers, and librarians to develop new article delivery options. Documents relating to the NAILDD project may be accessed online through ARL at URL gopher://arl.cni.org under the menu choices: Access to Research Resources/Interlibrary Loan and Document Delivery.

DIRECTORIES

IE1. Coffman, Steve, and Pat Wiedensohler, eds. *The Fiscal Directory of Fee-Based Research and Document Supply Services.* 4th ed. [N.p.]: County of Los Angeles Public Library; American Library Association, 1993.

A listing of over 500 information and document providers with information on what they provide, methods for making requests, delivery options, and charges. Includes geographic and subject-specialty indexes.

GENERAL WORKS

IE2. Cain, Mark. "Periodical Access in an Era of Change: Characteristics and a Model." *The Journal of Academic Librarianship* 21 (Sept. 1995): 365–70.

The author suggests a model for deciding when to buy and when to rely on remote access, either document delivery or traditional ILL. A case is made for emphasizing use of commercial document delivery and utilizing ILL only as a last resort.

IE3. Carrigan, Dennis P. "From Just-in-Case to Just-in-Time: Limits to the Alternative Library Service Model." *Scholarly Publishing* 26 (April 1995): 173–82.

Interesting article that reviews some of the restrictions imposed by publishers who limit or eliminate affordable document delivery options for their publications.

IE4. Coons, Bill, and Peter McDonald. "Implications of Commercial Document Delivery." *College & Research Libraries News* 56 (Oct. 1995): 626–31.

For annotation, *see* **GD38**.

IE5. Goodyear, Mary Lou, and Jane Dodd. "From the Library of Record to the Library as Gateway: An Analysis of Three Electronic Table-of-Contents Services." *Library Acquisitions: Practice & Theory* 18, no. 3 (1994): 253–64.

The authors compare title coverage of three commercial table-of-contents services that offer article delivery: *UnCover*, Faxon, and OCLC's *ArticleFirst*. Provides useful data analysis of subject overlap and areas of unique title holdings.

IE6. Gossen, Eleanor A., and Suzanne Irving. "Ownership Versus Access and Low-Use Periodical Titles." *Library Resources & Technical Services* 39, no. 1 (Jan. 1995): 43-52.

For annotation, *see* **GD40**.

IE7. Jackson, Mary E. "The Future of Resource Sharing: The Role of the Association of Research Libraries." *Journal of Library Administration* 21, no. 1/2 (1995): 193–202.

Provides an overview of the ARL projects (including NAILDD) designed to influence the evolution of ILL into a comprehensive program including not only ILL but also document delivery, on-site access, and local ownership. These projects play a major role in engaging not only librarians, but also library utilities and database vendors, in discussions to identify a model for integrated, one-stop shopping for the user to obtain needed documents.

IE8. Jaguszewski, Janice M., and Jody L. Kempf. "Four Current Awareness Databases: Coverage and Currency Compared." *Database* 18 (Feb./March 1995): 34–44.

The authors compare four databases for coverage, document delivery options, and attractive, unique features for users: *Current Contents on Diskette, UnCover, Inside Information,* and *ContentsFirst*. The focus is on the sciences with special attention given to journal coverage in mathematics and chemistry.

IE9. Kaser, Dick, ed. *Document Delivery in an Electronic Age*. Philadelphia: National Federation of Abstracting and Information Services, 1996.

Contributors to this book include attorneys, information brokers, and commercial document suppliers. Looks at traditional and nontraditional methods of document delivery such as subscriptions, photocopying, interlibrary loan, and microfiche. Other topics discussed include electronic document delivery systems, legal issues, and an AT&T case study. Fourteen document suppliers share their view on the future of document delivery. Includes directory of current suppliers and contact information for each.

IE10. Khalil, Mounir. "Document Delivery: A Better Option?" *Library Journal* 118 (Feb. 1, 1993): 43–47.

The author reviews the pros and cons of using commercial suppliers in place of traditional interlibrary loan. The article includes a list of commercial services with contact information and a description of services, coverage, fees, and strengths.

IE11. Kingma, Bruce R. "Access to Journal Articles: A Model of the Cost Efficiency of Document Delivery and Library Consortia." In *ASIS '94: Proceedings of the 57th ASIS Annual Meeting, Alexandria, Va., October 17–20, 1994,* edited by Bruce Maxian, 8–11. Medford, N.J.: Learned Information, 1994.

Outlines a project funded by CLR to collect empirical data to develop a cost-effective model for providing journal articles access in an academic library consortium.

IE12. Kohl, David F. "Revealing UnCover: Simple, Easy Article Delivery." *Online* 19, no. 3 (May/June 1995): 52–60.

The author describes mediated, subsidized article delivery from *UnCover* for faculty at University of Cincinnati (Ohio) following a major serial cancellation. *UnCover* service options are covered in detail with some mention of alternative article suppliers.

IE13. Kurosman, Kathleen, and Barbara Ammerman Durniak. "Document Delivery: A Comparison of Commercial Document Suppliers and Interlibrary Loan Services." *College & Research Libraries* 55 (1994): 129–39.

Librarians at Vassar College share data showing that traditional ILL was faster and less expensive than ordering articles from four commercial suppliers during a sample period in 1991/92.

IE14. Machovec, George S. "Criteria for Selecting Document Delivery Suppliers." *Online Libraries and Microcomputers* 12, no. 5 (1994): 1–5.

The author reviews important trends affecting traditional ILL and document delivery, emphasizing the need for offering the user flexibility and choice for obtaining material and requiring national standards for ILL, telecommunications, serial identification, patron record information, and circulation transactions.

IE15. Martin, Harry S. III, and Curtis L. Kendrick. "A User-Centered View of Document Delivery and Interlibrary Loan." *Library Administration and Management* 8, no. 4 (1994): 223–27.

Essential reading for librarians as they review options for integration of document delivery choices into library services. This paper presents the vision of scholars for locating and retrieving documents in a timely manner without mediated assistance from library staff. The scenario assumes the necessary technology and suggests fundamental changes in library philosophy and operations are needed to meet researchers' needs.

IE16. McDaniel, Elizabeth, and Ronald Epp. "Fee-based Information Services: The Promises and Pitfalls of a New Revenue Source in Higher Education." *Cause/Effect* 18 (Summer 1995): 35–39.

Insightful analysis of an unsuccessful attempt at the University of Hartford to establish a fee-based electronic information service for off-campus clients. This article reviews useful information for libraries considering utilization of local collections for a fee-based document delivery service to individuals and/or companies.

IE17. Mitchell, Eleanor, and Sheila Walters. *Document Delivery Services: Issues and Answers.* Medford, N.J.: Learned Information, 1995.

Handbook of practical information for integrating use of document delivery avenues into library services. Includes a comprehensive overview of the issues involved, covering traditional ILL, commercial document suppliers, electronic document retrieval and transmission technology, and end-user options. Well indexed, with an extensive bibliography.

IE18. Pedersen, Wayne, and David Gregory. "Interlibrary Loan and Commercial Document Supply: Finding the Right Fit." *Journal of Academic Librarianship* 20, no. 5/6 (1994): 263–72.

Analyzes six commercial suppliers accessible via OCLC and compared with ILL for cost, turnaround time, and fill rate. Advantages and disadvantages of specific suppliers are outlined in detail. Provides a useful framework for integrating the strengths of commercial suppliers into an interlibrary loan operation.

IE19. Roche, Marilyn M. *ARL/RLG Interlibrary Loan Cost Study: A Joint Effort by the Association of Research Libraries and the Research Libraries Group.* Washington, D.C.: Association of Research Libraries, June 1993.

This benchmark study provides cost information for both ILL borrowing and ILL lending activity, based on data collected in seventy-six U.S. and Canadian research libraries. It shows the average cost for a completed ILL transaction (incurred by both the lender and the borrower) to be close to $30—almost $19.00 for the borrower and $11.00 for the lender. This study provides important data for making decisions on whether to purchase or borrow material and when to use commercial suppliers.

IE20. Sellers, Minna, and Joan Beam. "Subsidizing Unmediated Document Delivery: Current Models and a Case Study." *Journal of Academic Librarianship* 21 (Nov. 1995): 459–66.

The authors report on a project at Colorado State University to test the viability of providing subsidized unmediated document delivery to the entire university community using UnCover. Comparison is made to mediated (e.g., ILL) document delivery for users, making a case for the need to have a single electronic interface for all user requests. Analysis of the project includes a user survey to evaluate the performance of the service.

IE21. Uses of Document Delivery Services. Mary E. Jackson and Karen Croneis, comps. SPEC Kit, no. 204. Washington, D.C.: Association of Research Libraries, Office of Management Services, 1994.

Includes a comprehensive bibliography and sample documentation, all pointing to a more user-centered environment for ordering documents. Includes material from pilot projects for document delivery; sample documents include selection policies, library flyers, document delivery service flyers, annual reports, evaluations, and statistics. Presents document delivery as a natural complement to onsite collections in response to shrinking library budgets, multiple serials cancellations, and perceived weaknesses in the traditional interlibrary loan system. Also covers reasons for use, charges and costs, service and quality, and end-user ordering as a method of reducing staff workloads.

IE22. Walters, Sheila. "The Direct Doc Pilot Project at Arizona State: User Behavior in a Non-mediated Document Delivery Environment." *Computers in Libraries* 15 (Oct. 1995): 22–26.

Analyzes faculty use of subsidized, unmediated document delivery services (UnCover, FirstSearch, and Eureka) to determine whether a user-initiated ordering service, with fast delivery direct to the user, could substitute for serials not held by the library. The project included educating faculty about the real cost of interlibrary loan in comparison to document delivery to gain acceptance for using library material funds for document delivery fees. Direct Doc services were implemented following the pilot with the modification of blocking titles owned by the library. With the addition of Direct Doc, Arizona State offers users an ideal model of access: nonmediated document delivery for most faculty, on-campus delivery of library-owned materials, and partially mediated delivery for less familiar document providers.

IE23. Wessling, Julie. "Document Delivery: A Primary Service for the Nineties." *Advances in Librarianship* 16 (1992): 1–31.

The author reviews the evolution of traditional interlibrary loan into an expanded document delivery service, concluding with the characteristics of a model library document delivery service.

IF
Library Security

Book theft is a growing problem in libraries. Publicized thefts in recent years, such as the case of Stephen Blumberg, show that thieves infiltrate both established systems in special collections departments and open stacks areas where valuable material may be shelved. Security is everybody's business and access services staffs often will play a daily role in maintaining and enforcing security measures in a library. All staff members should be trained fully in emergency procedures and know the resources available for assistance.

Library security is taking on a new dimension with the rising number of violent crimes committed in libraries. Libraries are not immune to violence—in 1994, there were eight homicides in U.S. libraries in the United States. Bomb threats are no longer unusual and arson and assault incidents are increasing.

IF1. Association of College & Research Libraries, Rare Books & Manuscripts Section, Security Committee. "Guidelines Regarding Thefts in Libraries." *College & Research Libraries News* 55 (Nov. 1994): 641–47.

These guidelines concern all library thefts, not just those that may occur in rare book or special collections departments. They cover what to do before a theft occurs, what to do after a theft occurs, and model legislation concerning theft, which librarians are urged to use to influence state laws for the prosecution and punishment of library thieves. Libraries are encouraged to appoint a Library Security Officer.

IF2. Chaney, Michael, and Ian MacDougall, eds. *Security and Crime Prevention in Libraries.* Aldershot, Hampshire, England: Gower, 1994.

This comprehensive review of security issues in English libraries provides an overall framework for identifying areas to be included in any library security policy. Some sections may be of limited use to American librarians, but valuable chapters make this book a good resource, especially those addressing crime prevention, library security audits, developing a security policy, and selecting a book theft detection system. Includes bibliographical references for further reading and index.

IF3. McNeal, Beth, and Denise J. Johnson, eds. *Patron Behavior in Libraries: A Handbook of Positive Approaches to Negative Situations.* Chicago: American Library Association, 1995.

Fifteen papers dealing with various aspects of negative patron behavior, ranging from disruptive to clear physical threat to patrons, staff, and collections. Recommends developing strategies for responding before problems occur. For all types and sizes of libraries.

IF4. St. Lifer, Evan, et al. "How Safe Are Our Libraries?" *Library Journal* 119 (Aug. 1994): 35–39.

The authors review recent cases of library theft and violence, offer practical tips on library security, and include a list of companies that handle book theft detection systems and other security aids.

IF5. *Stop Thief! Strategies for Keeping Your Collections from Disappearing.* [2 sound cassettes] (ALA-535) Chicago: American Library Association, 1995.

Many useful suggestions are presented. A program recorded at the 114th annual conference of the American Library Association, held June 22–28, 1995, in Chicago, Illinois. Sponsored by LAMA, Buildings and Equipment Section, Safety and Security of Library Buildings Committee and ACRL, Rare Books & Manuscripts Section, Security Committee.

IF6. Tehrani, Farideh, guest ed. "Library Security: Special Issue." *New Jersey Libraries* 27 (Fall 1994).

A special issue devoted to both practical approaches for library security and specific practices in a variety of public libraries and academic libraries in New Jersey. The problems and solutions are applicable in many locations. An annotated bibliography on items addresses security issues in all types of libraries.

IF7. Turner, Anne. *It Comes with the Territory: Handling Problem Situations in Libraries.* Jefferson, N.C.: McFarland & Co., 1993.

The author, using an engaging style, offers practical advice for handling difficult situations and difficult patrons in libraries. She covers writing rules and procedures for dealing with various situations and also addresses training staff to follow procedures. Appendixes include American Library Association policies as well as sample policies from the Santa Cruz, California, library where Turner is director.

IF8. *Violence in the Library: Prevention, Preparedness and Response.* Workshop. Layne Consultants International, Cultural Protection Specialists, P.O. Box 1, Dillon, CO 80435. Phone: 970-468-5522.

Workshop offered by certified protection professionals, who have experience with museums, historic sites, and libraries. This group has consulted for the American Library Association, offering workshops for librarians in cooperation with ALA's Library Administration and Management Association. Workshops deal with violent crimes in libraries and how staff members can protect themselves and their patrons.

ABOUT THE EDITORS

Peggy Johnson is Planning and Special Projects Officer at the University of Minnesota Libraries. Her previous positions include Interim Collection Development Officer, University Libraries, and Assistant Director, St. Paul Campus Libraries, University of Minnesota. She has consulted on library development in Morocco, Uganda, and Rwanda. Recent monographs include *Guide to Technical Services Resources* (American Library Association, 1994); *Collection Management and Development: Issues in an Electronic Era* with Bonnie MacEwan (American Library Association, 1994); *Recruiting, Educating, and Training Librarians for Collection Development* with Sheila S. Intner (Greenwood, 1994); and *The Searchable Internet Bibliography* with Lee English (American Library Association, 1996). She writes a bimonthly column on collection management for *Technicalities*. Johnson has an MA from the Graduate Library School, University of Chicago, and an MBA from Metropolitan (Minn.) State University.

Wesley L. Boomgaarden is Preservation Officer at the Ohio State University Libraries, Columbus, a position he has held since 1984. Prior to that, he was Head, Preservation Microfilming Office at the Research Libraries, New York Public Library, and served on the staffs of the Columbia University Libraries Preservation

Department, the Conservation Department of the Minnesota Historical Society, and Macalester College Library. Boomgaarden has numerous publications on the topic of preservation.

Deborah E. Burke is Head of Acquisitions and Serials Management for the St. Paul Campus Libraries, University of Minnesota. She previously held a variety of positions at the University of Minnesota, including Head of the Wilson Library Listening Room and Head of the Wilson Library Reserve Room. She is a Member-at-Large of the Minnesota Library Association's Academic and Research Libraries Division Board.

Janet Swan Hill is Associate Director for Technical Services at the University of Colorado Libraries in Boulder and has previously held positions as Head of Cataloging at Northwestern University Library and Head of Map Cataloging at the Library of Congress. An active writer and American Library Association member, she served on ALA's Committee on Cataloging: Description and Access in various capacities for eleven years, including six as the ALA representative to the Joint Steering Committee for the Revision of the Anglo-American Cataloguing Rules. Her concern for the education of catalogers in the mid-1980s led to the formation of the ALA Committee on Education, Training, and Recruitment of Catalogers.

Sheila S. Intner is Professor at the Graduate School of Library and Information Science, Simmons College, Boston. She taught at the University of Haifa and Hebrew University in Jerusalem during 1992-93 on a Fulbright grant. She is frequently a speaker at professional forums and widely published. Intner is editor of the journal *Technicalities* and of a monographic series published by ALA titled Frontiers of Access to Library Materials. Her recent monographs include *Intnerfaces: Relationships between Library Technical and Public Services* (Libraries Unlimited, 1993); *Recruiting, Educating, and Training Librarians for Collection Development* with Peggy Johnson (Greenwood, 1994); and *Standard Cataloging for School and Public Libraries*, 2nd ed. (Libraries Unlimited, 1996).

Susan Morris is Assistant to the Director for Cataloging at the Library of Congress. She was previously a monograph cataloger for the Library of Congress, specializing in religion and philosophy. She also cataloged at Harvard College Library and the Free Library of Philadelphia, where she was on the staff of the Edwin A.

Fleisher Collection of Orchestral Music. Morris is a member of the Association for Library Collections & Technical Services and serves on the editorial board of the *Journal of Internet Cataloging*.

Genevieve S. Owens is the Program Manager for Information Resources Selection at Bucknell University's Bertrand Library. Previously, she was a Reference Librarian and the Head of Collection Development at the University of Missouri–St. Louis. She is active in the Association for Library Collections & Technical Services, particularly its Library Materials Price Index Committee and Collection Management and Development Section.

Karen A. Schmidt is Acquisitions Librarian at the University of Illinois at Urbana-Champaign, where she has worked for the past fourteen years. She also serves as acting coordinator of collection development. Schmidt has written extensively on acquisitions, with a focus on education, as well as on collections and library history, and has won several awards for her research. Schmidt is a regular contributor to library scholarly journals and is active in the Association for Library Collections & Technical Services and the Association of College and Research Libraries.

Julie Wessling is Assistant Dean for Public Services at Colorado State University Libraries, where she was previously Head of Interlibrary Loan Services. Under her leadership, Colorado State developed a patron-initiated ILL electronic request system, ZAP, which has been extended for use by libraries throughout Colorado. ZAP served as a beta test program for linking a local request program with the OCLC ILL subsystem package. She has written numerous articles and given presentations on automated ILL and document delivery issues.

Nancy J. Williamson is Professor Emerita at the University of Toronto, where she continues to teach a course in the subject approach to information and to supervise PhD students at the Faculty of Information Studies. She chairs the FID/CR (the Committee on Classification Research for Knowledge Organization of the International Federation for Information and Documentation [FID]), and has written extensively on the topic of subject access. Williamson is currently conducting research on the feasibility of converting the Universal Decimal Classification into a fully faceted system. She holds degrees in Library and Information Science from the University of Toronto and Case Western Reserve.

AUTHOR/TITLE INDEX

The index is arranged according to the ALA filing rules except that numbers are filed as if spelled out, and & is filed as if spelled *and*.

Some persons and organizations listed in the index have served as authors, compilers, and editors. The titles by these entities are sub-arranged alphabetically without regard to the "comp." and "ed." designations. However, titles for which there are joint authors are filed after those of single authorship (or compilership or editorship), arranged alphabetically by the surname of the second author (or compiler or editor). Titles by the same group of joint authors (or compilers or editors) are arranged alphabetically.

For specific divisions, sections, or committees within the American Library Association, see that specific name. Organization names are spelled out; see pp. xii-xviii for a list of acronyms.

A

AACR2 Decisions & Rule Interpretations, L. C. Howarth, comp., CCC1
"AACR2R: Dissemination and Use in Canadian Libraries," L. C. Howarth and J. Weihs, CCA4
"AACR2R Use in Canadian Libraries and Implications for Bibliographic Databases," L. C. Howarth and J. Weihs, CCA5
AAT-L: Art & Architecture Thesaurus Discussion List, EAD1
Abridged Dewey Decimal Classification and Relative Index, M. Dewey, DBB2

"The Academic Library: A Time of Crisis, Change, and Opportunity," R. M. Dougherty and A. P. Dougherty, AHA4
"Academic Library Collection Development and Management Positions: Announcements in *College and Research Libraries News* from 1980–1991," W. C. Robinson, GB4
"The Academic Library Collection in an On-Line Environment," R. Atkinson, GE1
"Access Blues: A Song We Are All Singing," K. Kennedy, FBB3

"Access, Ownership, and the Future of Collection Development," R. Atkinson, GE2
Access Services in Libraries: New Solutions for Collection Management, G. Sapp, IA2
"Access to Journal Articles: A Model of the Cost Efficiency of Document Delivery and Library Consortia," B. R. Kingman, IE11
"Access to Library Materials in Remote Storage," C. Q. Bellanti, IB2
"Access to Nonbook Materials: The Limits of Subject Indexing for Visual and Aural Languages," E. Svenonius, ED5
"Access to Serials: National and International Cooperation," A. A. Mullis, FF1
Access versus Assets: A Comprehensive Guide to Resource Sharing for Academic Librarians, B. B. Higginbotham and S. Bowdoin, GH5
"Achieving Success as a Serials Librarian: Some Advice," H. M. Grochmal, FBA4
ACQNET, BAC1, FAD1
Acquisitions Course Syllabus, ALCTS Acquisitions Section, BAA1
"Acquisitions Ethics: The Evolution of Models for Hard Times," M. C. Bushing, BI2
"Acquisitions Principles and the Future of Acquisitions: Information Soup, the Soup-Hungry, and Libraries' Five Dimensions," J. W. Barker, BA1
AcqWeb, BAD4
Ad Hoc Subcommittee on Copyright, American Library Association, IAC1
"Adult Fiction in Medium-Sized U.S. Public Libraries: A Survey," J. H. Sweetland, GE18
Advances in Preservation and Access, B. B. Higginbotham, ed., HA1

"After Cutter: Authority Control in the Twenty-first Century," J. Younger, CBF15
After the Electronic Revolution, Will You Be the First to Go?, A. Hirshon, ed., AHA7
Against the Grain, BAC2
"Agreement and Disagreement among Fiction Reviews in *Library Journal, Booklist* and *Publishers Weekly*," J. W. Palmer, GD48
"An Aid for Total Quality Searching: Developing a Hedge Book," M. J. Klatt, EC5
AIIM Buying Guide, HN24
Albrechtsen, Hanne, and Susanne Oernager, eds. *Knowledge Organization and Quality Management: Proceedings of the Third International ISKO Conference*, DAA1
"ALCTS Commercial Technical Services Committee, June 28, 1994," S. Flood, AHA5
"ALCTS Creative Ideas in Technical Services Discussion Group," A. McGreer, AHC6
"ALCTS/Role of the Professional in Academic Research Technical Services Departments Discussion Group," N. J. Gibbs, AHB4
Alexander, Adrian W. "Periodical Prices, 1990–1992," FE1
———. "Periodical Prices, 1991–1993," FE2
———. "Periodical Prices, 1992–1994," FE3
Alexander, Adrian W., and Kathryn Hammell Carpenter. "U.S. Periodical Price Index for 1993," FE4
———. "U.S. Periodical Price Index for 1994," FE5
———. "U.S. Periodical Price Index for 1995," FE6
Alexander, Michael. "Digital Data Retrieval: Testing Excalibur," EF1

Aliprand, Joan M. "Linking of Alternate Graphic Representation in USMARC Authority Records," CBF1

Al-Kharashi, Ibrahim A., and Martha W. Evens. "Comparing Words, Stems, and Roots as Index Terms in an Arabic Information Retrieval System," EAB1

Allen, Barbara McFadden. *See* Olson, Georgine N.

Allen, Bryce. "Improved Browsable Displays: An Experimental Test," EC1

Allen, Nancy H., and James F. Williams II. "The Future of Technical Services: An Administrative Perspective," AHA1

Allerton Park Institute, AGB1

Allerton Park Institute. "New Roles for Classification in Libraries and Information Networks: Presentations and Reports from the Thirty-Sixth Allerton Institute," DBA2

"Allocating the Materials Funds Using Total Cost of Materials," C. Cubberly, GF1

Allocation Formulas in Academic Libraries, J. H. Tuten and B. Jones, comps., GF4

Aluri, Rao, Alasdair Kemp, and John J. Boll. *Subject Analysis in Online Catalogs*, DA1

Alvarado, Rubén Urbizagástegui. "Cataloging Pierre Bourdieu's Books," EE1

Alvey, Christine E. *See* Scott, Mona L.

Amalgamations & the Centralisation of Technical Services: Profit or Loss, J. Thawley and P. G. Kent, eds., AHA13

"Ambiguities in the Use of Certain Library of Congress Subject Headings for Form and Genre Access to Moving Image Materials," D. Miller, DCB7

American Library Association, CAE17, IAC1

American Library Association. Intellectual Freedom Committee. *Guidelines for the Development of Policies and Procedures Regarding User Behavior and Library Usage*, ICI1

American Library Association. Office for Intellectual Freedom. *Intellectual Freedom Manual*, GD13

American Library Association, Subcommittee on Guide for Training Collection Development Librarians. *Guide for Training Collection Development Librarians*, GB5

American National Standard for Imaging Media (Film): Thermally Processed Silver Microfilm: Specifications for Stability, ANSI, HN4

American National Standard for Imaging Media: Photographic Activity Test, ANSI, HN5

American National Standard for Imaging Media: Photographic Films, Papers, and Plates: Glossary of Terms Pertaining to Stability, ANSI, HN6

American National Standard for Imaging Media (Photography): The Effectiveness of Chemical Conversion of Silver Images against Oxidation: Methods for Measuring, ANSI, HN7

American National Standard for Imaging Media: Processing Safety Photographic Film: Storage, ANSI, HN8

American National Standard for Permanence of Paper for Publications and Documents in Libraries and Archives, NISO, HN2

American National Standard for Photography: Determination of Residual Thiosulfate and Other Related Chemicals in Processed Photographic Materials, ANSI, HN9

American National Standard for Photography: Film Dimensions: Film for Documentary Reproduction, ANSI, HN10

American National Standard for Photography: Processed Vesicular Photographic Film—Specifications for Stability, ANSI, HN12

American National Standard for Photography: Specifications for Safety Film, ANSI, HN11

American National Standards Institute. *American National Standard for Imaging Media (Film): Thermally Processed Silver Microfilm: Specifications for Stability*, HN4

———. *American National Standard for Imaging Media: Photographic Activity Test*, HN5

———. *American National Standard for Imaging Media: Photographic Films, Papers, and Plates: Glossary of Terms Pertaining to Stability*, HN6

———. *American National Standard for Imaging Media (Photography): The Effectiveness of Chemical Conversion of Silver Images against Oxidation: Methods for Measuring*, HN7

———. *American National Standard for Imaging Media: Processing Safety Photographic Film: Storage*, HN8

———. *American National Standard for Photography: Determination of Residual Thiosulfate and Other Related Chemicals in Processed Photographic Materials*, HN9

———. *American National Standard for Photography: Film Dimensions: Film for Documentary Reproduction*, HN10

———. *American National Standard for Photography: Processed Vesicular Photographic Film—Specifications for Stability*, HN12

———. *American National Standard for Photography: Specifications for Safety Film*, HN11

———. *Guidelines for the Construction, Format, and Management of Monolingual Thesauri*, DCE3, EB1

American Society of Indexers, EAD7

AMIA Newsletter, HAA1

AMIA-L, HAB2

"The Anabasis from Analog to Digital Escalates: The Year's Work in the Reproduction of Library Materials, 1992," T. A. Bourke, HAC1

"An Analysis of Courses in Cataloging and Classification and Related Areas Offered in Sixteen Graduate Library Schools . . .," D. McAllister-Harper, CBA8

Analytical Review of the Library of the Future, K. M. Drabenstott, AD1

"Anatomy of a Small Step Forward: The Electronic Reserve Book Room at San Diego State University," D. L. Bosseau, IC1

Anderson, Beth G., et al., comps. *Curriculum Materials Center Collection Development Policy*, GC4

Anderson, James D. "Standards for Indexing: Revising the American National Standard Guidelines Z39.4," EB2

Anderson, Joanne S. *Guide for Written Collection Policy Statements*, GC1

"Anglo-American Cataloguing Rules in the Online Environment: A Literature Review," R. Fattahi, CCB2

Anglo-American Cataloguing Rules, Second Edition, 1988 Revision Amendments 1993, CCA1

Anthes, Mary A. *See* Chrzastowski, Tina E.

"The Ant*ill: Using Perl to Automate the ILL Lending Process," A. Toyofuku and C. Riggs, IDA18

The Application of Expert Systems in Libraries and Information Centres, A. Morris, AIB3

"The Approval Plan Profiling Session," R. F. Nardini, GD4
"Approval Plans: Politics and Performance," R. F. Nardini, BB7, GD5
"Approval Slips and Faculty Participation in Book Selection at a Small University Library," A. E. Arnold, BB1, GD1
Archer, John. "Give Me Barcodes, or Give Me Carpal Tunnel!," AIC1
ARCHIVES, HAB3
"Are We on Equal Terms Yet? Subject Headings Concerning Women in *LCSH*, 1975–1991," M. N. Rogers, DCB8
Ariel, IDE1
Ariel Address List and Directory, IDC1
ARIEL-L, IAC4
ARL/RLG Interlibrary Loan Cost Study: A Joint Effort by the Association of Research Libraries and the Research Libraries Group, M. M. Roche, IE19
Arnold, Amy E. "Approval Slips and Faculty Participation in Book Selection at a Small University Library," BB1, GD1
Arseneau, Marjo. *See* Perrault, Anna H.
"The Art of Projecting: The Cost of Keeping Periodicals," L. Ketcham and K. Born, FE9
The Artist's Complete Health and Safety Guide, M. Rossol, HC11
ASIS SIG/CR Classification Research Workshops, DAA2
ASIS Thesaurus of Information Science and Librarianship, J. L. Milstead, DCE4
ASIS-L, EAD2
Association for Educational Communications and Technology; Association for College and Research Libraries. "Standards for Community, Junior, and Technical College Learning Resources Programs," GAC3

Association for Information and Image Management (U.S.). *AIIM Buying Guide*, HN24
———. *Electronic Imaging Output Printers*, HN25
———. *Glossary of Imaging Technology*, HN20
———. *Performance Guideline for the Legal Acceptance of Records Produced by Information Technology Systems*, HN21
———. *Recommended Practice for Microfilming Printed Newspapers on 35mm Roll Microfilm*, HN13
———. *Recommended Practice for Quality Control of Image Scanners*, HN26
———. *Resolution as It Relates to Photographic and Electronic Imaging*, HN22
———. *Standard for Information and Image Management: Microfiche*, HN14
———. *Standard for Information and Image Management; Microfilm Package Labeling*, HN15
———. *Standard for Information and Image Management: Micrographics—Splices for Image Microfilm—Dimensions and Operational Constraints*, HN16
———. *Standard Recommended Practice—File Format for Storage and Exchange of Images—Bi-level Image File Format, Part 1*, HN17
———. *Standard Recommended Practice: Identification of Microforms*, HN18
———. *Standard Recommended Practice: Monitoring Image Quality of Roll Microfilm and Microfiche Scanners*, HN27
Association for Library Collections & Technical Services, CAE18, FAE3
———. "Guidelines for ALCTS Members to Supplement the ALA Code of Ethics," AE1
———, Acquisitions Section, BAD1

Association for Library Collections & Technical Services, Acquisitions Section. *Book and Serial Vendors for Asia and the Pacific*, FAB1

———, ———, Education Committee. *Acquisitions Course Syllabus*, BAA1

Association for Library Collections & Technical Services, Commercial Technical Services Committee. *Outsourcing Cataloging, Authority Work and Physical Processing: A Checklist of Considerations*, CBG5

———, Committee on Cataloging: Description and Access. *Guidelines for Bibliographic Description of Reproductions*, CCD28

———, Interactive Multimedia Guidelines Review Task Force. *Guidelines for Bibliographic Description of Interactive Multimedia*, CCD13

———, Preservation and Reformatting Section, HAB15

———, Publisher/Vendor-Library Relations Committee. *Principles & Standards of Acquisitions Practice*, BI1

———, Reproduction of Library Materials Section, Subcommittee on Preservation Photocopying Guidelines. "Guidelines for Preservation Photocopying," HL1

———, Serials Section. *Serials Acquisitions Glossary*, FAA2

Association for Library Collections & Technical Services. *See also* American Library Association, Subcommittee on Guide for Training Collection Development Librarians

Association of College and Research Libraries, FAE2. *See also* Association for Educational Communications and Technology

———. "Guidelines for University Undergraduate Libraries: A Draft," GAC1

———. "Standards for College Libraries, 1995 Edition," GAC2

———, Copyright Committee, IAC1

———, Rare Books and Manuscripts Section. "Guidelines for the Loan of Rare and Unique Materials," IDD1

———, ———. "Guidelines Regarding Thefts in Libraries," IF1

———, ———. "Selection of General Collection Materials for Transfer to Special Collection," GE28

Association of Moving Image Archivists. *AMIA Newsletter*, HAA1

Association of Research Libraries, GAD1, IAA1

———. *Transborder Interlibrary Loan: Shipping Interlibrary Loan Materials from the U.S. to Canada*, IDD4

———, Office of Management Services, AGA1

Association of Specialized and Cooperative Library Agencies, Interlibrary Cooperation and Networking Section, Interlibrary Cooperation Discussion Group, IAC1

———, Libraries Serving Special Populations Section, IAC1

Astle, Deanna L. "Staff Involvement: The Key to the Successful Merger of Monograph and Serial Acquisitions Functions at Clemson University Library," BG1

Atkinson, Ross. "The Academic Library Collection in an On-Line Environment," GE1

———. "Access, Ownership, and the Future of Collection Development," GE2

———. "Crisis and Opportunity: Reevaluating Acquisitions Budgeting in an Age of Transition," GF5

Auburn University Libraries Cataloging Department, CAE1

"Authority Files in Online Catalogs Revisited," N. S. Bangalore, CBF2
AUTOCAT: Library Cataloging and Authorities Discussion Group, EAD3
"Automated Acquisitions and Serials Control," W. Saffady, FDF9
"Automated Mapping of Topical Subject Headings into Faceted Index Strings Using the *Art & Architecture Thesaurus* as a Machine Readable Dictionary," J. A. Brusch and T. Petersen, EF2
Automated Support to Indexing, G. M. Hodge, EF6
"Automatic Thesaurus Generation for an Electronic Community System," H. Chen et al., DCE6, EE5
Automating Preservation Management in ARL Libraries, P. Brennan and J. Reed-Scott, comps., HI1
"Automation and Technical Services Organization," R. Bazirjian, AHC1
"Automation: The Bridge between Technical Services and Government Documents," D. M. Pierce and E. Theodore-Shusta, AHA9
AV in Public and School Libraries: Selection and Policy Issues, M. J. Hughes and B. Katz, eds., GD31
AVISO, IDE2
AVISO-L Mailing List: AVISO Inter-Library Loan Software Mailing List, IAC5
Ayres, F. H. "Bibliographic Control at the Cross Roads," CAC1

B

BACKSERV, FDE1
Baer, William M.. *See* Courtois, Martin P.
Bagnall, Roger. *Digital Imaging of Papyri*, HH1
Baird, Brian J. "Motivating Student Employees: Examples from Collections Conservation," HI2, IB1

Baker, Barry B., ed. *Cooperative Cataloging: Past, Present and Future*, CBI1
———, ed. "Technical Services Report," AG1
Baker, Nicholson. "Discards," CAC2
Baker, Sharon L. *The Responsive Public Library Collection: How to Develop and Market It*, GE14
Baker, Shirley K., and Mary E. Jackson. *The Future of Resource Sharing*, IDA1
———. *Maximizing Access, Minimizing Cost: A First Step toward the Information Access Future*, IDA2
"Balancing Act for Library Materials Budgets: Use of a Formula Allocation," M. Niemeyer et al., GF2
Banach, Patricia. "Migration from an In-House Serials System to INNOPAC at the University of Massachusetts at Amherst," FDF1
Bangalore, Nirmala S. "Authority Files in Online Catalogs Revisited," CBF2
Banks, Jennifer. *Options for Replacing and Reformatting Deteriorated Materials*, HI3
Barker, Joseph W. "Acquisitions Principles and the Future of Acquisitions: Information Soup, the Soup-Hungry, and Libraries' Five Dimensions," BA1
Barnett, Judith B. "OCLC Cataloging Peer Committees: An Overview," CBI2
Barnett, Patricia J. *See* Petersen, Toni
Barnum, George. *See* Walsh, Jim
"The Battelle Mass Deacidification Process: A New Method for Deacidifying Books and Archival Materials," J. Wittekind, HC14
Bazirjian, R. "Automation and Technical Services Organization," AHC1
Beam, Joan. *See* Sellers, Minna

Beatty, Sue. "Subject Enrichment Using Contents or Index Terms: The Australian Defense Force Academy Experience," DA2

Beehler, Sandra A., and Patricia G. Court. "Speaking in Tongues: Communications between Technical Services and Public Services in an Online Environment," AA1

Beghtol, Clare. *The Classification of Fiction: The Development of a System Based on Theoretical Principles,* DBA7

――――. " 'Facets' as Interdisciplinary Undiscovered Public Knowledge: S. R. Ranganathan in India and L. Guttman in Israel," DBD3

A Beginner's Guide to Copy Cataloging on OCLC/Prism, L. M. Schultz, CBE9

Bell, Timothy A. H. *See* Moffatt, Alistair

Bellanti, Claire Q. "Access to Library Materials in Remote Storage," IB2

Benchmarking Interlibrary Loan: A Pilot Project, IDA3

"The Benefits of Online Series Authority Control," H. H. McCurley, CBF9

Benemann, William E. "The Cathedral Factor: Excellence and the Motivation of Cataloging Staff," CBA1

Bennett, Scott, Kenneth Frazier, and Laura Gasaway. "Special Section: Fair Use and Copyright," IAA2

Berman, Sanford, ed. *Prejudices and Antipathies: A Tract on the LC's Subject Headings Concerning People,* DCA1

"Between a Rock and a Hard Place: The Future of the Subscription Agent," FBB1

Bevis, Mary D., and Sonja L. McAbee. "NOTIS as an Impetus for Change in Technical Services Departmental Staffing," AHC2, FDB1

Beyer, Carrie, ed. *Preservation Research and Development: Round Table Proceedings,* HA8

"Beyond Access: New Concepts, New Tensions for Collection Development in a Digital Environment," W. P. Lougee, GE4

"Beyond the Fringe: Administratively Decentralized Collections at the University of Michigan," C. E. Reinke, GD52

Biblarz, Dora. "Richard Abel," BB2

The Bibliographic Control and Preservation of Latin Americanist Library Resources: A Status Report with Suggestions, D. C. Hazen, HD3

"Bibliographic Control at the Cross Roads," F. H. Ayres, CAC1

Bibliographic Guide to Microform Publications: 1992, HK1

"The Bibliographic Utilities in 1993: A Survey of Cataloging Support and Other Services," W. Saffady, AIB5, CBI11

"Bibliography of Articles Related to Electronic Journal Publications and Publishing," D. F. W. Morrison, FAC4

Biblo, Lisa. *See* Watt, Marcia

Bielefield, Arlene, and Lawrence Cheeseman. *Maintaining the Privacy of Library Records: A Handbook and Guide,* IA1

Bierbaum, Esther Green. "A Modest Proposal: No More Main Entry," CCB1

――――. "Searching for the Human Good: Some Suggestions for a Code of Ethics for Technical Services," AA2

"Binding Conventions for Music Materials," E. Tibbits, HB2

"Biology Journal Use at an Academic Library: A Comparison of Use Studies," D. E. Schmidt et al., GD44

Biosonnas, Christian M. "Darwinism in Technical Services: Natural Selection in an Evolving Information Delivery Environment," AA3

Black, Leah, and Colleen Hyslop. "Telecommuting for Original Cataloging at the Michigan State University Libraries," CBD1

Black, William K., ed. *Libraries and Student Assistants: Critical Links*, HI4

Blackwell Group, FDD1

Blake, Virgil L. P. and Renee Tjoumas. "The Conspectus Approach to Collection Evaluation: Panacea or False Prophet?," GG7

Blakeslee, Jan. "Indexing Encyclopedia and Multivolume Works," EAB2

Bloomfield, Masse. "A Look at Subject Headings: A Plea for Standardization," DCA4

Bloss, Alex. "The Value-Added Acquisitions Librarian: Defining Our Role in a Time of Change," BA2

"Blurring the Lines in Technical Services," T. L. Davis, AHC5

Boakye, G. "Challenges and Frustrations of an Acquisitions Librarian in a Developing Country: The Case of Balme Library," BG2

Boardman, Edna M. "How to Run a Tight Ship in the Magazine Stacks," FBA1

Boll, John J. *See* Aluri, Rao

Bonario, Steve, and Ann Thornton. "Library-Oriented Lists and Electronic Serials," FAC1

"Book and Periodical Indexing," H. W. Wellisch, EAB25

Book and Serial Vendors for Asia and the Pacific, ALCTS Acquisitions Section, FAB1

"Book Deterioration and Loss: Magnitude and Characteristics in Ohio Libraries," E. T. O'Neill and W. L. Boomgaarden, HI14

BOOK_ARTS-L, HAB4

Boomgaarden, Wesley L. *See* O'Neill, Edward T.

Born, Kathleen. *See* Ketcham, Lee

Bosch, Stephen, Patricia Promis, and Chris Sugnet. *Guide to Selecting and Acquiring CD-ROMs, Software, and Other Electronic Publications*, BAA2, GD22

Boss, Richard W. "Client/Server Technology for Libraries with a Survey of Vendor Offerings," AID1

———. "Technical Services Functionality in Integrated Library Systems," AID2

Bosseau, Don L. "Anatomy of a Small Step Forward: The Electronic Reserve Book Room at San Diego State University," IC1

———, Beth Shapiro, and Jerry Campbell. "Digitizing the Reserve Function: Steps toward Electronic Document Delivery," IC2

"Bottoming Out the Bottomless Pit with the Journal Usage/Cost Relational Index," C. Francq, GE35

Boucher, Virgina. *Interlibrary Loan Practices Handbook*, IDB1

Bourdon, Francoise. *International Cooperation in the Field of Authority Data: An Analytical Study with Recommendations*, CBF3

Bourke, Thomas A. "The Anabasis from Analog to Digital Escalates: The Year's Work in the Reproduction of Library Materials, 1992," HAC1

Bovey, J. D. "Building a Thesaurus for a Collection of Cartoon Drawings," EE2

Bowdoin, Sally. *See* Higginbotham, Barbra Buckner

Boxes for the Protection of Books: Their Design and Construction, L. Carlson et al., comps., HC1

Boyarski, Jennie S., and Kate Hickey, eds. *Collection Management in the Electronic Age: A Manual for Creating Community College Collection Development Policy Statements*, GC3

Boyle, Deirdre. *Video Preservation: Securing the Future of the Past*, HJ1

Brancolini, Kristine, and Rick E. Provine, comps. *Video Collections and Multimedia in ARL Libraries*, GD35

Braun, Janice, and Lola Raykovic Hopkins. "Collection-Level Cataloging, Indexing, and Preservation of the Hoover Institution Pamphlet Collection on Revolutionary Change in Twentieth Century Europe," CBE1

"Breaking New Ground in Fostering Preservation: The Society of American Archivists' Preservation Management Training Program," T. O. Walters, HM7

Brennan, Patricia, and Jutta Reed-Scott, comps. *Automating Preservation Management in ARL Libraries*, HI1

———, comps. *Cooperative Strategies in Foreign Acquisitions*, GE8

Brennan, Patricia. *See also* Mathews, Michael

Bricks and Mortar for the Mind: Statewide Preservation Program for Rhode Island, Rhode Island Council for the Preservation of Research Resources, HD16

"A Brief History of Library-Vendor Relations since 1950," W. Fisher, BF2

"A Brief Survey of ARL Libraries' Cataloging of Instructional Materials," L. Varughese and G. Poirier, CCD24

Bril, Patricia L. *See* Budd, John M.

Brisson, Roger. "The Cataloger's Workstation and the Continuing Transformation of Cataloging," CBC1

Broadway, Rita. *See* Coulter, Cynthia

Brooke, F. Dixon Jr., and Allen Powell. "EBSCO 1995 Serials Price Projections," FE7

Brooks, Connie. *See* Reich, Vicky

Brooks, Terence A. "People, Words, and Perception: A Phenomenological Investigation of Textuality," EC2

Brower, Carol. *See* Wilhelm, Henry

Brown, Lynne C. Branche. "An Expert System for Predicting Approval Plan Receipts," BB3

———. "Vendor Evaluation," BF1

Brown, Lynne C. Branche. *See also* Stanley, Nancy Marke

Brugger, Judith M., Michael Kaplan, and Joseph A. Kiegel, comps. *Technical Services Workstations*, AIC4

Brusch, Joseph A., and Toni Petersen. "Automated Mapping of Topical Subject Headings into Faceted Index Strings Using the *Art & Architecture Thesaurus* as a Machine Readable Dictionary," EF2

Bruwelheide, Janis H. *The Copyright Primer for Librarians and Educators*, HE1, IAA3

———. *Copyright Issues for the Electronic Age*, HE2

Bryant, Philip. "Quality of a National Bibliographic Service: In the Steps of John Whytefeld—An Admirable Cataloguer," CBI3

Buckland, Michael K. "What Will Collection Developers Do?," GAE1

———, Barbara A. Norgard, and Christian Plaunt. "Filing, Filtering, and the First Few Found," EAA1

Budd, John M., and Patricia L. Bril. "Education for Collection Management Results of a Survey of Educators and Practitioners," GB6

Budd, John M., and Karen A. Williams. "CD-ROMs in Academic Libraries: A Survey," GD23

Budd, John M. *See also* Harloe, Bart

Bugs, Mold & Rot. [Proceedings of] A Workshop on Residential Moisture Problems, Health Effects, Building Damage, and Moisture Control, HG1

Bugs, Mold & Rot II. [Proceedings of] A Workshop on Control of Humidity for Health, Artifacts, and Buildings, HG2

"Building a Better Mousetrap: Enhanced Cataloging and Access for the Online Catalog," S. A. Wittenbach, DA10

"Building a Thesaurus for a Collection of Cartoon Drawings," J. D. Bovey, EE2

"Building and Managing an Acquisitions Program," C. P. Hawks, BA3

"Building Racially Diverse Collections: An Afrocentric Approach," G. Johnson-Cooper, GD7

Bukoff, Ronald N. "Censorship and the American College Library," GD14

Burrows, Toby, and Philip G. Kent, eds. *Serials Management in Australia and New Zealand: Profile of Excellence,* FBA2

Bush, Carmel C., Margo Sassé, and Patricia Smith. "Toward a New World Order: A Survey of Outsourcing Capabilities of Vendors for Acquisitions, Cataloging and Collection Development Services," AHB1

Bushing, Mary C. "Acquisitions Ethics: The Evolution of Models for Hard Times," BI2

Bustos, Roxann, comp. *Interlibrary Loan in College Libraries,* IDA4

"But Is It an Online Shelflist? Classification Access on Eight OPACs," D. Kneisner and C. Willman, DBA8

Buyers Laboratory, Inc., "Test Reports on 15 Photocopiers," HL2

Byrd, Jacqueline, and Kathryn Sorury. "Cost Analysis of NACO Participation at Indiana University," CBF4

C

Cain, Mark. "Periodical Access in an Era of Change: Characteristics and a Model," IE2

"Cait—Computer-Assisted Indexing Tutor: Implemented for Training at NAL," H. Irving, EAB9

Calhoun, John C. "Serials Citations and Holdings Correlation," GD36

The California Preservation Program, California State Library, HD9

California State Library. *The California Preservation Program,* HD9

CALIPR: An Automated Tool to Assess Preservation Needs of Books and Document Collections for Institutional or Statewide Planning, B. W. Ogden and M. Jones, HI13

Callahan, Daren, and Judy MacLeod. "Recruiting and Retention Revisited: A Study of Entry Level Catalogers," CBA2

Callahan, Daren. *See also* MacLeod, Judy

Camden, Beth Picknally, and Jean L. Cooper. "Controlling a Cataloging Backlog; or Taming the Bibliographical Zoo," CBH1

Campbell, Jerry. *See* Bosseau, Don L.

Canadian Conservation Institute. *CCI Newsletter,* HAA2

Canadian Cooperative Preservation Project: Final Summary Report, R. W. Manning, HD6

Canadian MARC Communications Format: Classification Data, DBA4

"Canon Formation, Library Collections, and the Dilemma of Collection Development," M. Myzyk, GD2

Caplan, Priscilla. "Providing Access to Online Information Resources: A Paper for Discussion," CCD8
——. "You Call It Corn, We Call It Syntax-Independent Metadata for Document-Like Objects," CAC3
Carey, Kjestine R. *See* Price, Anna L.
Caring for Your Collections: Preserving & Protecting Your Art & Other Collectibles, HA2
"A CARL Model for Cooperative Collection Development in a Regional Consortium," D. Cochenour and J. S. Rutstein, GH1
Carlo, Paula Wheeler, and Allen Natowitz. "*Choice* Book Reviews in American History, Geography and Area Studies: An Analysis of 1988–1993," GD46
Carlson, Lage, et al., comps. *Boxes for the Protection of Books: Their Design and Construction*, HC1
Carpenter, David, and Malcolm Getz. "Evaluation of Library Resources in the Field of Economics: A Case Study," GG2
Carpenter, Eric. *See* Pankake, Marcia
Carpenter, Kathryn Hammell. *See* Alexander, Adrian W.
Carrigan, Dennis P. "From Just-in-Case to Just-in-Time: Limits to the Alternative Library Service Model," IE3
——. "Toward a Theory of Collection Development," GAE2
Carter, Ruth C., and Paul B. Kohberger, Jr. "Using SPSS/PC+ and NOTIS Downloaded Files of Current Subscription Records at the University of Pittsburgh," FDF2
Cassel, Rachel. "Selection Criteria for Internet Resources," GD24
Casserly, Mary F., and Judith L. Hegg. "A Study of Collection Development Personnel Training and Evaluation in Academic Libraries," GB7

"Catalog Record Contents Enhancement," T. S. Weintraub and W. Shimoguchi, CCB8
"Catalogers and Workstations: A Retrospective and Future View," H. R. Lange, CBC3
Cataloger's Desktop, Library of Congress, CBC5
"The Cataloger's Workstation and the Continuing Transformation of Cataloging," R. Brisson, CBC1
"A Cataloger's Workstation: Using a NeXT Computer and Digital Librarian Software to Access the Anglo-American Cataloguing Rules," J. Gomez, CBC2
Cataloging and Classification: An Introduction, L. M. Chan, CAA1, DAB1
Cataloging and Classification for Library Technicians, M. L. Kao, DAB2
Cataloging & Classification Quarterly, CAD1
Cataloging and the Small Special Library, J. W. Palmer, CBA9
"Cataloging Collection-Level Records for Archival Video and Audio Recordings," K. J. M. Haynes et al., CBE4
Cataloging Concepts: Descriptive Cataloging, M. W. Cundiff, CBB3
"Cataloging Electronic Texts: The University of Virginia Experience," E. Gaynor, CCD12
Cataloging Government Publications Online, C. C. Sheryko, ed., CCD27
"Cataloging in the 1990s: Managing the Crisis (Mentality)," J. D. LeBlanc, CAC6
Cataloging Internet Resources: A Manual and Practical Guide, N. B. Olson, ed., CCD17
Cataloging Nonbook Resources: A How-to-Do-It Manual for Librarians, M. B. Fecko, CCD22
"The Cataloging of Primary State Legal Material," M. Maben, CCD26

"Cataloging Pierre Bourdieu's Books," R. U. Alvarado, EE1
Cataloging Unpublished Nonprint Materials: A Manual of Suggestions, Comments and Examples, V. Urbanski et al., CCD25
"Cataloging with Copy: Methods for Increasing Productivity," S. J. Smith, CBE10
Cataloguer's Toolbox, CAE2
Cataloguing and Indexing of Electronic Resources, CCD9
"The Cathedral Factor: Excellence and the Motivation of Cataloging Staff," W. E. Benemann, CBA1
Catriona: Cataloguing and Retrieval of Information over Networks, CCD10
CatSkill, An Interactive Multimedia Training Package to Teach AACR2 and MARC, CBB1
CCI Newsletter, HAA2
"CD-ROM Longevity: A Select Bibliography," M. Watt and L. Biblo, HJ19
"CD-ROMs in Academic Libraries: A Survey," J. M. Budd and K. A. Williams, GD23
Censorship and Selection: Issues and Answers for Schools, H. Reichman, GD20
"Censorship and the American College Library," R. N. Bukoff, GD14
Center for Electronic Texts in the Humanities. *ETEXTCTR: Electronic Text Center Discussion Group,* HAB9
Center for Research Libraries, GAD2
"Challenges and Frustrations of an Acquisitions Librarian in a Developing Country: The Case of Balme Library," G. Boakye, BG2
Chan, Lois Mai. *Cataloging and Classification: An Introduction,* CAA1, DAB1
_____. *Library of Congress Subject Headings: Principles and Application,* DCB2, EE3
_____, John P. Comaromi, Joan S. Mitchell, and Mohinder P. Satija. *Classification Decimale de Dewey: Guide pratique,* DBB6
_____, John P. Comaromi, Joan S. Mitchell, and Mohinder P. Satija. *Dewey Decimal Classification: A Practical Guide,* DBB7
Chaney, Michael, and Ian MacDougall, eds. *Security and Crime Prevention in Libraries,* IF2
Chang, B. C. *See* Urbanski, Verna
"Changing Relationships in the Acquisition and Delivery of Library Materials: A Survey," J. Ogburn, BG5
The Changing Role of Book Repair in ARL Libraries, R. Silverman and M. Gradinette, eds., HC2
Chaplan, Margaret A. "Mapping LaborLine Thesaurus Terms to Library of Congress Subject Headings: Implications for Vocabulary Switching," EE4
"Characteristics of Duplicate Records in OCLC's Online Union Catalog," E. T. O'Neill et al., CBI7
Cheeseman, Lawrence. *See* Bielefield, Arlene
Chen, Chiou-Sen Dora. *Serials Management: A Practical Guide,* FAA1
Chen, Hsinchun, et al. "Automatic Thesaurus Generation for an Electronic Community System," DCE6, EE5
Cherry, Joan M. "Improving Subject Access in OPACs: An Exploratory Study on User's Queries," DA3
_____, et al. "OPACs in Twelve Canadian Academic Libraries: An Evaluation of Functional Capabilities and Interface Features," EC3

Chiang, Belinda. "Migration from Microlinx to NOTIS: Expediting Serials Holdings Conversion through Programmed Function Keys," FDF3

Child, Margaret S. *Directory of Information Sources on Scientific Research Related to the Preservation of Sound Recordings, Still and Moving Images and Magnetic Tape*, HJ2

"*Choice* Book Reviews in American History, Geography and Area Studies: An Analysis of 1988–1993," P. W. Carlo and A. Natowitz, GD46

Christensen, Peter G. *See* Sweetland, James H.

Chrzastowski, Tina E., and Mary A. Anthes. "Seeking the 99% Chemistry Library: Extending the Serials Collection through the Use of Decentralized Document Delivery," GD37

Chrzastowski, Tina E., and Karen A. Schmidt. "Surveying the Damage: Academic Library Serial Cancellations 1987–88 through 1989–90," GE33

Chu, Clara M., and Ann O'Brien. "Subject Analysis: The Critical First Stage in Indexing," EAB3

Cimino, James J. "Vocabulary and Health Care Information Technology: State of the Art," DCC1

CircPlus, IAC6

"Citation as a Form of Library Use," J. L. Kelland and A. P. Young, GG5

Citations for Serial Literature, FAD2

Clack, Doris H. "Education for Cataloging: A Symposium Paper," CBA3

———. "Subject Access to African American Studies Resources," DCB4

Clack, Mary Elizabeth. "The Role of Training in the Reorganization of Cataloging Services," CBB2

Clark, Tom. "On the Cost Differences between Publishing a Book in Paper and in the Electronic Medium," BH1

Class FC: A Classification for Canadian History, DBC4

Class PS8000: A Classification for Canadian Literature, DBC5

"Classification and Shelflisting as Value Added: Some Remarks on the Relative Worth and Price of Predictability, Serendipity, and Depth of Access," J. D. LeBlanc, DA7

Classification Decimale de Dewey: Guide practique, L. M. Chan et al., DBB6

Classification Issues for Knowledge Organization, DAC1

Classification: Its Kinds, Systems, Elements and Applications, D. W. Langridge, DBA1

The Classification of Fiction: The Development of a System Based on Theoretical Principles, C. Beghtol, DBA7

Classification: Options and Opportunities, A. Thomas, DBA3

Classification Plus, DBC2

Classification Plus & Cataloger's Desktop, DBC3

Classification Research for Knowledge Representation and Organization: Proceedings of the Fifth International Study Conference on Classification Research, N. J. Williamson and M. Hudon, DAA3

"Client/Server Technology for Libraries with a Survey of Vendor Offerings," R. W. Boss, AID1

Cline, Nancy M. "Staffing: The Art of Managing Change," GB1

Clinic on Library Applications of Data Processing, AGB3

Cloonan, Michèle Valerie. *Global Perspectives on Preservation Education*, HM1

"A Closer World: A Review of Acquisitions Literature, 1992," L. German, BAB1

"Closing the Loop: How Did We Get Here and Where Are We Going?" M. T. Reid, BF3

CNI-Copyright, HAB5, IAC7

Coalition for Networked Information. *CNI-Copyright*, HAB5

Cochenour, Donnice, and Joel S. Rutstein. "A CARL Model for Cooperative Collection Development in a Regional Consortium," GH1

Coffey, James R. "Competency Modelling for Hiring in Technical Services: Developing a Methodology," AHA2

Coffman, Steve, and Pat Wiedensohler, eds. *The Fiscal Directory of Fee-Based Research and Document Supply Services*, IE1

Cohen, Jonathan D. "Highlights: Language- and Domain-Independent Automatic Indexing Terms for Abstracting," EF3

Coleman, Christopher D. G., comp. *Preservation Education Directory*, HM2

Collantes, Lourdes Y. "Degrees of Agreement in Naming Objects and Concepts for Information Retrieval," EE6

"Collection Assessment in Academic Libraries: Institutional Effectiveness in Microcosm," W. A. Henderson et al., GG1

Collection Assessment in Music Libraries, J. Gottlieb, GE23

Collection Conservation Treatment: A Resource Manual for Program Development and Conservation Technician Training, including "Report on Training the Trainers," M. Jones, HC7

"Collection Development," B. Morton, GE11

"Collection Development and Acquisitions in a Changing University Environment," K. Flowers, BG3

Collection Development and Collection Evaluation: A Sourcebook, M. R. Gabriel, GAB1

Collection Development and Finance: A Guide to Strategic Library-Materials Budgeting, M. S. Martin, GF7

"Collection Development and Scholarly Communication in the Era of Electronic Access," B. Harloe and J. M. Budd, GE3

"Collection Development for the Electronic Media: A Conceptual and Organizational Model," S. G. Demas, GD25

"Collection Development Guidelines for Selective Federal Depository Libraries," L. A. Rosenblatt, GE12

"Collection Development Issues of Academic and Public Libraries: Converging or Diverging?" K. S. Nilsen, GAE4

"Collection Development of Genre Literature," E. Futas, GE15

"Collection Development Policies: A Cunning Plan," P. Johnson, GC8

Collection Development Policies and Procedures, E. Futas, ed., GC6

Collection Development Policies Committee, Collection Development & Evaluation Section, Reference & Adult Services Division. "The Relevance of Collection Development Policies: Definition, Necessity, and Applications," GC2

"Collection Development Policies in the Information Age," D. C. Hazen, GC7

"Collection Development Strategies for a University Center Library," C. S. Hurt et al., GD50

"Collection-Level Cataloging, Indexing, and Preservation of the Hoover Institution Pamphlet Collection on Revolutionary Change in Twentieth Century Europe," J. Braun and L. R. Hopkins, CBE1

"Collection Management," R. S. Karp, GAB3

Collection Management and Development: Issues in an Electronic Era, P. Johnson and B. MacEwan, eds., GAE3

Collection Management in the Electronic Age: A Manual for Creating Community College Collection Development Policy Statements, J. Boyarski and K. Hickey, eds., GC3

"Collection- or Archival-Level Description for Monographic Collections," R. Saunders, CBE8

Collection Policies, McGill University Libraries and E. V. Silvester, ed., GC11

Collections Conservation, R. C. DeCandido, HC3

The College of Charleston Conference: Issues in Book and Serial Acquisitions, AGB2, BAD2

Colorado Preservation Alliance. *On the Road to Preservation: A State-Wide Preservation Action Plan for Colorado*, HD10

Comaromi, John P. *See* Chan, Lois Mai

"Commentaries on Collection Bias," M. Pankake et al., GD19

Commerce Department's White Paper on National and Global Information Infrastructure. Executive Summary for the Library and Educational Community, A. P. Lutzker, IAA6

Commission on Preservation and Access, HAB16

——. *Newsletter*, HAA3

"Compact Shelving of Circulating Collections," S. Sam and J. A. Major, IB5

"Comparing Web Browsers: Mosaic, Cello, Netscape, WinWeb and InternetWorks Lite," G. R. Notess, EF10

"Comparing Words, Stems, and Roots as Index Terms in an Arabic Information Retrieval System," I. A. Al-Kharashi and M. W. Evens, EAB1

"A Comparison between the Online Catalog and the Card Catalog: Some Considerations for Redesigning Bibliographic Standards," R. Fattahi, CAC4

"A Comparison of *AACR2R* and French Cataloging Rules," N. A. Jacobowitz, CCA6

"A Comparison of OCLC and WLN Hit Rates for Monographs and an Analysis of the Types of Records Retrieved," R. E. Ross, CBI10

"Comparison of Out-of-Print Searching Methods," M. Eldredge and W. Ludington, BD1

"Competency Modelling for Hiring in Technical Services: Developing a Methodology," J. R. Coffey, AHA2

"Computer-Assisted Database Indexing: The State-of-the-Art," G. M. Hodge, EF7

Computer Images for Research, Teaching, and Publication in Art History and Related Disciplines, C. S. Rhyne, HH21

"Computers and Technical Services," P. F. Philips, AA12

"The Concept of Inadequacy in Uniform Titles," D. Nelson and J. Marner, CCB5

"The Concept of *Work* for Moving Image Materials," M. M. Yee, CCD5

"Confessions of a Videotape Restorer; Or, How Come These Tapes All Need to Be Cleaned Differently?" J. Lindner, HJ6

Connaway, Lynn Silipigni. "An Examination of the Inclusion of a Sample of Selected Women Authors in *Books for College Libraries*," GD6
CONSER, CAE12
CONSER Cataloging Manual, CCD19
CONSER Editing Guide, 1994 Edition, CCD20
CONSERline: Newsletter of the CONSER (Cooperative Online Serials) Program, Library of Congress and OCLC, Inc., CAD2
Conservation DistList, HAB6
Conservation OnLine, HAB17
"The Conspectus Approach to Collection Evaluation: Panacea or False Prophet?" V. L. P. Blake and R. Tjoumas, GG7
"Contract Acquisitions: Change, Technology, and the New Library/Vendor Partnership," G. M. Shirk, BG8
"Contract Negotiations for the Commercial Microforms Publishing of Library and Archival Materials: Guidelines for Librarians and Archivists," Subcommittee on Contract Negotiations for Commercial Reproduction of Library & Archival Materials, HK3
"Controlling a Cataloging Backlog; or Taming the Bibliographical Zoo," B. P. Camden and J. L. Cooper, CBH1
Conversion Tables: LCC-Dewey, Dewey-LCC, M. L. Scott and C. E. Alvey, DBA6
Conway, Paul. "Digitizing Preservation," HH2
_____. *Preservation in the Digital World*, HH3
_____. "Selecting Microfilm for Digital Preservation: A Case Study from Project Open Book," HH4

_____, and Shari Weaver. *The Setup Phase of Project Open Book: A Report to the Commission on Preservation and Access on the Status of an Effort to Convert Microfilm to Digital Imagery*, HH5
Cook, Eleanor L., and Pat Farthing. "A Technical Services Perspective of Implementing an Organizational Review while Simultaneously Installing an Integrated Library System," AHC3, FBA3
"Cool Tools for Searching the Web: A Performance Evaluation," M. P. Courtois et al., EF4
Coons, Bill, and Peter McDonald. "Implications of Commercial Document Delivery," GD38, IE4
COOPCAT: Cooperative Cataloging, CBI4
Cooper, Jean L. *See* Camden, Beth Picknally
Cooper, Mary Campbell. *See* Schrock, Nancy Carlson
Cooper, Michael D., and George F. McGregor. "Using Article Photocopy Data in Bibliographic Models for Journal Collection Management," GD39
Cooperative Cataloging: Past, Present and Future, B. B. Baker, CBI1
"A Cooperative Cataloging Project between Two Large Academic Libraries," J. Kiegel and M. Schellinger, CBI6
"Cooperative Collection Development at the Research Triangle University Libraries: A Model for the Nation," P. B. Dominguez and L. Swindler, GH3
"Cooperative Collection Management: The Conspectus Approach," G. N. Olson and B. F. Allen, GG8
"Cooperative Preservation of State-Level Publications: Preserving the Literature of New York State Agriculture and Rural Life," D. Wright et al., HD8

289

Author/Title Index

Cooperative Strategies in Foreign Acquisitions, P. Brennan and J. Reed-Scott, comps., GE8
"Copyright and Digital Libraries," P. Samuelson, IAA8
Copyright and Intellectual Property Resources, IAD1
Copyright Clearance Center, IAA4
Copyright Office, Library of Congress, IAA9
The Copyright Primer for Librarians and Educators, J. H. Bruwelheide, HE1, IAA3
Copyright, Public Policy, and the Scholarly Community, M. Mathews and P. Brennan, eds., IAA7
"The Core Record: A New Bibliographic Standard," W. Cromwell, CBE2
Cornell University Library Technical Services Manual, CAE3, FDA1
Cornish, Graham. "Training Opportunities for Interlibrary Loan and Document Supply Staff," IDA5
"The Corruption of Cataloging," M. Gorman, CAC5
"Cost Analysis of NACO Participation at Indiana University," J. Byrd and K. Sorury, CBF4
"Cost-Benefit Analysis for B/W Acetate: Cool/Cold Storage vs. Duplication," S. Puglia, HG7
Coulter, Cynthia, and Lola Halpin, with Rita Broadway. "Who Needs to Know What? Essential Communication for Automation Implementation and Effective Reorganization," AHC4
Court, Patricia G. *See* Beehler, Sandra A.
Courtois, Martin P., William M. Baer, and Marcella Stark. "Cool Tools for Searching the Web: A Performance Evaluation," EF4
Cox, John. "EDI: The Modern Way to Do Business Together," BC1

Coyle, Karen, ed. *Format Integration and Its Effect on Cataloging, Training and Systems*, CCE1
Cramer, Michael D. "Licensing Agreements: Think before You Act," FCB1
"Creating a World Wide Web Resource Collection," S. Pointek and K. Garlock, GD34
"Crisis and Opportunity: Reevaluating Acquisitions Budgeting in an Age of Transition," R. Atkinson, GF5
"Criteria for Selecting Document Delivery Suppliers," G. S. Machovec, IE14
Cromwell, Willy. "The Core Record: A New Bibliographic Standard," CBE2
Cromwell, Willy. *See also* Reich, Vicky
Croneis, Karen. *See* Jackson, Mary E.
"Crossing Subject Boundaries: Collection Management of Environmental Studies in a Multi-Library System," B. Defilice and C. Rinaldo, GE7
Crump, Michele J., and LeiLani Freund. "Serial Cancellations and Interlibrary Loan: The Link and What It Reveals," GE34
Crystal, David. "Some Indexing Decisions in the Cambridge Encyclopedia Family," EF5
Cubberly, Carol. "Allocating the Materials Funds Using Total Cost of Materials," GF1
Cuestra, Emerita M. *See* Meiseles, Linda
Cundiff, Margaret Welk. *Cataloging Concepts: Descriptive Cataloging*, CBB3
Curl, Margo Warner. "Enhancing Subject and Keyword Access to Periodical Abstracts and Indexes," DA4
CurrentCites, FAC2
Curriculum Materials Center Collection Development Policy, B. G. Anderson, et al., comps., GC4

Customer Service: A How-to-Do-It Manual for Librarians, S. Walters, IA5
Cybulski, Walter. *See* Wright, Dorothy

D

Daniel, Rodney. *See* Weibel, Stuart
Dannelly, Gay. "Resource Sharing in the Electronic Era: Potentials and Paradoxes," GH2
"Darwinism in Technical Services: Natural Selection in an Evolving Information Delivery Environment," C. M. Biosonnas, AA3
Data Elements for Binding of Library Materials, NISO, HN1
"Dates in Added Entries: An Analysis of an AUTOCAT Discussion," D. Nelson and J. Marner, CCB6
Davis, Elisabeth B. *See* Schmidt, Diane E.
Davis, Frances. "A Plan for Evaluating a Small Library Collection," GG3
Davis, Susan, Deanna Iltis, and Judy Chandler Irvin. "Integrating Documents Processing into Traditional Technical Services," AHA3
Davis, Trisha L. "Blurring the Lines in Technical Services," AHC5
———, and James Huesmann. *Serials Control Systems for Libraries*, FDF4
Dean, Barbara. "Toward a Code of Ethics for Acquisitions Librarians," BI3
DeCandido, Robert C. *Collections Conservation*, HC3
Decimal Classification System: A Bibliography for the Period 1876–1994, S. Gupta, DAD1
Defilice, Barbara, and Constance Rinaldo. "Crossing Subject Boundaries: Collection Management of Environmental Studies in a Multi-Library System," GE7

"Degrees of Agreement in Naming Objects and Concepts for Information Retrieval," L. Y. Collantes, EE6
Delaney, Thomas. "Electronic Reserve: The Library Goes to the People," IC3
"Demand for Document Delivery and Interlibrary Loan in Academic Settings," M. T. Kinnucan, IDA11
Demas, Samuel G. "Collection Development for the Electronic Media: A Conceptual and Organizational Model," GD25
———, Peter McDonald, and Gregory Lawrence. "The Internet and Collection Development: Mainstreaming Selection of Internet Resources," GD26
Demas, Samuel G. *See also* Wright, Dorothy
DeMiller, Anna L. *See* Rutstein, Joel S.
den Beyker, Karin. *See* Russell, Gordon
"Departmental Profiles: A Collection Development Aid," R. E. Stelk et al., GD54
"Designing an Expert System for Classifying Office Documents," D. Savic, EAB20
DeStefano, Paula. "Use-Based Selection for Preservation Microfilming," HI5
Developing Library and Information Center Collections, G. E. Evans, GAA1
Dewey Decimal Classification: A Practical Guide, L. M. Chan et al., DBB7
Dewey Decimal Classification and Relative Index, M. Dewey, DBB3
Dewey for Windows: DDC 21, DBB5
Dewey, Melvil. *Abridged Dewey Decimal Classification and Relative Index*, DBB2
———. *Dewey Decimal Classification and Relative Index*, DBB3
———. *Sistema de Clasificacion Decimal Dewey*, DBB4

291

Author/Title Index

Dick, Gerald K. *LC's Author Numbers,* DBD1
Dickinson, Gail K. *Selection and Evaluation of Electronic Resources,* GD27
Diedrichs, Carol Pitts. *See* Hawks, Carol Pitts
"Difficult Choices": How Can Scholars Help Save Endangered Research Resources?, G. George, HM4
Digital Collections Inventory Report, P. A. McClung, HH18
"Digital Data Retrieval: Testing Excalibur," M. Alexander, EF1
Digital Imaging of Papyri, R. Bagnall, HH1
Digital Imaging Technology for Preservation: Proceedings from an RLG Symposium . . ., N. E. Elkington, ed., HH6
The Digital Preservation Consortium: Mission and Goals, D. J. Waters and A. Kenney, HH29
"Digital-to-Microfilm Conversion: An Interim Preservation Solution," A. R. Kenney, HH15
The Digitization of Primary Textual Sources, P. Robinson, HH23
"Digitizing Preservation," P. Conway, HH2
"Digitizing the Reserve Function: Steps toward Electronic Document Delivery," D. L. Bosseau, IC2
Dillon, Martin. *Measuring the Impact of Technology on Libraries: A Discussion Paper,* AA4
"The Direct Doc Pilot Project at Arizona State: User Behavior in a Non-Mediated Document Delivery Environment," S. Walters, IE22
Directory of Information Sources on Scientific Research Related to the Preservation of Sound Recordings, Still and Moving Images and Magnetic Tape, M. S. Child, HJ2
Disaster Response and Prevention for Computers and Data, M. B. Kahn, HF2

"Discarding the Main Entry in an Online Cataloging Environment," R. C. Winke, CCB9
"Discards," N. Baker, CAC2
"The Dis-Integrating Library System: Effects of New Technologies in Acquisitions," R. Ray, BG6
"The Distribution of Information: The Role for Online Public Access Catalogs," J. R. Matthews, EC6
DocDel-L, IAC8
"Document Delivery: A Better Option?" M. Khalil, IE10
"Document Delivery: A Comparison of Commercial Document Suppliers and Interlibrary Loan Services," K. Kurosman and B. A. Durniak, IE13
"Document Delivery: A Primary Service for the Nineties," J. Wessling, IE23
Document Delivery in an Electronic Age, D. Kaser, ed., IE9
"Document Delivery over the Internet: Electronic Document Delivery Systems Used by Libraries," M. E. Jackson, IDA9
Document Delivery Services: Issues and Answers, E. Mitchell and S. Walters, IE17
"Document Ordering Standards: The ILL Protocol and Z39.50 Item Order," F. Turner, IDD6
Dodd, Jane. *See* Goodyear, Mary Lou
"Does Outsourcing Mean 'You're Out'?" J. Dwyer, AHB2
Dole, Wanda V. "Myth and Reality: Using the OCLC/AMIGOS Collection Analysis CD to Measure Collections against Peer Collections and against Institutional Priorities," GG9
Doll, Carol A. "School Media Center and Public Library Collections and the High School Curriculum," GE20

Dominguez, Patricia Buck, and Luke Swindler. "Cooperative Collection Development at the Research Triangle University Libraries: A Model for the Nation," GH3

Dougherty, Ann P. *See* Dougherty, Richard M.

Dougherty, Richard M., and Ann P. Dougherty. "The Academic Library: A Time of Crisis, Change, and Opportunity," AHA4

Down, Nancy. "Subject Access to Individual Works of Fiction: Participating in the OCLC/Fiction Project," DA5

Drabenstott, Karen M. *Analytical Review of the Library of the Future*, AD1

———, and Diane Vizine-Goetz. *Using Subject Headings for Online Retrieval: Theory and Practice and Potential*, DCA2

Drabenstott, Karen M. *See also* Franz, Lori

Drewes, Jeanne M. "A Widening Circle: Preservation Literature Review, 1992," HAC2

Duke, John K. "Slow Revolution: The Electronic AACR2," CCA2

Dunn, Pam. *See* McMahon, Suzanne

Dunshire, G. "The Potential of the Internet and Networks for Library Acquisitions," BC2

Duranceau, Ellen. "Vendors and Librarians Speak on Outsourcing, Cataloging and Acquisitions," CBG1, FBB2

Durniak, Barbara Ammerman. *See* Kurosman, Kathleen

Dworaczek, Marian, and Victor G. Wiebe. "E-Journals: Acquisition and Access," FCB2

Dwyer, Jim. "Does Outsourcing Mean 'You're Out'?," AHB2

Dykeman, Amy. "Faculty Citations: An Approach to Assessing the Impact of Diminishing Resources on Scientific Research," GG4

E

East, Dennis. "User Views of Compact Shelving in an Open Access Library," IB3

"EBSCO 1995 Serials Price Projections," F. D. Brooke and A. Powell, FE7

EBSCO Information Services, FDD2

Edelbute, Thomas. "A Pro-Cite Authority File on a Network," CBF5

"EDI: The Modern Way to Do Business Together," J. Cox, BC1

"EDI/EDIFACT," S. K. Paul, FDF7

"Education for Cataloging: A Symposium Paper," D. H. Clack, CBA3

"Education for Collection Management Results of a Survey of Educators and Practitioners," J. M. Budd and P. L. Bril, GB6

"The Education of Catalogers: The View of the Practitioner/Educator," A. F. Evans, CBA4

The Education of Staff and Users for the Proper Handling of Archival Materials, H. Ford, HM3

"Educators and Practitioners Reply: An Assessment of Cataloging Education," J. MacLeod and D. Callahan, CBA7

Edwards, F. "Licence to Kill?," BH2

"Effective Liaison Relationships in an Academic Library," C. Wu et al., GD55

Effective Library Signage, K. W. Ragsdale and D. J. Kenney, comps., ICI2

"Effectiveness of Surname-Title-Words Searches by Scholars," F. G. Kilgour, CBF7, EC4

"E-Journals: Acquisition and Access," M. Dworaczek and V. G. Wiebe, FCB2

Eldredge, Mary. "United Kingdom Approval Plans and United States Academic Libraries: Are They Necessary and Cost Effective?," BB4

Eldredge, Mary, and William Ludington. "Comparison of Out-of-Print Searching Methods," BD1
"An Electrifying Year: A Year's Work in Serials, 1992," J. F. Riddick, FAC6
"Electronic and Print Information: Active Distribution and Passive Retention in Relation to a Murder—A Case Study," F. K. Groen, GD15
"Electronic Data Interchange (EDI): The Exchange of Ordering, Claiming, and Invoice Information from a Library Perspective," G. Kelly, BC5
"Electronic Discussion Lists and Journals: A Guide for Technical Services Staff," V. Reich et al., AC1, HA9
"Electronic ILL at Colorado State University," J. Smith, IDA16
Electronic Imaging Output Printers, AIIM, HN25
"Electronic Information and Acquisitions," R. Heseltine, BC3
"Electronic Journal Subscriptions," L. R. Keating et al., FCB6
Electronic Journals in ARL Libraries: Issues and Trends, E. Parang and L. Saunders, comps., FCB3, GD28
Electronic Journals in ARL Libraries: Policies and Procedures, E. Parang and L. Saunders, comps., FCB4
"Electronic Reserve and Copyright," M. B. Jensen, IC7
"Electronic Reserve: The Library Goes to the People," T. Delaney, IC3
Elkington, Nancy E., ed. *Digital Imaging Technology for Preservation: Proceedings from an RLG Symposium . . .*, HH6
———, ed. *RLG Archives Microfilming Manual*, HH7
EMEDIA: Electronic Cataloging Issues in Libraries, CCD11
Emergency Planning and Management in College Libraries, S. C. George, HF1

Encyclopedia of Recorded Sound in the United States, G. A. Marco, HJ3
"End-User Understanding of Subdivided Subject Headings," L. Franz et al., DCB5
"Enhancing Subject and Keyword Access to Periodical Abstracts and Indexes," M. W. Curl, DA4
"Enhancing USMARC Records with Tables of Contents," DA6
Enssle, Halcyon R. "Reserve On-line: Bringing Reserve into the Electronic Age," IC4
"Ensuring the Longevity of Digital Documents," J. Rothenberg, HH24
Environmental Guidelines for the Storage of Paper Records, NISO, HN3
EPIC: European Preservation Information Center, HAB18
"The Epic Struggle: Subject Retrieval from Large Bibliographic Databases," H. R. Tibbo, EC10
EPIC-LST, HAB7
Epp, Ronald. See McDaniel, Elizabeth
Erbolato-Ramsey, Christiane, and Mark L. Grover. "Spanish and Portuguese Online Cataloging: Where Do You Start from Scratch?" CBE3
ERECS-L: Management & Preservation of Electronic Records, HAB8
ERIC Identifier Authority List (IAL) 1992, C. R. Weller and J. E. Houston, eds., EE16
Erikson, Rodney. See Reed, Lawrence
Ester, Michael. "Image Quality and Viewer Perception," HH8
ETEXTCTR: Electronic Text Center Discussion Group, HAB9
"Ethical Considerations in Decision Making," P. Johnson, BI4
"Ethics in Cataloging," S. S. Intner, CAC7
European Commission on Preservation and Access, HAB1, HAB18
European Commission on Preservation and Access. *EPIC-LST*, HAB7

The European Register of Microform Masters—Supporting International Cooperation, W. Schwartz, HK2
"Evaluating Adult Fiction in the Smaller Public Library," J. J. Senkevitch and J. H. Sweetland, GG6
"Evaluating Electronic Texts in the Humanities," S. Hockey, GD30
"Evaluation of Library Resources in the Field of Economics: A Case Study," D. Carpenter and M. Getz, GG2
"The Evaluation, Selection, and Acquisition of Legal Looseleaf Publications," M. J. Petit, BH6
Evans, Anaclare F. "The Education of Catalogers: The View of the Practitioner/Educator," CBA4
Evans, G. Edward. *Developing Library and Information Center Collections,* GAA1
———, and Sandra M. Heft. *Introduction to Technical Services,* AB1
Evens, Martha W. *See* Al-Kharashi, Ibrahim A.
"The Evolution of Approval Services," M. Warzala, BB9
The Evolving National Information Network: Background and Challenges, D. E. Van Houwelling and M. J. McGill, HH28
"An Examination of the Inclusion of a Sample of Selected Women Authors in *Books for College Libraries,*" L. S. Connaway, GD6
Examples to Accompany "Descriptive Cataloging of Rare Books," CCD1
EXLIBRIS: Rare Books and Special Collections Forum, CCD2, HAB10
"The Expert Cataloging Assistant Project at the National Library of Medicine," P. Weiss, CBF14
"Expert System Applications in Cataloging, Acquisitions, and Collection Development: A Status Review," J. Jeng, AIB2

"An Expert System for Predicting Approval Plan Receipts," L. C. B. Brown, BB3
"An Expert System for Quality Control and Duplicate Detection in Bibliographic Databases," M. J. Ridley, AIB4
"Expert Systems in Technical Services and Collection Management," C. P. Hawks, AIB1, GAF1
Extension and Corrections to the UDC, DBD2

F
" 'Facets' as Interdisciplinary Undiscovered Public Knowledge: S. R. Ranganathan in India and L. Guttman in Israel," C. Beghtol, DBD3
"Faculty Citations: An Approach to Assessing the Impact of Diminishing Resources on Scientific Research," A. Dykeman, GG4
Fair Use Guidelines for Electronic Reserve Systems, ICI3
Farrell, David. "Fundraising for Collection Development Librarians," GF9
Farthing, Pat. *See* Cook, Eleanor L.
Fattahi, Rahmatollah. "Anglo-American Cataloguing Rules in the Online Environment: A Literature Review," CCB2
———. "A Comparison between the Online Catalog and the Card Catalog: Some Considerations for Redesigning Bibliographic Standards," CAC4
"Favorable and Unfavorable Book Reviews: A Quantitative Study," R. J. Greene and C. D. Spornick, GD47
Faxon Company Home Page, FDD3
Faxon Source Online, FDD4
The Feather River Institute on Acquisitions and Collection Development, AGB4, BAD3

Fecko, Mary Beth. *Cataloging Nonbook Resources: A How-to-Do-It Manual for Librarians*, CCD22

"Fee-based Information Services: The Promises and Pitfalls of a New Revenue Source in Higher Education," E. McDaniel and R. Epp, IE16

Feng, Suliang. *See* Williamson, Nancy J.

Fidel, Raya. "User-Centered Indexing," EAB4

Fidel, Raya. *See also* Lunin, Lois F.

File Management and Information Retrieval Systems: A Manual for Managers and Technicians, S. L. Gill, EAA2

"Filing, Filtering, and the First Few Found," M. K. Buckland et al., EAA1

Film Preservation 1993: A Study of the Current State of American Film Preservation, A. Melville and S. Simmon, comps., HJ4

"Financial Issues for Collection Managers in the 1990s," E. L. Wiemers, GF8

Findley, Marcia. "Using the OCLC/AMIGO Collection Analysis Compact Disk to Evaluate Art and Art History Collections," GG10

"Firing an Old Friend, Painful Decisions: The Ethics between Librarians and Vendors," R. L. Presley, BI5, FDC3

"A Firm Order Vendor Evaluation Using a Stratified Sample," S. M. Rouzer, BF4

First Steps for Handling & Drying Water Damaged Materials, M. B. Kahn, HF3

The Fiscal Directory of Fee-Based Research and Document Supply Services, S. Coffman and P. Wiedensohler, IE1

Fiscella, Joan B., and Nancy Sack. "Independent Office Collections and the Evolving Role of Academic Librarians," GD49

Fisher, Janet H., and John Tagler, eds. "Perspectives on Firm Serials Prices," FE8

Fisher, William. "A Brief History of Library-Vendor Relations since 1950," BF2

"The Floating Standard: One Answer to Cataloging Schizophrenia," S. S. Intner, CBE5

Flood, Susan. "ALCTS Commercial Technical Services Committee, June 28, 1994," AHA5

Florida State Historical Records Advisory Board. *Historical Records Advisory Board Strategic Plan*, HD11

Flowers, Janet L. "Systems Thinking about Acquisitions and Serials Issues and Trends: A Report on the 1993 Charleston Conference," FAF1

Flowers, Kay. "Collection Development and Acquisitions in a Changing University Environment," BG3

Folcarelli, Ralph J. *See* Gillespie, John T.

Fons, Theodore A., and Wendy Sistrunk. "The Future of Technical Services: A Report on the NETSL Spring Conference," AA5

Ford, Helen. *The Education of Staff and Users for the Proper Handling of Archival Materials*, HM3

"Foreign Concepts: Indexing and Indexes on the Continent," M. Robertson, EAB17

Format Integration and Its Effect on Cataloging, Training and Systems, K. Coyle, ed., CCE1

"Formula-Based Subject Allocation: A Practical Approach," L. O. Rein, GF3

Fountain, Joanna F. *Headings for Children's Materials: An LCSH/Sears Companion*, DCC2

"Four Current Awareness Databases: Coverage and Currency Compared," J. M. Jaguszewski and J. L. Kempf, IE8

Fouty, Kathleen G. *Implementing an Automated Circulation System: A How-to-Do-It Manual,* IC5

Fox, Lisa L., ed. *Preservation Microfilming: A Guide for Librarians and Archivists,* HH9

Francq, Carole. "Bottoming Out the Bottomless Pit with the Journal Usage/Cost Relational Index," GE35

Franklin, Hugh L. "Sci/Tech Book Approval Plans Can Be Effective," BB5

Franklin, Jonathan A. "Once Pieces of the Collection Development Puzzle: Issues in Drafting Format Selection Guidelines," GC5

Franz, Lori, John Powell, Suzann Jude, and Karen M. Drabenstott. "End-User Understanding of Subdivided Subject Headings," DCB5

Frase, Robert W. "Permanent Paper: Progress Report II," HC4

Frazier, Kenneth. *See* Bennett, Scott

"From Just-in-Case to Just-in-Time: Limits to the Alternative Library Service Model," D. P. Carrigan, IE3

"From Smart Guesser to Smart Navigator: Changes in Collection Development for Research Libraries in a Network Environment," Y. Zhou, GE6

"From Text to Hypertext by Indexing," A. Salminen et al., EAB19

"From the Library of Record to the Library as Gateway: An Analysis of Three Electronic Table-of-Contents Services," M. L. Goodyear and J. Dodd, IE5

Frost, Carolyn O. "Quality in Technical Services: A User-Centered Definition for Future Information Environments," AHA6

Fugmann, Robert. "Representational Predictability: Key to the Resolution of Several Pending Issues in Indexing and Information Supply," EAB5

"Fundraising for Collection Development Librarians," D. Farrell, GF9

Fuseler, Elizabeth A. *See* Rutstein, Joel S.

Futas, Elizabeth. "Collection Development of Genre Literature," GE15

_____, ed. *Collection Development Policies and Procedures,* GC6

"A Future for Technical Services," V. Reich, AA13

The Future Is Now: The Changing Face of Technical Services, AHB3

The Future of Resource Sharing, S. K. Baker and M. E. Jackson, IDA1

"The Future of Resource Sharing: The Role of the Association of Research Libraries," M. E. Jackson, IE7

"The Future of Technical Services: An Administrative Perspective," N. H. Allen and J. F. Williams II, AHA1

"The Future of Technical Services: A Report on the NETSL Spring Conference," T. A. Fons and W. Sistrunk, AA5

"The Future Using an Integrated Approach: The OhioLINK Experience," P. O'Connor et al., IC9

G

Gabriel: Gateway to Europe's National Libraries, IDC2

Gabriel, Michael R. *Collection Development and Collection Evaluation: A Sourcebook,* GAB1

Gans, Alfred. *Serials Publishing and Acquisitions in Australia,* FCA1

Garlock, Kristen. *See* Pointek, Sherry

Gasaway, Laura N. and Sarah K. Wiant. *Libraries and Copyright: A Guide to Copyright Law in the 1990s,* IAA5

Gasaway, Laura N. *See also* Bennett, Scott

Gascon, Pierre. "Le Répertoire de vedettes-matière de la Bibliothèque de l'Université Laval: sa génè et son évolution," EE7

"Gay, Lesbian, and Bisexual Titles: Their Treatment in the Review Media and Their Selection by Libraries," J. H. Sweetland and P. G. Christensen, GD9

Gaylord Preservation Pathfinder series, HA3

Gaynor, Edward. "Cataloging Electronic Texts: The University of Virginia Experience," CCD12

George, Gerald. *"Difficult Choices":* *How Can Scholars Help Save Endangered Research Resources?,* HM4

George, Susan C., comp. *Emergency Planning and Management in College Libraries,* HF1

Gerhard, Kristin H., Trudi E. Jacobson, and Susan G. Williamson. "Indexing Adequacy and Interdisciplinary Journals: The Case of Women's Studies," EAB6, GE26

Germain, J. Charles. "Publishing in the International Marketplace," BH3, FCB5

German, Lisa. "A Closer World: A Review of Acquisitions Literature, 1992," BAB1

Gertz, Janet. *Oversize Color Images Project, 1994–1995. Final Report of Phase I,* HH10

———, ed. *Proceedings of the New York State Seminar on Mass Deacidification,* HC10

———. "Selection for Preservation: A Digital Solution for Illustrated Texts," HH11

———. "Ten Years of Preservation in New York State: The Comprehensive Research Libraries," HD1

———, et al. "Preservation Analysis and the Brittle Book Problem in College Libraries: The Identification of Research-Level Collections and Their Implications," GE32, HI6

Getty Art History Information Program, HAB19

Getz, Malcolm. *See* Carpenter, David

Gibbs, Nancy J. "ALCTS/Role of the Professional in Academic Research Technical Services Departments Discussion Group," AHB4

Gill, Suzanne L. *File Management and Information Retrieval Systems: A Manual for Managers and Technicians,* EAA2

Gillespie, John T., and Ralph J. Folcarelli. *Guides to Library Collection Development,* GAB2

Gilmer, Lois C. *Interlibrary Loan: Theory and Management,* IDB2

"GIPSY: Automated Geographic Indexing of Text Documents," A. G. Woodruff and C. Plaunt, EF12

Giral, Angela, and Arlene G. Taylor. "Indexing Overlap and Consistency between the *Avery Index to Architectural Periodicals* and the *Architectural Periodicals Index,"* EAB7

Gitistan, Darrin. "Subjects of Concern: Selected Examples Illustrating Problems Affecting Information Retrieval on Iran and Related Subjects Using *LCSH,"* DCB6

"Give Me Barcodes, or Give Me Carpal Tunnel!" J. Archer, AIC1

Global Perspectives on Preservation Education, M. V. Cloonan, HM1

Glossary of Imaging Technology, AIIM, HN20

Godby, Jean. *See* Weibel, Stuart

Gomez, Joni. "A Cataloger's Workstation: Using a NeXT Computer and Digital Librarian Software to Access the *Anglo-American Cataloguing Rules,"* CBC2

Goodman, Judi A. *See* Keating, Lawrence R.
Goodyear, Mary Lou, and Jane Dodd. "From the Library of Record to the Library as Gateway: An Analysis of Three Electronic Table-of-Contents Services," IE5
"The Gordon and Breach Litigation: A Chronology and Summary," A. L. O'Neill, BH5
Gorman, Michael. "The Corruption of Cataloging," CAC5
Gossen, Eleanor A., and Suzanne Irving. "Ownership Versus Access and Low-Use Periodical Titles," GD40, IE6
Gottlieb, Jane, ed. *Collection Assessment in Music Libraries*, GE23
Gottlieb, Jane. *See also* Roosa, Mark
Gould, S. "A Voucher Scheme to Simplify Payment for International Interlibrary Transactions," IDA6
Gradinette, Maria, and Randy Silverman. "The Library Collections Conservation Discussion Group: Taking a Comprehensive Look at Book Repair," HC5
———. "Who, What, and Where in Book Repair: Institutional Profiles of the LCCDG," HC6
Gradinette, Maria. *See also* Silverman, Randy
Graham, Peter S. *Intellectual Preservation: Electronic Preservation of the Third Kind*, HH12
Greenberg, Jane. "Intellectual Control of Visual Archives: A Comparison between the *Art & Architecture Thesaurus* and the *Library of Congress Thesaurus for Graphic Materials*," DCC3
Greene, Robert J., and Charles D. Spornick. "Favorable and Unfavorable Book Reviews: A Quantitative Study," GD47

Gregory, David. *See* Pedersen, Wayne
Grochmal, Helen M. "Achieving Success as a Serials Librarian: Some Advice," FBA4
———. "Selecting Electronic Journals," GD41
Groen, Frances K. "Electronic and Print Information: Active Distribution and Passive Retention in Relation to a Murder—A Case Study," GD15
Grover, Mark L. *See* Erbolato-Ramsey, Christiane
Grund, Angelika. "ICONCLASS: On Subject Analysis of Iconographic Representations of Works of Art," DBD4
"Guerrilla Collection Development: Time-Saving Tactics for Busy Librarians," B. Quinn, GAE6
Guide for Training Collection Development Librarians, ALA Subcommittee on Guide for Training Collection Development Librarians, GB5
Guide for Written Collection Policy Statements, J. S. Anderson, GC1
Guide to Centres of International Lending, IFLA, IDB3
Guide to Cooperative Collection Development B. Harloe, ed., GH4
Guide to Document Delivery Centres, IFLA, IDB4
Guide to Indexing and Cataloging with the Art & Architecture Thesaurus, T. Petersen and P. J. Barnett, eds., DCE5, EE14
Guide to Selecting and Acquiring CD-ROMs, Software, and Other Electronic Publications, S. Bosch et al., BAA2, GD22
Guide to Technical Services Resources, P. Johnson, ed., AD2
"Guidelines and Procedures for Telefacsimile and Electronic Delivery of Interlibrary Loan Requests," RASD Management of Public Services Section, IDD2

Author/Title Index

"Guidelines for ALCTS Members to Supplement the ALA Code of Ethics," ALCTS, AE1

Guidelines for Bibliographic Description of Interactive Multimedia, ALCTS Interactive Multimedia Guidelines Review Task Force, CCD13

Guidelines for Bibliographic Description of Reproductions, ALCTS Committee on Cataloging: Description and Access, CCD28

Guidelines for Cataloging Monographic Electronic Texts at the Center for Electronic Texts in the Humanities, A. Hoogscarspel, CCD14

Guidelines for Electronic Preservation of Visual Materials, HH13

"Guidelines for Preservation Photocopying," ALCTS Reproduction of Library Materials Section, HL1

Guidelines for the Construction, Format, and Management of Monolingual Thesauri, ANSI, DCE3, EB1

Guidelines for the Development of Policies and Procedures Regarding User Behavior and Library Usage, ALA Intellectual Freedom Committee, ICI1

"Guidelines for the Loan of Rare and Unique Materials," ACRL Rare Books & Manuscripts Section, IDD1

"Guidelines for University Undergraduate Libraries: A Draft," ACRL, GAC1

"Guidelines Regarding Thefts in Libraries," ACRL Rare Books & Manuscripts Section, IF1

Guides to Library Collection Development, J. T. Gillespie and R. J. Folcarelli, GAB2

Gupta, Sushima. *Decimal Classification System: A Bibliography for the Period 1876–1994,* DAD1

Gwinn, Nancy E. *A National Preservation Program for Agricultural Literature,* HD2

H

Haar, John. "Scholar or Librarian? How Academic Libraries' Dualistic Concept of the Bibliographer Affects Recruitment," GB2

Halpin, Lola. *See* Coulter, Cynthia

Hancock-Beaulieu, Micheline. *See* Jones, Susan

Handman, Gary, ed. *Video Collection Development in Multi-Type Libraries: A Handbook,* GD29

Harloe, Bart, ed. *Guide to Cooperative Collection Development,* GH4

———, and John M. Budd. "Collection Development and Scholarly Communication in the Era of Electronic Access," GE3

Harmeyer, Dave. "Potential Collection Development Bias: Some Evidence on a Controversial Topic in California," GD16

Harmon, James D. *Integrated Pest Management in Museum, Library, and Archival Facilities: A Step by Step Approach for the Design, Development, Implementation, & Maintenance of an Integrated Pest Management Program,* HG3

Harrassowitz Online, FDD5

Harri, Wilbert. "Implementing Electronic Data Interchange in the Library Acquisitions Environment," FDF5

Harsock, Ralph. *Notes for Music Catalogers: Examples Illustrating AACR2 in the Online Bibliographic Record,* CCD4

Harvey, Ross. *Preservation in Libraries: A Reader,* HA4

———. *Preservation in Libraries: Principles, Strategies and Practices for Librarians,* HA5

Hawks, Carol Pitts. "Building and Managing an Acquisitions Program," BA3

———. "Expert Systems in Technical Services and Collection Management," AIB1, GAF1

Haynes, Craig, comp. *Providing Public Service to Remote Users*, IC16

Haynes, Douglas. "Pro-Cite for Library and Archival Condition Surveys," HI7

Haynes, Kathleen J. M., Jerry D. Saye, and Lynda Lee Kaid. "Cataloging Collection-Level Records for Archival Video and Audio Recordings," CBE4

Hazen, Dan C. *The Bibliographic Control and Preservation of Latin Americanist Library Resources: A Status Report with Suggestions*, HD3

———. "Collection Development Policies in the Information Age," GC7

———. *Preservation Priorities in Latin America: A Report from the Sixtieth IFLA Meeting*, HD4

———. *The Production and Bibliographic Control of Latin American Preservation Microforms in the United States*, HD5

Headings for Children's Materials: An LCSH/Sears Companion, J. F. Fountain, DCC2

Headings for Tomorrow: Public Access Display of Subject Headings, M. M. Yee, ed., DCA3, EC12

Health Hazards Manual for Artists, M. McCann, HC8

Heaney, Michael. "Object-Oriented Cataloging," CCA3

Hebert, Francoise. "Service Quality: An Unobtrusive Investigation of Interlibrary Loan in Large Public Libraries," IDA7

Heft, Sandra M. *See* Evans, G. Edward

Hegg, Judith L. *See* Casserly, Mary F.

Hemmasi, Harriette. "The Music Thesaurus Project at Rutgers University," EE8

Henderson, William Abbot, William J. Hubbard, and Sonja L. McAbee. "Collection Assessment in Academic Libraries: Institutional Effectiveness in Microcosm," GG1

Henigman, Barbara. *See* Preece, Barbara G.

Heseltine, R. "Electronic Information and Acquisitions," BC3

Hewitt, Joe. "On the Nature of Acquisitions," BA4

Hickey, Kate. *See* Boyarski, Jennie S.

Higginbotham, Barbra Buckner, ed. *Advances in Preservation and Access*, HA1

———, and Sally Bowdoin. *Access versus Assets: A Comprehensive Guide to Resource Sharing for Academic Librarians*, GH5

"Highlights: Language- and Domain-Independent Automatic Indexing Terms for Abstracting," J. D. Cohen, EF3

Hirshon, Arnold, ed. *After the Electronic Revolution, Will You Be the First to Go?*, AHA7

———, and Barbara Winters. *Outsourcing Technical Services: A How-to-Do-It Manual for Librarians*, CBG2

Historical Records Advisory Board Strategic Plan, Florida State Historical Records Advisory Board, HD11

Hitchcock, Steve, Leslie Carr, and Wendy Hall. *A Survey of STM Online Journals 1990–1995: The Calm before the Storm*, FAC3

Hockey, Susan. "Evaluating Electronic Texts in the Humanities," GD30

Hodge, Gail M. *Automated Support to Indexing*, EF6

———. "Computer-Assisted Database Indexing: The State-of-the-Art," EF7

Holley, Beth, and Mary Ann Sheble. *A Kaleidoscope of Choices: Reshaping Roles and Opportunities for Serialists*, FAF2

Holley, Beth. *See also* Sheble, Mary Ann

Hong Yi. "Indexing Languages, New Progress in China," EAB8

Hoogscarspel, Annelies. *Guidelines for Cataloging Monographic Electronic Texts at the Center for Electronic Texts in the Humanities,* CCD14

Hopkins, Lola Raykovic. *See* Braun, Janice

Horny, Karen L. "Taking the Lead: Catalogers Can't Be Wallflowers," CBA5

Houston, J. E. *See* Weller, Carolyn R.

"How Safe Are Our Libraries?" E. St. Lifer, IF4

"How the Richard Abel Co., Inc. Changed the Way We Work," A. L. O'Neill, BB8

"How to Distinguish and Catalog Chinese Personal Names," Q. Hu, CCB3

"How to Evaluate Serials Suppliers," P. G. Kent, FDC1

"How to Run a Tight Ship in the Magazine Stacks," E. M. Boardman, FBA1

"How to Survive the Present While Preparing for the Future: A Research Library Strategy," E. Smith and P. Johnson, GH7

Howarth, Lynne C., comp. *AACR2 Decisions & Rule Interpretations,* CCC1

———. "Modelling Technical Services in Libraries: A Microanalysis Employing Domain Analyis and Ishikawa ('Fishbone') Diagrams," AA6

———, and Jean Weihs. "AACR2R: Dissemination and Use in Canadian Libraries," CCA4

———, and Jean Weihs. "AACR2R Use in Canadian Libraries and Implications for Bibliographic Databases," CCA5

Howarth, Lynne C. *See also* Weihs, Jean

Hu, Qianli. "How to Distinguish and Catalog Chinese Personal Names," CCB3

Hubbard, William J. *See* Henderson, William Abbot

Hudon, Michèle. *Le Thésaurus: Conception, elaboration, gestion,* DCE7

Hudon, Michèle. *See also* Williamson, Nancy J.

Hudson, Gary A. "The MSUS/PALS Acquisitions Subsystem Vendor File," BC4

Huesmann, James. *See* Davis, Trisha

Hughes, Janet. "Use of Faculty Publication Lists and ISI Citation Data to Identify a Core List of Journals with Local Importance," GD42

Hughes, Margaret J., and Bill Katz, eds. *AV in Public and School Libraries: Selection and Policy Issues,* GD31

Hukyk, Barbara. *See* Walsh, Jim

Humphrey, Susanne. "The MedIndEx® Prototype for Computer Assisted MEDLINE® Database Indexing," EF8

Hunt, Caroline C. "Technical Services and the Faculty Client in the Digital Age," AHA8

Hupp, Stephen L. "The Reserve Department Revisited: A Study of Faculty and Student Use of the Capital University Library Reserve Department," IC6

Hurt, Charlene S., et al. "Collection Development Strategies for a University Center Library," GD50

Hyslop, Colleen. *See* Black, Leah

I

IAML-L, HAB11

"ICONCLASS: On Subject Analysis of Iconographic Representations of Works of Art," A. Grund, DBD4

If We Build It: Scholarly Communications and Networking Technologies, S. McMahon et al., eds., FAF3

"ILL Staffing: A Survey of Michigan Academic Libraries," J. L. Kimmel, IDA10

ILL-L, IAC9
Iltis, Deanna. *See* Davis, Susan
"The Image Database: A Need for Innovative Indexing and Retrieval," S. Oernager, ED3
"Image Quality and Viewer Perception," M. Ester, HH8
Imagelib, HAB12
"Impact of Basic Books v. Kinko's Graphics on Reserve Services at the University of Colorado, Boulder," S. Seaman, IC10
"Implementation of an Online Series Authority File at Auburn University," H. H. McCurley, CBF10
Implementing an Automated Circulation System: A How-to-Do-It Manual, K. G. Fouty, IC5
"Implementing Electronic Data Interchange in the Library Acquisitions Environment," W. Harri, FDF5
"Implications and Rules in Thesauri," R. E. Kent, EE11
"Implications of Commercial Document Delivery," B. Coons and P. McDonald, GD38, IE4
"Improved Browsable Displays: An Experimental Test," B. Allen, EC1
"Improving Subject Access in OPACs: An Exploratory Study on User's Queries," J. M. Cherry, DA3
"In Situ Generation of Compressed Inverted Files," A. Moffatt and T. A. H. Bell, EF9
"Independent Office Collections and the Evolving Role of Academic Librarians," J. B. Fischella and N. Sack, GD49
The Index and Abstract Directory: An International Guide to Services and Serials Coverage, EAC1
"Indexing Adequacy and Interdisciplinary Journals: The Case of Women's Studies," K. H. Gerhard et al., EAB6, GE26
Indexing and Abstracting Society of Canada, EAD8
"Indexing and Retrieval Performance: The Logical Evidence," D. Soergel, EAB21
Indexing Books, N. C. Mulvany, EAB14
"Indexing Encyclopedia and Multi-volume Works," J. Blakeslee, EAB2
"Indexing Film and Video Images for Storage and Retrieval," J. Turner, ED6
"Indexing for Information," H. R. Tibbo, EAB23
"Indexing Languages, New Progress in China," Y. Hong, EAB8
"Indexing Overlap and Consistency between the *Avery Index to Architectural Periodicals* and the *Architectural Periodicals Index*," A. Giral and A. G. Taylor, EAB7
Indexing, Providing Access to Information: Looking Back, Looking Ahead, N. C. Mulvany, ed., EAB15
"Indexing the Internet," T. G. McFadden, ED2
Index-L, EAD4
"Indoor Environment Standards: A Report on the NYU Symposium," E. McCrady, HG5
Information and Documentation—Guidelines for the Content, Organization, and Presentation of Indexes: Draft International Standard, ISO/TC46/SC9, ISO, EB3
Information on Microfiche Headers: A Draft American National Standard, NISO, HN19
"Information-Retrieval Systems: Systems Analysis of Problems of Quality Management," E. Sukiasyan, EAB22
Information Services for People with Developmental Disabilities, L. L. Walling and M. M. Irwin, IA4
"The Information Universe: Will We Have Chaos or Control?," A. G. Taylor, AA14

"Integrated Library Systems for Minicomputers and Mainframes: A Vendor Study," W. Saffady, AID6

Integrated Pest Management in Museum, Library, and Archival Facilities: A Step by Step Approach for the Design, Development, Implementation, & Maintenance of an Integrated Pest Management Program, J. D. Harmon, HG3

"Integrating Documents Processing into Traditional Technical Services," S. Davis et al., AHA3

"Intellectual Control of Visual Archives: A Comparison between the *Art & Architecture Thesaurus* and the *Library of Congress Thesaurus for Graphic Materials,*" J. Greenberg, DCC3

Intellectual Freedom Manual, ALA Office for Intellectual Freedom, GD13

Intellectual Preservation: Electronic Preservation of the Third Kind, P. S. Graham, HH12

Intellectual Property and the National Information Infrastructure: The Report of the Working Group on Intellectual Property Rights, HE3

Interactive Electronic Serials Cataloging Aid (IESCA), CCD15

"Interactive Thesaurus Navigation: Intelligence Rules OK?" S. Jones et al., EE10

INTERCAT, EAD5

"Interfaces" [column], S. S. Intner, AA8

"Interfaces" [column], J. Weihs, AA15

Interfaces: Relationships between Library Technical and Public Services, S. S. Intner, AA7

"Interlibrary Loan and Commercial Document Supply: Finding the Right Fit," W. Pedersen and D. Gregory, IE18

Interlibrary Loan Data Elements, NISO, IDD5

Interlibrary Loan in College Libraries, R. Bustos, comp., IDA4

"Interlibrary Loan of Alternative Materials: A Balanced Sourcebook," B. Massis and W. Vitzansky, eds., IDB5

Interlibrary Loan Policies Directory, L. H. Morris, IDC4

Interlibrary Loan Practices Handbook, V. Boucher, IDB1

Interlibrary Loan: Theory and Management, L. C. Gilmer, IDB2

Interloc: The Electronic Marketplace for Books, BD2

International Association of Music Libraries. *IAML-L,* HAB11

International Conference on Interlending & Document Supply, IAC3

"International Cooperation in Subject Analysis," D. McGarry, EC7

International Cooperation in the Field of Authority Data: An Analytical Study with Recommendations, F. Bourdon, CBF3

International Directory of Serials Specialists, J. I. Whiffin, FAB2

International Federation of Library Associations and Institutions. *Guide to Centres of International Lending,* IDB3

―――. *Guide to Document Delivery Centres,* IDB4

―――, Section on Document Delivery and Interlending, IAC2

International Federation of Television Archives, HAB20

International Organization for Standardization. *Information and Documentation—Guidelines for the Content, Organization, and Presentation of Indexes: Draft International Standard, ISO/TC46/SC9,* EB3

The International Serials Industry, H. Woodward and S. Pilling, eds., FCA3

International Subscription Agents, L. R. Wilkas, FAB3

"The International Use of the *Dewey Decimal Classification*," R. Sweeney, DBB1
"The Internet and Collection Development: Mainstreaming Selection of Internet Resources," S. G. Demas et al., GD26
"The Internet and Technical Services: A Point Break Approach," G. M. McCombs, AA9, FDF6
"The Internet in Acquisitions Work: A Status Report," J. G. Montgomery, BC6
"The Internet in Technical Services: The Impact for Acquiring Resources and Providing Bibliographic Access on Technical Services," J. A. Mumm and A. Sitkin, AA11
Internet Resources for Cataloging, V. T. Sha, CAA4, DAB3
Intner, Sheila S. "Ethics in Cataloging," CAC7
———. "The Floating Standard: One Answer to Cataloging Schizophrenia," CBE5
———. "Interfaces" [column], AA8
———. *Interfaces: Relationships between Library Technical and Public Services*, AA7
———. "Outsourcing: What Does It Mean for Technical Services?," AHB5, FDB2
———. "Preservation Training for Library Users," HM5
———. "The Relationship of Acquisitions to Resource Sharing: An Informal Analysis," BG4
———. "The Re-Professionalization of Cataloging," CBA6
———. "Taking Another Look at Minimal Level Cataloging," CBE6
———. "Training Staff for Preservation," HM6
Intner, Sheila S. *See also* Johnson, Peggy
Introduction to Technical Services, G. E. Evans and S. M. Heft, AB1

"Invisible Collections within Women's Studies: Practical Suggestions for Access and Assessment," C. Paries and P. A. Scott, GE25
"Invisible Thesauri: The Year 2000," J. L. Milstead, DCE1
IPI Storage Guide for Acetate Film, HG4
Irvin, Judy Chandler. *See* Davis, Susan
Irvine, Ann. "Is Centralized Collection Development Better? The Results of a Survey," GE16
Irving, Holly. "Cait—Computer-Assisted Indexing Tutor: Implemented for Training at NAL," EAB9
Irving, Suzanne. *See* Gossen, Eleanor A.
Irwin, Dale. "Local Systems and Authority Control," CBF6
Irwin, Marilyn. *See* Walling, Linda Lucas
"Is Centralized Collection Development Better? The Results of a Survey," A. Irvine, GE16
"Is Keyword Searching the Answer?," J. Tillotson, EC11
Ison, Jan. "Patron-Initiated Interlibrary Loan," IDA8
Isoperms: An Environmental Management Tool, D. K. Sebera, HG9
It Comes with the Territory: Handling Problem Situations in Libraries, A. Turner, IF7
ITS for Windows: Integrated Technical Services Workstation, The Library Corporation, CBC4

J

Jackson, Mary E. "Document Delivery over the Internet: Electronic Document Delivery Systems Used by Libraries," IDA9
———. "The Future of Resource Sharing: The Role of the Association of Research Libraries," IE7

Jackson, Mary E., and Karen Croneis, comps. *Uses of Document Delivery Services,* IE21

Jackson, Mary E. *See also* Baker, Shirley K.

Jacobowitz, Neil A. "A Comparison of AACR2R and French Cataloging Rules," CCA6

Jacobs, Dorothy S. "The Vertical File: An Overview and Guide," GD11

Jacobson, Trudi E. *See* Gerhard, Kristin H.

Jaguszewski, Janice M., and Jody L. Kempf. "Four Current Awareness Databases: Coverage and Currency Compared," IE8

Jahr, Rudy. *See* Schmidt, Diane E.

Jeng, Judy "Expert System Applications in Cataloging, Acquisitions, and Collection Development: A Status Review," AIB2

Jensen, Mary Brandt. "Electronic Reserve and Copyright," IC7

Joachim, Martin D., ed. *Languages of the World: Cataloging Issues and Problems,* CAA2

Jobe, Peggy. *See* Lombardo, Nancy

Johns, Cecily, and William Z. Schenck, eds. *Selection of Library Materials for Area Studies: Part II. Australia, Canada, and New Zealand,* GE9

Johnson, Denise J. *See* McNeal, Beth

Johnson, Marda L. "Technical Services Productivity Alternatives," AHB6

Johnson, Peggy. "Collection Development Policies: A Cunning Plan," GC8

———. "Ethical Considerations in Decision Making," BI4

———, ed. *Guide to Technical Services Resources,* AD2

———. "Writing Collection Development Policy Statements: Format, Content, Style," GC9

———. "Writing Collection Development Policy Statements: Getting Started," GC10

Johnson, Peggy, and Sheila S. Intner, eds. *Recruiting, Educating, and Training Librarians for Collection Development,* GB8

Johnson, Peggy, and Bonnie MacEwan, eds. *Collection Management and Development: Issues in an Electronic Era,* GAE3

Johnson, Peggy. *See also* Smith, Eldred

Johnson-Cooper, Glendora. "Building Racially Diverse Collections: An Afrocentric Approach," GD7

Jones, Beverly. *See* Tuten, Jane H.

Jones, C. Lee. *Preservation Film: Platform for Digital Access Systems,* HH14

Jones, D. Lee, ed. *The Library Fax/Ariel Directory: More than 10,560 Fax Sites and 555 Ariel Sites,* IDC3

Jones, Maralyn. *Collection Conservation Treatment: A Resource Manual for Program Development and Conservation Technician Training, including "Report on Training the Trainers,"* HC7

Jones, Maralyn. *See also* Ogden, Barkley W.

Jones, Susan, and Micheline Hancock-Beaulieu. "Support Strategies for Interactive Thesaurus Navigation," EE9

Jones, Susan, et al. "Interactive Thesaurus Navigation: Intelligence Rules OK?," EE10

Jones, Wayne, ed. *Serials Canada: Aspects of Serials Work in Canadian Libraries,* FBA5

Joswick, Kathleen E., and Jeanne Koekkoek Stierman. "Perceptions vs. Use: Comparing Faculty Evaluations of Journal Titles with Faculty and Student Usage," GD43

Joswick, Kathleen E., and John P. Stierman. "Systematic Reference Weeding: A Workable Model," GE29

"The Journal Deselection Project: The LSUMC-S Experience," B. E. Tucker, GE36

Journal of Interlibrary Loan, Document Delivery & Information Supply, IAB1

Jude, Suzann. *See* Franz, Lori

K

Kachel, Debra E. "Look Inward before Looking Outward: Preparing the School Library Media Center for Cooperative Collection Development," GE21

Kahn, Miriam B. *Disaster Response and Prevention for Computers and Data,* HF2

———. *First Steps for Handling & Drying Water Damaged Materials,* HF3

Kaid, Lynda Lee. *See* Haynes, Kathleen J. M.

A Kaleidoscope of Choices: Reshaping Roles and Opportunities for Serialists, B. Holley and M. A. Sheble, FAF2

Kaneko, Hideo. "RLIN CJK and East Asian Library Community," CBI5

Kao, Mary L. *Cataloging and Classification for Library Technicians,* DAB2

Kaplan, Michael. "Technical Services Workstations Improve Productivity," AIC2

Kaplan, Michael. *See also* Brugger, Judith M.

Karon, B. L. *See* Urbanski, Verna

Karp, Rashelle S. "Collection Management," GAB3

———. "Technical Services," AD3

Kascus, Marie, and Faith Merriman. "Using the Internet in Serials Management," FDE2

Kaser, Dick, ed. *Document Delivery in an Electronic Age,* IE9

Kathman, Jane McGurn. *See* Kathman, Michael

Kathman, Michael, and Jane McGurn Kathman, comps. *Managing Student Employees in College Libraries,* HI8, IC8

Katz, Bill. *See* Hughes, Margaret J.; Parrish, Karen

Keating, Lawrence R. II, Christa Easton Reinke, and Judi A. Goodman. "Electronic Journal Subscriptions," FCB6

Kelland, John Laurence, and Arthur P. Young. "Citation as a Form of Library Use," GG5

Kellerman, Lydia Suzanne. "Moving Fragile Materials: Shrink-Wrapping at Penn State," HI9

Kelly, Glen "Electronic Data Interchange (EDI): The Exchange of Ordering, Claiming, and Invoice Information from a Library Perspective," BC5

Kemp, Alasdair. *See* Aluri, Rao

Kempf, Jody L. *See* Jaguszewksi, Janice M.

Kendrick, Curtis L. *See* Martin, Harry S. III

Kennedy, Kit. "Access Blues: A Song We Are All Singing," FBB3

Kenney, Anne R. "Digital-to-Microfilm Conversion: An Interim Preservation Solution," HH15

Kenney, Anne R., and Stephen Chapman. *Tutorial: Digital Resolution Requirements for Replacing Text-Based Material: Methods for Benchmarking Image Quality,* HH16

Kenney, Anne R., and Lynn K. Personius. *A Testbed for Advancing the Role of Digital Technologies for Library Preservation and Access: Final Report by Cornell University to the Commission on Preservation and Access,* HH17

Kenney, Anne R. *See also* Waters, Donald J.

Kenney, Donald J. *See* Ragsdale, Kate W.

Kent, Philip G. "How to Evaluate Serials Suppliers," FDC1

Kent, Philip G. *See also* Burrows, Toby

Kent, Robert E. "Implications and Rules in Thesauri," EE11
Ketcham, Lee, and Kathleen Born. "The Art of Projecting: The Cost of Keeping Periodicals," FE9
———. "Projecting Serials Costs: Banking on the Past to Buy for the Future," FE10
———. "Serials *vs.* the Dollar Dilemma: Currency Swings and Rising Costs Play Havoc with Prices," FE11
Keys, Marshall. "On the Future of the OCLC Regional Networks," AIA1
Khalil, Mounir. "Document Delivery: A Better Option?," IE10
Kiegel, Joseph A., and Merry Schellinger. "A Cooperative Cataloging Project between Two Large Academic Libraries," CBI6
Kiegel, Joseph A. *See also* Brugger, Judith M.
Kilgour, Frederick G. "Effectiveness of Surname-Title-Words Searches by Scholars," CBF7, EC4
Kimmel, Janice L. "ILL Staffing: A Survey of Michigan Academic Libraries," IDA10
Kingma, Bruce R. "Access to Journal Articles: A Model of the Cost Efficiency of Document Delivery and Library Consortia," IE11
Kinnucan, Mark T. "Demand for Document Delivery and Interlibrary Loan in Academic Settings," IDA11
Klatt, Mary J. "An Aid for Total Quality Searching: Developing a Hedge Book," EC5
Kleinberg, Ira. "Making the Case for Professional Indexers: Where Is the Proof?," EAB10
Kneisner, Dan, and Carrie Willman. "But Is It an Online Shelflist? Classification Access on Eight OPACs," DBA8
Knowing the Score: Preserving Collections of Music, M. Roosa and J. Gottleib, comps., HJ5

Knowledge Organization, DAC2
Knowledge Organization and Quality Management: Proceedings of the Third International ISKO Conference, H. Albrechtsen and S. Oernager, eds., DAA1
Knutson, Gunnar. "The Year's Work in Descriptive Cataloging, 1992," CAB1
Kohberger, Paul B., Jr. *See* Carter, Ruth C.
Kohl, David F. "Revealing UnCover: Simple, Easy Article Delivery," IE12
Koltay, Zsuza. "Multimedia in the Research Library: Collection and Services," GD32
Krikos, Linda. "Women's Studies Periodical Indexes: An In-Depth Comparison," GE27
Kruger, Betsy. "U.K. Books and Their U.S. Imprints: A Cost and Duplication Study," BB6
Kurosman, Kathleen, and Barbara Ammerman Durniak. "Document Delivery: A Comparison of Commercial Document Suppliers and Interlibrary Loan Services," IE13

L

Labaree, Robert V. "The Regulation of Hate Speech on College Campuses and the 'Library Bill of Rights,'" GD17
LaGuardia, Cheryl. "Virtual Dreams Give Way to Digital Reality," GD33
Lange, Holley R. "Catalogers and Workstations: A Retrospective and Future View," CBC3
Langridge, Derek W. *Classification: Its Kinds, Systems, Elements and Applications*, DBA1
Languages of the World: Cataloging Issues and Problems, M. D. Joachim, ed., CAA2
Lawrence, Gregory. *See* Demas, Samuel G.

Layne, Sara Shatford. "Some Issues in the Indexing of Images," ED1
LC Subject Headings: Weekly Lists, Library of Congress, Office for Subject Cataloging Policy, DCB3
LCCN: LC Cataloging Newsline, CAD3, EAD6
LC's Author Numbers, G. K. Dick, DBD1
"LCSH Online at the State Library of Queensland," H. Thurlow, DCB1
Leazer, Gregory H. *See* Smiraglia, Richard P.
LeBlanc, James D. "Cataloging in the 1990s: Managing the Crisis (Mentality)," CAC6
_____. "Classification and Shelflisting as Value Added: Some Remarks on the Relative Worth and Price of Predictability, Serendipity, and Depth of Access," DA7
_____. "Towards Finding More Catalog Copy: The Possibility of Using OCLC and the Internet to Supplement RLIN Searching," CBE7
Lee, Hur-Li. "The Library Space Problem, Future Demand, and Collection Control," GE30
Lehmann, Stephen, and James H. Spohrer. "The Year's Work in Collection Development, 1992," GAB4
Leonard, Barbara G. "The Metamorphosis of the Information Resources Budget," GF6
Lesher, Marcella. *See* Sylvia, Margaret
The Librarian's Yellow Pages, HAA4
Libraries and Copyright: A Guide to Copyright Law in the 1990s, L. N. Gasaway and S. K. Wiant, IAA5
Libraries and Student Assistants: Critical Links, W. K. Black, ed., HI4
Library Acquisitions: Practice & Theory, AF1
Library Administration and Management Association, Building and Equipment Section, Safety and Security of Library Buildings Committee, IAC1

_____, Library Storage Discussion Group, IAC1
_____, Systems and Services Section, Circulation/Access Services Committee, IAC1
Library and Information Science, Mansfield University, CAE4
"The Library Collections Conservation Discussion Group: Taking a Comprehensive Look at Book Repair," M. Gradinette and R. Silverman, HC5
The Library Corporation. *ITS for Windows: Integrated Technical Services Workstation,* CBC4
The Library Fax/Ariel Directory: More than 10,560 Fax Sites and 555 Ariel Sites, D. L. Jones, IDC3
Library Mosaics, AF2
Library of Congress, CAE13, GAD3
Library of Congress. *Cataloger's Desktop,* CBC5
"The Library of Congress and the 'Bibliography' Note," A. J. Schimizzi, CCB7
Library of Congress Cataloging Directorate, CAE14
The Library of Congress Classification: A Content Analysis of the Schedules in Preparation for Their Conversion into Machine-Readable Form, N. J. Williamson et al., DBC1
Library of Congress, Copyright Office, IAA9
Library of Congress, MARC Editorial Division. *MARC Conversion Manual–Authorities (Names). Content Designation Conventions and Online Procedures,* CCE2
Library of Congress, Office for Subject Cataloging Policy. *LC Subject Headings: Weekly Lists,* DCB3
Library of Congress, Prints and Photographs Division. *Thesaurus for Graphic Materials,* EE12
Library of Congress Subject Headings: Principles and Application, L. M. Chan, DCB2, EE3

Author/Title Index

Library Research Models: A Guide to Classification, Cataloging, and Computers, T. Mann, CAA3
"Library Security: Special Issue," F. Tehrani, ed., IF6
Library Services for Non-Affiliated Patrons, E. S. Mitchell, comp., ICI4
"A Library Shelver's Performance Evaluation as It Relates to Reshelving Accuracy," S. C. Sharp, IB6
"The Library Space Problem, Future Demand, and Collection Control," H. L. Lee, GE30
"Library-Oriented Lists and Electronic Serials," S. Bonario and A. Thornton, FAC1
"Licence to Kill?" F. Edwards, BH2
"Licensing Agreements: Think before You Act," M. D. Cramer, FCB1
Liddy, Elizabeth D., and Corinne L. Jörgensen. "Reality Check! Book Index Characteristics that Facilitate Information Access," EAB11
Lin, Joseph C. "Undifferentiated Names: A Cataloging Rule Overlooked by Catalogers, Reference Librarians, and Library Users," CBF8
Lindner, Jim. "Confessions of a Videotape Restorer; Or, How Come These Tapes All Need to Be Cleaned Differently?," HJ6
"Linking of Alternate Graphic Representation in USMARC Authority Records," J. M. Aliprand, CBF1
LIS-FID, DAE1
"Loading the GPO MARC Tapes: 1992 Preconference of ALA/Government Documents Round Table," AID3
"Local Area Networks and Wide Area Networks for Libraries," J. R. Matthews and M. R. Parker, AIC3
"Local Systems and Authority Control," D. Irwin, CBF6

Lombardo, Nancy, and Peggy Jobe. "What's MIME Is Yours," IDA12
"A Look at Subject Headings: A Plea for Standardization," M. Bloomfield, DCA4
"Look Inward before Looking Outward: Preparing the School Library Media Center for Cooperative Collection Development" D. E. Kachel, GE21
"Looking Trojan Gift Horses in the Mouth: Are Special Issues Special Enough to Pay Extra Money for and Bind?" T. Stankus, FCA2
Lougee, Wendy P. "Beyond Access: New Concepts, New Tensions for Collection Development in a Digital Environment," GE4
Lu, Min-Huei. "A Study of OP and OSI Cancellations in an Academic Library," BH4
Ludington, William. *See* Eldredge, Mary
Lui-Palmer Thesaurus Construction System, DCE8
Lunin, Lois F. and Raya Fidel, eds. "Perspectives on . . . Indexing," EAB12
Lutzker, Arnold P. *Commerce Department's White Paper on National and Global Information Infrastructure. Executive Summary for the Library and Educational Community*, IAA6

M

Maben, Michael. "The Cataloging of Primary State Legal Material," CCD26
MacDougall, Ian. *See* Chaney, Michael
MacEwan, Bonnie. *See* Johnson, Peggy
Machovec, George S. "Criteria for Selecting Document Delivery Suppliers," IE14
MacLennan, Birdie. *Serials in Cyberspace: Collections, Resources, and Services*, FDA2

MacLeod, Judy, and Daren Callahan. "Educators and Practitioners Reply: An Assessment of Cataloging Education," CBA7
MacLeod, Judy. *See also* Callahan, Daren
Magnetic Tape Storage and Handling: A Guide for Libraries and Archives, J. W. C. Van Bogart, HJ17
Maintaining the Privacy of Library Records: A Handbook and Guide. A. Bielefield and L. Cheeseman, IA1
"Making Hard Choices: Canceling Print Indexes," M. Sylvia and M. Lesher, GE19
"Making Lemonade: The Challenges and Opportunities of Forced Reference Serials Cancellations: One Academic Library's Experiences," S. Wise, GE37
"The Making of a Standard," E. U. Mangan, CAC8
"Making the Case for Professional Indexers: Where Is the Proof?" I. Kleinberg, EAB10
Managing a Mold Invasion: Guidelines for Disaster Response, L. O. Price, HG6
Managing Acquisitions and Vendor Relations: A How-to-Do-It Manual, H. S. Miller, BAA3
Managing Performing Arts Collections in Academic and Public Libraries, C. A. Sheehy, ed., GE13
Managing Preservation: A Guidebook, HA6, HI10
Managing Student Employees in College Libraries, M. Kathman and J. M. Kathman, HI8, IC8
Managing the Preservation of Serial Literature: An International Symposium, M. A. Smith, ed., HI11
"A Mandate for Change in the Library Environment," R. M. Silberman, AHA11
Mangan, Elizabeth U. "The Making of a Standard," CAC8

"Manifestations and Near-Equivalents of Moving-Image Works: A Research Project," M. M. Yee, CCD7
"Manifestations and Near-Equivalents: Theory, with Special Attention to Moving-Image Materials," M. M. Yee, CCD6
Mann, Thomas. *Library Research Models: A Guide to Classification, Cataloging, and Computers,* CAA3
Manning, Ralph W. *Canadian Cooperative Preservation Project: Final Summary Report,* HD6
Mansfield University, CAE4
"Mapping LaborLine Thesaurus Terms to Library of Congress Subject Headings: Implications for Vocabulary Switching," M. A. Chaplan, EE4
MARC Conversion Manual–Authorities (Names). Content Designation Conventions and Online Procedures, Library of Congress, MARC Editorial Division, CCE2
Marcella, Rita, and Robert Newton. *A New Manual of Classification,* DBA5
Marco, Guy A., ed. *Encyclopedia of Recorded Sound in the United States,* HJ3
Marker, Rhonda, and Melinda Ann Reagor. "Variation in Place of Publication: A Model for Cataloging Simplification," CCB4
"Marketing to Libraries: What Works and What Doesn't," V. Reich, FBB4
Marner, Jonathan. *See* Nelson, David
Martin, Harry S. III, and Curtis L. Kendrick. "A User-Centered View of Document Delivery and Interlibrary Loan," IE15
Martin, Murray S. *Collection Development and Finance: A Guide to Strategic Library-Materials Budgeting,* GF7

311

Author/Title Index

Massachusetts Task Force on Preservation and Access to the Library, Archives, Public Records, and Governmental Communities and the Citizens of Massachusetts. *Preserved to Serve: The Massachusetts Preservation Agenda*, HD12

Massimiliano, Giurelli, and Giliola Negrini. "A Tool to Guide the Logical Process of Conceptual Structuring," EE13

Massis, Bruce, and Winnie Vitzansky, eds. "Interlibrary Loan of Alternative Materials: A Balanced Sourcebook," IDB5

Material Published by Members of the Library of Congress Preservation Directorate: A Bibliography, C. Zimmerman, HAC3

Mathews, Michael, and Patricia Brennan, eds. *Copyright, Public Policy, and the Scholarly Community*, IAA7

Matthews, Joseph R. "The Distribution of Information: The Role for Online Public Access Catalogs," EC6

———, and Mark R. Parker. "Local Area Networks and Wide Area Networks for Libraries," AIC3

———, and Mark R. Parker. "Microcomputer-Based Automated Library Systems: New Series, 1993," AID4

Maximizing Access, Minimizing Cost: A First Step toward the Information Access Future, S. K. Baker and M. E. Jackson, IDA2

McAbee, Sonja L. *See* Bevis, Mary D.; Henderson, William Abbot

McAllister-Harper, Desretta. "An Analysis of Courses in Cataloging and Classification and Related Areas Offered in Sixteen Graduate Library Schools . . . ," CBA8

McCann, Michael. *Health Hazards Manual for Artists*, HC8

McClellan, Charles. *See* Salminen, Airi

McClung, Patricia A. *Digital Collections Inventory Report*, HH18

———, ed. *RLG Digital Image Access Project: Proceedings from an RLG Symposium*, HH22

McColl, Amy M., comp. *NACO Participants' Manual*, CBF11

McCombs, Gillian M. "The Internet and Technical Services: A Point Break Approach," AA9, FDF6

McCoy, Patricia Sayre. "Technical Services and the Internet," AA10

McCrady, Ellen. "Indoor Environment Standards: A Report on the NYU Symposium," HG5

McCue, Janet, with Dongming Zhang. "Technical Services and the Electronic Library: Defining Our Roles and Divining the Partnership," AID5

McCurley, Henry H., Jr. "The Benefits of Online Series Authority Control," CBF9

———. "Implementation of an Online Series Authority File at Auburn University," CBF10

McDaniel, Elizabeth, and Ronald Epp. "Fee-based Information Services: The Promises and Pitfalls of a New Revenue Source in Higher Education," IE16

McDonald, Peter. *See* Coons, Bill; Demas, Samuel G.

McFadden, Thomas G. "Indexing the Internet," ED2

McGarry, Dorothy. "International Cooperation in Subject Analysis," EC7

McGill, Michael J. *See* Van Houwelling, Douglas E.

McGill University Libraries, and Elizabeth V. Silvester, ed. *Collection Policies*, GC11

McGreer, Anne. "ALCTS Creative Ideas in Technical Services Discussion Group," AHC6

McGregor, George F. *See* Cooper, Michael D.

McMahon, Suzanne, Miriam Palm, and Pam Dunn, eds. *If We Build It: Scholarly Communications and Networking Technologies,* FAF3

McMillan, Gail, and Marilyn L. Norstedt, eds. *New Scholarship—New Serials,* FAF4

McNeal, Beth, and Denise J. Johnson, eds. *Patron Behavior in Libraries: A Handbook of Positive Approaches to Negative Situations,* IF3

"Measuring Diversity of Opinion in Public Library Collections," J. Serebnick and F. Quinn, GD21

Measuring the Impact of Technology on Libraries: A Discussion Paper, M. Dillon, AA4

Media Stability Studies: Final Report, J. W. C. Van Bogart, HJ18

"The MedIndEx® Prototype for Computer Assisted MEDLINE® Database Indexing," S. Humphrey, EF8

Meiseles, Linda, and Emerita M. Cuestra. "Multiformat Periodicals: A New Challenge for the Periodicals Manager," FDE3

Melville, Annette, and Scott Simmon, comps. *Film Preservation 1993: A Study of the Current State of American Film Preservation,* HJ4

———, comps. *Redefining Film Preservation: A National Plan,* HJ10

Memorial University of Newfoundland. Cataloguers Toolbox, CAE2

Merriman, Faith. *See* Kascus, Marie

Merriman, John. "The Work of Subscription Agents," FDC2

"The Metamorphosis of the Information Resources Budget," B. G. Leonard, GF6

Metz, Paul, and Bruce Obenhaus. "Virginia Tech Sets Policy on Controversial Materials," GD18

Metz, Paul. *See also* Stelk, Roger Edward

Micco, Mary. "Subject Authority Control on the World of the Internet, Part 1," DD1

———. "Subject Authority Control on the World of the Internet, Part 2," DD2

Michigan. State Historical Records Advisory Board. *Strategies to Preserve Michigan's Historical Records,* HD13

"Microcomputer-Based Automated Library Systems: New Series, 1993," J. R. Matthews and M. R. Parker, AID4

"Migration from an In-House Serials System to INNOPAC at the University of Massachusetts at Amherst," P. Banach, FDF1

"Migration from Microlinx to NOTIS: Expediting Serials Holdings Conversion through Programmed Function Keys," B. Chiang, FDF3

Miller, David. "Ambiguities in the Use of Certain Library of Congress Subject Headings for Form and Genre Access to Moving Image Materials," DCB7

———. "Outsourcing Cataloging: The Wright Experience," CBG4

Miller, Eric. *See* Weibel, Stuart

Miller, Heather S. *Managing Acquisitions and Vendor Relations: A How-to-Do-It Manual,* BAA3

Milstead, Jessica L., ed. *ASIS Thesaurus of Information Science and Librarianship,* DCE4

———. "Invisible Thesauri: The Year 2000," DCE1

———. "Needs for Research in Indexing," EAB13

"Minnesota Opportunities for Technical Services Excellence (MOTSE): An Innovative CE Program for Technical Services Staff," S. Nevin, AB2

MIT Cataloging Oasis, CAE5

313

Mitchell, Eleanor, and Sheila Walters. *Document Delivery Services: Issues and Answers*, IE17

Mitchell, Eugene S., comp. *Library Services for Non-Affiliated Patrons*, ICI4

Mitchell, Joan S. *See* Chan, Lois Mai

"MLA's Statement on the Significance of Primary Records," HI12

"Modelling Technical Services in Libraries: A Microanalysis Employing Domain Analyis and Ishikawa ('Fishbone') Diagrams," L. C. Howarth, AA6

"A Modest Proposal: No More Main Entry," E. G. Bierbaum, CCB1

Moffatt, Alistair, and Timothy A. H. Bell. "In Situ Generation of Compressed Inverted Files," EF9

Mohlhenrich, Janice, ed. *Preservation of Electronic Formats & Electronic Formats for Preservation*, HH19

Mohr, Deborah A. *See* Schuneman, Anita

Molholt, Pat, and Toni Petersen. "The Role of the *Art & Architecture Thesaurus* in Communicating about Visual Art," DCE2

Montgomery, J. G. "The Internet in Acquisitions Work: A Status Report," BC6

"The More Things Change, the More They Stay the Same: East-West Exchanges 1960-1993," M. S. Olsen, BE1

Morris, Anne, ed. *The Application of Expert Systems in Libraries and Information Centres*, AIB3

Morris, Leslie H. *Interlibrary Loan Policies Directory*, IDC4

Morrison, David F. W. "Bibliography of Articles Related to Electronic Journal Publications and Publishing," FAC4

Morton, Bruce. "Collection Development," GE11

"Motivating Student Employees: Examples from Collections Conservation," B. J. Baird, HI2, IB1

"Moving Fragile Materials: Shrink-Wrapping at Penn State," L. S. Kellerman, HI9

"The MSUS/PALS Acquisitions Subsystem Vendor File," G. A. Hudson, BC4

Mullis, Albert A. "Access to Serials: National and International Cooperation," FF1

Multicultural Acquisitions, K. Parrish and B. Katz, eds., GD8

"Multiformat Periodicals: A New Challenge for the Periodicals Manager," L. Meiseles and E. M. Cuestra, FDE3

"Multimedia in the Research Library: Collection and Services," Z. Koltay, GD32

MultiTes 5.2, DCE9

Mulvany, Nancy C. *Indexing Books*, EAB14

———, ed. *Indexing, Providing Access to Information: Looking Back, Looking Ahead*, EAB15

Mumm, James A., and Ann Sitkin. "The Internet in Technical Services: The Impact for Acquiring Resources and Providing Bibliographic Access on Technical Services," AA11

MUSEUM-L, HAB13

"The Music Thesaurus Project at Rutgers University," H. Hemmasi, EE8

"Myth and Reality: Using the OCLC/AMIGOS Collection Analysis CD to Measure Collections against Peer Collections and against Institutional Priorities," W. V. Dole, GG9

Myzyk, Mark. "Canon Formation, Library Collections, and the Dilemma of Collection Development," GD2

N

NACO Participants' Manual, A. M. McColl, comp., CBF11

Nardini, Robert F. "The Approval Plan Profiling Session," GD4
———. "Approval Plans: Politics and Performance," BB7, GD5
National Film Preservation Board Home Page, HAB21
National Information Standards Organization. *American National Standard for Permanence of Paper for Publications and Documents in Libraries and Archives*, HN2
———. *Data Elements for Binding of Library Materials*, HN1
———. *Environmental Guidelines for the Storage of Paper Records*, HN3
———. *Information on Microfiche Headers: A Draft American National Standard*, HN19
———. *Interlibrary Loan Data Elements*, IDD5
———. *Proposed American National Standard Format for Circulation Transactions Standards*, ICI5
———. *Proposed American National Standard Guidelines for Indexes and Related Information Retrieval Devices*, EB4
"National Interlibrary Loan Code for the United States 1993," RASD Management of Public Services Section, IDD3
National Library of Canada, CAE16
National Library of Canada Electronic Publications Pilot Project, CCD16
National Library of Medicine (U.S.), CAE15
National Media Lab. *NML Bits: Newsletter of the National Media Lab*, HAA5
A National Preservation Program for Agricultural Literature, N. E. Gwinn, HD2
"National Series Authority File Derailed?" J. J. Reimer, CBF13
"Native American Literature for Young People: A Survey of Collection Development Methods in Public Libraries," R. Tjoumas, GD10

Natowitz, Allen. See Carlo, Paula Wheeler
"Needs for Research in Indexing," J. L. Milstead, EAB13
Negrini, Giliola. *See* Massimiliano, Giurelli
Nelson, Antoinette. *See* Wilkes, Adeline
Nelson, David, and Jonathan Marner. "The Concept of Inadequacy in Uniform Titles," CCB5
———. "Dates in Added Entries: An Analysis of an AUTOCAT Discussion," CCB6
NetFirst, DD3
"Networking: An Overview for Leaders of Academic Medical Centers," W. B. Panko, AIA3
Nevin, Susanne. "Minnesota Opportunities for Technical Services Excellence (MOTSE): An Innovative CE Program for Technical Services Staff," AB2
A New Manual of Classification, R. Marcella and R. Newton, DBA5
"New Roles for Classification in Libraries and Information Networks: Presentations and Reports from the Thirty-Sixth Allerton Institute," DBA2
New Scholarship—New Serials, G. McMillan and M. L. Norstedt, eds., FAF4
New Tools for Preservation: Assessing Long-Term Environmental Effects on Library and Archives Collections, J. M. Reilly et al., HG8
New York State Education and Research Network (NYSERNet) Conference, AGB5
NewJour: Electronic Journals & Newsletters, FAC5
Newsletter on Serials Pricing Issues, FAD3
"Newspaper Indexing: The San Antonio Express-News," C. Peterson, ED4
Newton, Robert. *See* Marcella, Rita

315

Niemeyer, Mollie, et al. "Balancing Act for Library Materials Budgets: Use of a Formula Allocation," GF2

Nilsen, Kirsti S. "Collection Development Issues of Academic and Public Libraries: Converging or Diverging?," GAE4

Nishimura, Douglas W. *See* Reilly, James M.

Nissley, Meta. "Rave New World: Librarians and Electronic Acquisitions," BA5

NML Bits: Newsletter of the National Media Lab, HAA5

"Nonbook Materials: Their Occurrence and Bibliographic Description in Canadian Libraries," J. Weihs and L. C. Howarth, CCD23

Norgard, Barbara A. *See* Buckland, Michael K.

Norstedt, Marilyn L. *See* McMillan, Gail

North American Permanent Papers, HC9

North American Serials Interest Group, FAE4

Northwestern University Library Technical Services, CAE6

Notes for Music Catalogers: Examples Illustrating AACR2 in the Online Bibliographic Record, R. Harsock, CCD4

Notess, Greg R. "Comparing Web Browsers: Mosaic, Cello, Netscape, WinWeb and InternetWorks Lite," EF10

———. "Searching the World-Wide Web: Lycos, WebCrawler and More," EF11

"NOTIS as an Impetus for Change in Technical Services Departmental Staffing," M. D. Bevis and S. L. McAbee, AHC2, FDB1

O

Obenhaus, Bruce. *See* Metz, Paul

Oberlin LCRI Cumulated: Prototype Edition, CCC2

"Object-Oriented Cataloging," M. Heaney, CCA3

O'Brien, Ann. "Online Catalogs: Enhancements and Developments," EC8

"OCLC Affiliated U. S. Regional Networks: A Special Partnership," AIA2

"OCLC Cataloging Peer Committees: An Overview," J. B. Barnett, CBI2

OCLC Forest Press DDC Web Site, DAE2

OCLC Online Computer Library Center, Inc., CAE19, IDE3

OCLC Systems & Services, AF3, IAB2

"OCLC/NCSA Metadata Workshop Report," S. Weibel et al., CAC9

OCLC's Resource Sharing Strategy, IDA13

O'Connor, Phyllis. "Remote Storage Facilities: An Annotated Bibliography," IB4

———, Susan Wehmeyer, and Susan Weldon. "The Future Using and Integrated Approach: The OhioLINK Experience," IC9

Oernager, Susanne. "The Image Database: A Need for Innovative Indexing and Retrieval," ED3

Oernager, Susanne. *See also* Albrechtsen, Hanne

Ogburn, Joyce L. "Changing Relationships in the Acquisition and Delivery of Library Materials: A Survey," BG5

———. "Special Section on Outsourcing," AHB7

Ogden, Barkley W., and Maralyn Jones. *CALIPR: An Automated Tool to Assess Preservation Needs of Books and Document Collections for Institutional or Statewide Planning,* HI13

Ogden, Barkley W. *See also* Strauss, Robert J.

Ogden, Sherelyn, ed. *Preservation of Library and Archival Materials: A Manual,* HA7

Ohio Historical Records Advisory Board. *The Ohio 2003 Plan: A Statement of Priorities and Preferred Approaches to Historical Records Programs in Ohio;* and *To Outwit Time: Preserving Materials in Ohio's Libraries and Archives,* HD14

The Ohio 2003 Plan: A Statement of Priorities and Preferred Approaches to Historical Records Programs in Ohio, Ohio Historical Records Advisory Board, HD14

"OhioLINK Inter-Institutional Lending Online: The Miami University Experience," J. Sessions et al., IDA15

Oklahoma Historical Records Advisory Board. *To Save Our Past: A Strategic Plan for Preserving Oklahoma's Documentary Heritage,* HD15

Olsen, Margaret S. "The More Things Change, the More They Stay the Same: East-West Exchanges 1960–1993," BE1

Olson, Georgine N., and Barbara McFadden Allen, eds. "Cooperative Collection Management: The Conspectus Approach," GG8

Olson, Nancy B., ed. *Cataloging Internet Resources: A Manual and Practical Guide,* CCD17

"On the Cost Differences between Publishing a Book in Paper and in the Electronic Medium," T. Clark, BH1

"On the Future of the OCLC Regional Networks," M. Keys, AIA1

"On the Nature of Acquisitions," J. Hewitt, BA4

On the Road to Preservation: A State-Wide Preservation Action Plan for Colorado, Colorado Preservation Alliance, HD10

"On the Subject of Subjects," A. G. Taylor, DA9

"Once Pieces of the Collection Development Puzzle: Issues in Drafting Format Selection Guidelines," J. A. Franklin, GC5

O'Neill, Ann L. "The Gordon and Breach Litigation: A Chronology and Summary," BH5

———. "How the Richard Abel Co., Inc. Changed the Way We Work," BB8

O'Neill, Edward T., and Wesley L. Boomgaarden. "Book Deterioration and Loss: Magnitude and Characteristics in Ohio Libraries," HI14

O'Neill, Edward T., Sally Rogers, and W. Michael Oskins. "Characteristics of Duplicate Records in OCLC's Online Union Catalog," CBI7

"Ongoing Changes in Stanford University Libraries Technical Services," S. W. Propas, AHB8

"Online Catalogs: Enhancements and Developments," A. O'Brien, EC8

OPAC Directory: An Annual Guide to OnLine Public Access Catalogs and Databases, M. Schuyler, IDC5

"OPACs in Twelve Canadian Academic Libraries: An Evaluation of Functional Capabilities and Interface Features," J. M. Cherry et al., EC3

Options for Replacing and Reformatting Deteriorated Materials, J. Banks, HI3

Organization of Collection Development, G. Rowley, comp., GB3

Organizing Knowledge, J. Rowley, EAB18

"Original Cataloging Errors: A Comparison of Errors Found in Entry-Level Cataloging with Errors Found in OCLC and RLIN," L. Romero, CBI9

Oskins, W. Michael. *See* O'Neill, Edward T.

317

Otero-Boisvert, Maria. "The Role of the Collection Development Librarian in the 90s and Beyond," GAE5

An Ounce of Preservation: A Guide to the Care of Papers and Photographs, C. A. Tuttle, HJ16

Outsourcing Cataloging, Authority Work and Physical Processing: A Checklist of Considerations, ALCTS Commercial Technical Services Committee, CBG5

"Outsourcing Cataloging: The Wright Experience," D. Miller, CBG4

"Outsourcing Copy Cataloging and Physical Processing: A Review of Blackwell's Outsourcing Services for the J. Hugh Jackson Library at Stanford University," K. A. Wilson, CBG8

"Outsourcing Library Production: The Leader's Role," C. H. Varner, AHB9

Outsourcing Technical Services: A How-to-Do-It Manual for Librarians, A. Hirshon and B. Winters, CBG2

"Outsourcing: What Does It Mean for Technical Services?" S. S. Intner, AHB5, FDB2

Overexposure: Health Hazards in Photography, S. D. Shaw and M. Rossol, HC13

Oversize Color Images Project, 1994–1995. Final Report of Phase I, J. Gertz, HH10

"Ownership Versus Access and Low-Use Periodical Titles," E. A. Gossen and S. Irving, GD40, IE6

"Ownership Versus Access: Shifting Perspectives for Libraries," J. S. Rutstein et al., GE5

P

Palm, Miriam. *See* McMahon, Suzanne

Palmer, Joseph W. "Agreement and Disagreement among Fiction Reviews in *Library Journal, Booklist* and *Publishers Weekly,*" GD48

———. *Cataloging and the Small Special Library,* CBA9

———. "Saving Those Historic Videotapes: It May Already Be Too Late," HJ7

Palmetto Reflections: A Plan for South Carolina's Documentary Heritage, South Carolina. State Historical Records Advisory Board, HD17

Pankake, Marcia, Karen Wittenborg, and Eric Carpenter. "Commentaries on Collection Bias," GD19

Panko, W. B., et al. "Networking: An Overview for Leaders of Academic Medical Centers," AIA3

Parang, Elizabeth, and Laverna Saunders. *Electronic Journals in ARL Libraries: Issues and Trends,* FCB3, GD28

———. *Electronic Journals in ARL Libraries: Policies and Procedures,* FCB4

Paries, Cindy, and Patricia A. Scott. "Invisible Collections within Women's Studies: Practical Suggestions for Access and Assessment," GE25

Parker, Mark R. *See* Matthews, Joseph R.

Parrish, Karen, and Bill Katz, eds. *Multicultural Acquisitions,* GD8

Patron Behavior in Libraries: A Handbook of Positive Approaches to Negative Situations, B. McNeal and D. J. Johnson, IF3

"Patron-Initiated Interlibrary Loan," J. Ison, IDA8

Patron Request System, IDE4

Paul, Sandra K. "EDI/EDIFACT," FDF7

Payson, Evelyn. "The Vertical File: Retain or Discard?," GD12

Pedersen, Wayne, and David Gregory. "Interlibrary Loan and Commercial Document Supply: Finding the Right Fit," IE18

Pelzer, N. L. "Veterinary Subject Headings and Classification: A Critical Analysis," DCC4

Pennsylvania State University Libraries, Acquisitions Services, BAD5

"People, Words, and Perception: A Phenomenological Investigation of Textuality," T. A. Brooks, EC2

"Perceptions vs. Use: Comparing Faculty Evaluations of Journal Titles with Faculty and Student Usage," K. E. Joswick and J. K. Stierman, GD43

Performance Guideline for the Legal Acceptance of Records Produced by Information Technology Systems, AIIM, HN21

"Periodical Access in an Era of Change: Characteristics and a Model," M. Cain, IE2

"Periodical Prices, 1990–1992," A. W. Alexander, FE1

"Periodical Prices, 1991–1993," A. W. Alexander, FE2

"Periodical Prices, 1992–1994," A. W. Alexander, FE3

"Periodical Records Conversion: From Union List to Statewide Network," S. L. Tsui, FF3

The Permanence and Care of Color Photographs: Traditional and Digital Color Prints, Color Negatives, Slides, and Motion Pictures, H. Wilhelm and C. Brower, HJ20

Permanence, Care, and Handling of CDs, HJ8

"Permanent Paper: Progress Report II," R. W. Frase, HC4

Perrault, Anna H. "The Shrinking National Collection: A Study of the Effects of the Diversion of Funds from Monographs to Serials on the Monograph Collections of Research Libraries," GD51

———, and Marjo Arseneau. "User Satisfaction and Interlibrary Loan Service: A Study at Louisiana State University," IDA14

"Perspectives on Firm Serials Prices," J. H. Fisher and J. Tagler, eds., FE8

"Perspectives on . . . Indexing," L. F. Lunin and R. Fidel, EAB12

"Perspectives on the Cataloging Literature: Cataloging's Prospects: Responding to Austerity with Innovation," C. Ruschoff, CBD2

"Perspectives on the Future of Union Listing," A. C. Schaffner, FF2

Petersen, Toni, and Patricia J. Barnett, eds. *Guide to Indexing and Cataloging with the* Art & Architecture Thesaurus, DCE5, EE14

Petersen, Toni. *See also* Brusch, Joseph A.; Molholt, Pat

Peterson, Candace. "Newspaper Indexing: The San Antonio Express-News," ED4

Petit, Michael J. "The Evaluation, Selection, and Acquisition of Legal Looseleaf Publications," BH6

Philips, Phoebe F. "Computers and Technical Services," AA12

Pierce, Darlene M., and Eileen Theodore-Shusta. "Automation: The Bridge between Technical Services and Government Documents," AHA9

Pilling, Stella. *See* Woodward, Hazel

PLA Handbook for Writers of Public Library Policies, PLA Policy Manual Committee, GC12

"A Plan for Evaluating a Small Library Collection," F. Davis, GG3

Plaunt, Christian. *See* Buckland, Michael K.; Woodruff, Allison Gyle

Pointek, Sherry, and Kristen Garlock. "Creating a World Wide Web Resource Collection," GD34

"Points of View: ILL PRISM Transfer," J. Smith, IDA17

Poirier, Gayle. *See* Varughese, Lola

Polyvinyl Acetate Adhesives for Double-Fan Adhesive Binding: Report on a Review and Specification Study, R. J. Strauss and B. W. Ogden, HB1

"Postmodern Acquisitions," S. Propas and V. Reich, BA6

"Potential Collection Development Bias: Some Evidence on a Controversial Topic in California," D. Harmeyer, GD16

"The Potential of the Internet and Networks for Library Acquisitions," G. Dunshire, BC2

Powell, Allen. *See* Brooke, F. Dixon Jr.

Powell, John. *See* Franz, Lori

"A Practitioner's View of the Education of Catalogers," J. J. Reimer, CBA10

Preece, Barbara G., and Barbara Henigman. "Shared Authority Control: Governance and Training," CBF12

Prejudices and Antipathies: A Tract on the LC's Subject Headings Concerning People, S. Berman, ed., DCA1

PRESED-L, HAB14

PRESED-X: Preservation Education Exchange, HAB22

Preservation Activities in Canada: A Unifying Theme in a Decentralized Country, K. Turko, HD7

"Preservation Analysis and the Brittle Book Problem in College Libraries: The Identification of Research-Level Collections and Their Implications," J. Gertz et al., GE32, HI6

Preservation Appraisal of Sound Recordings Collections, S. Smolian, HJ15

Preservation Concerns in Construction and Remodeling of Libraries: Planning for Preservation, M. Trinkley, HG10

Preservation Education Directory, C. D. G. Coleman, comp., HM2

Preservation Film: Platform for Digital Access Systems, C. L. Jones, HH14

Preservation in Libraries: A Reader, R. Harvey, HA4

Preservation in Libraries: Principles, Strategies and Practices for Librarians, R. Harvey, HA5

Preservation in the Digital World, P. Conway, HH3

Preservation Microfilming: A Guide for Librarians and Archivists, L. L. Fox, HH9

"Preservation Needs of Oversized Illustrations in Geology Master's Theses," S. J. Scott, HC12

Preservation of Archival Materials: A Report of the Task Force on Archival Selection to the Commission on Preservation and Access, HJ9

Preservation of Archival Records: Holdings Maintenance at the National Archives, M. L. Ritzenthaler, HJ11

Preservation of Electronic Formats & Electronic Formats for Preservation, J. Mohlhenrich, HH19

Preservation of Library and Archival Materials: A Manual, S. Ogden, ed., HA7

Preservation Priorities in Latin America: A Report from the Sixtieth IFLA Meeting, D. C. Hazen, HD4

Preservation Research and Development: Round Table Proceedings, C. Beyer, ed., HA8

"Preservation Technologies: Photocopies, Microforms, and Digital Imaging—Pros and Cons," M. P. Trader, HH27

"Preservation Training for Library Users," S. S. Intner, HM5

Preserved to Serve: The Massachusetts Preservation Agenda, Massachusetts Task Force on Preservation and Access to the Library, Archives, Public Records, and Governmental Communities and the Citizens of Massachusetts, HD12

Preserving Archives and Manuscripts, M. L. Ritzenthaler, HJ12

Preserving Digital Information, Task Force on Archiving of Digital Information, HH26
Preserving Library Materials: A Manual, S. G. Swartzburg, HA10, IB7
Preserving Scientific Data on Our Physical Universe: A New Strategy for Archiving the Nation's Scientific Information Resources, HH20
Presley, Roger L. "Firing an Old Friend, Painful Decisions: The Ethics between Librarians and Vendors," BI5, FDC3
Price, Anna L., and Kjestine R. Carey. "Serials Use Study Raises Questions about Cooperative Ventures," GH6
Price, Lois Olcott. *Managing a Mold Invasion: Guidelines for Disaster Response,* HG6
"The Price of Materials and Collection Development in Larger Public Libraries," W. C. Robinson, GE17
Princeton University Libraries Technical Services Department, CAE7
Principles & Standards of Acquisitions Practice, ALCTS, Publisher/Vendor-Library Relations Committee, BI1
Proceedings of the New York State Seminar on Mass Deacidification, J. Gertz, ed., HC10
Proceedings of the Seminar on Cataloging Digital Documents, October 12–14, 1994, CCD18
"A Pro-Cite Authority File on a Network," T. Edelbute, CBF5
"Pro-Cite for Library and Archival Condition Surveys," D. Haynes, HI7
The Production and Bibliographic Control of Latin American Preservation Microforms in the United States, D. C. Hazen, HD5
Program for Cooperative Cataloging, CBI8
"Projecting Serials Costs: Banking on the Past to Buy for the Future," L. Ketcham and K. Born, FE10

Promis, Patricia. *See* Bosch, Stephen
"PromptCat: A Projected Service for Automatic Cataloging—Results of a Study at the Ohio State University Libraries," M. M. Rider, CBG6
Propas, Sharon W. "Ongoing Changes in Stanford University Libraries Technical Services," AHB8
———, and Vicky Reich. "Postmodern Acquisitions," BA6
Proposed American National Standard Format for Circulation Transactions Standards, NISO, ICI5
Proposed American National Standard Guidelines for Indexes and Related Information Retrieval Devices, NISO, EB4
Protecting Your Collections: A Manual of Archival Security, G. Trinkaus-Randall, HF4
"Providing Access to Online Information Resources: A Paper for Discussion," P. Caplan, CCD8
Providing Public Service to Remote Users, C. Haynes, ICI6
Provine, Rick E. *See* Brancolini, Kristine
Public & Access Services Quarterly, IAB3
Public Library Association, Policy Manual Committee. *PLA Handbook for Writers of Public Library Policies,* GC12
"Publishing in the International Marketplace," J. C. Germain, BH3, FCB5
Puglia, Steven. "Cost-Benefit Analysis for B/W Acetate: Cool/Cold Storage vs. Duplication," HG7

Q
QTECH Web, CAE8
"The Quality and Timeliness of Chinese and Japanese Monographic Records in the RLIN Database," J. H. Tsao, CBI14

"Quality in Technical Services: A User-Centered Definition for Future Information Environments," C. O. Frost, AHA6
"Quality of a National Bibliographic Service: In the Steps of John Whytefeld—An Admirable Cataloguer," P. Bryant, CBI3
Queens University Libraries. QTECH Web, CAE8
Quinn, Brian. "Guerrilla Collection Development: Time-Saving Tactics for Busy Librarians," GAE6
———. "Recent Theoretical Approaches in Classification and Indexing," DA8, EAB16
———. "Some Implications of the Canon Debate for Collection Development," GD3
Quinn, Frank. *See* Serebnick, Judith

R

Ragsdale, Kate W., and Donald J. Kenney, comps. *Effective Library Signage*, ICI2
Rare and Valuable Government Documents: A Resource Packet on Identification, Preservation and Security Issues for Government Documents Collections, J. Walsh et al., comps., HA11
Rasmussen, Lane. *See* Stelk, Roger Edward
"Rave New World: Librarians and Electronic Acquisitions," M. Nissley, BA5
Ray, Ron. "The Dis-Integrating Library System: Effects of New Technologies in Acquisitions," BG6
ReadiCat, FDD6
Readmore Webserver, FDD7
Reagor, Melinda Ann. *See* Marker, Rhonda
"Reality Check! Book Index Characteristics that Facilitate Information Access," E. D. Liddy and C. L. Jörgensen, EAB11

Recent Setbacks in Conservation, HAA6
"Recent Theoretical Approaches in Classification and Indexing," B. Quinn, DA8, EAB16
Recommended Practice for Microfilming Printed Newspapers on 35mm Roll Microfilm, AIIM, HN13
Recommended Practice for Quality Control of Image Scanners, AIIM, HN26
Records in Architectural Offices: Suggestion for the Organization, Storage and Conservation of Architectural Office Archives, N. C. Schrock and M. C. Cooper, HJ14
"Recruiting and Retention Revisited: A Study of Entry Level Catalogers," D. Callahan and J. MacLeod, CBA2
Recruiting, Educating, and Training Librarians for Collection Development, P. Johnson and S. S. Intner, eds., GB8
Redefining Film Preservation: A National Plan, A. Melville and S. Simmon, HJ10
"Redesigning Technical Services Work Areas for the 21st Century," K. A. Wilson, AHA14
Reed, Lawrence L., and Rodney Erikson. "Weeding: A Quantitative and Qualitative Approach," GE31
Reed-Scott, Jutta. *See* Brennan, Patricia
Reference & Adult Services Division, Collection Development & Evaluation Section, Collection Development Policies Committee, "The Relevance of Collection Development Policies: Definition, Necessity, and Applications," GC2
———, Management of Public Services Section. "Guidelines and Procedures for Telefacsimile and Electronic Delivery of Interlibrary Loan Requests," IDD2

———, ———. "National Interlibrary Loan Code for the United States 1993," IDD3

Reference & Adult Services Division. *See also* American Library Association, Subcommittee on Guide for Training Collection Development Librarians

Reference and User Services Association, Management and Operation of Public Services Section, Interlibrary Loan Committee, IAC1

"Reflecting the Maturation of a Profession: Thirty-Five Years of *Library Resources & Technical Services*," R. P. Smiraglia and G. H. Leazer, AD4

Register of Indexers Available, EAC2

"The Regulation of Hate Speech on College Campuses and the 'Library Bill of Rights,'" R. V. Labaree, GD17

Reich, Vicky. "A Future for Technical Services," AA13

———. "Marketing to Libraries: What Works and What Doesn't," FBB4

———, Connie Brooks, Willy Cromwell, and Scott Wicks. "Electronic Discussion Lists and Journals: A Guide for Technical Services Staff," AC1, HA9

Reich, Vicky. *See also* Propas, Sharon W.

Reichman, Henry. *Censorship and Selection: Issues and Answers for Schools*, GD20

Reid, Marion T. "Closing the Loop: How Did We Get Here and Where Are We Going?," BF3

Reilly, James M., Douglas W. Nishimura, and Edward Zinn. *New Tools for Preservation: Assessing Long-Term Environmental Effects on Library and Archives Collections*, HG8

Reimer, John J. "National Series Authority File Derailed?," CBF13

———. "A Practitioner's View of the Education of Catalogers," CBA10

Rein, Laura O., et al. "Formula-Based Subject Allocation: A Practical Approach," GF3

Reinke, Christa Easton. "Beyond the Fringe: Administratively Decentralized Collections at the University of Michigan," GD52

Reinke, Christa Easton. *See also* Keating, Lawrence R. II

"Reinvent Catalogers!" E. J. Waite, CBG7

"The Relationship of Acquisitions to Resource Sharing: An Informal Analysis," S. S. Intner, BG4

"The Relevance of Collection Development Policies: Definition, Necessity, and Applications" Collection Development Policies Committee, RASD, CODES Collection Development Policies Committee, GC2

"Remote Storage Facilities: An Annotated Bibliography," P. O'Connor, IB4

"Reorganization of Technical Services Staff in the 90s," K. A. Wilson, AHC9

"Reorganizing Acquisitions at the Pennsylvania State University Libraries: From Work Units to Teams," N. M. Marke, and L. B. Brown, BG9

"Le Répertoire de vedettes-matière de la Bibliothèque de l'Université Laval: sa génè et son évolution," P. Gascon, EE7

Répertoire de vedettes-matière (RVM), DCC5

"Representational Predictability: Key to the Resolution of Several Pending Issues in Indexing and Information Supply," R. Fugmann, EAB5

Reproduction of Copyrighted Works by Educators and Librarians, HE4

"The Re-Professionalization of Cataloging," S. S. Intner, CBA6

Research Libraries Group, Inc., CAE20, IDE5

Reser, David W., comp. *Towards a New Beginning in Cooperative Cataloging: The History, Progress, and Future of the Cooperative Cataloging Council*, CBI13

"The Reserve Department Revisited: A Study of Faculty and Student Use of the Capital University Library Reserve Department," S. L. Hupp, IC6

"Reserve On-line: Bringing Reserve into the Electronic Age," H. R. Enssle, IC4

"Reshaping the Serials Vendor Industry," D. Tonkery, FBB5

Resolution as It Relates to Photographic and Electronic Imaging, AIIM, HN22

"Resource Sharing in the Electronic Era: Potentials and Paradoxes," G. Dannelly, GH2

The Responsive Public Library Collection: How to Develop and Market It, S. L. Baker, GE14

Retrospective Conversion: History, Approaches, Considerations, B. Schottlaender, AID7

Rettig, James. *Rettig on Reference*, GAD4

Rettig on Reference, J. Rettig, GAD4

"Revealing UnCover: Simple, Easy Article Delivery," D. F. Kohl, IE12

Rhode Island Council for the Preservation of Research Resources. *Bricks and Mortar for the Mind: Statewide Preservation Program for Rhode Island*, HD16

Rhyne, Charles S. *Computer Images for Research, Teaching, and Publication in Art History and Related Disciplines*, HH21

"Richard Abel," D. Biblarz, BB2

Richter, Linda, and Joan Roca. "An X12 Implementation in Serials: MSUS/PALS and Faxon," FDF8

Riddick, John F. "An Electrifying Year: A Year's Work in Serials, 1992," FAC6

Rider, Mary M. "PromptCat: A Projected Service for Automatic Cataloging—Results of a Study at the Ohio State University Libraries," CBG6

Ridley, M. J. "An Expert System for Quality Control and Duplicate Detection in Bibliographic Databases," AIB4

Rinaldo, Constance. *See* Defilice, Barbara

Ritzenthaler, Mary Lynn. *Preservation of Archival Records: Holdings Maintenance at the National Archives*, HJ11

———. *Preserving Archives and Manuscripts*, HJ12

RLG Archives Microfilming Manual, N. E. Elkington, ed., HH7

RLG Digital Image Access Project: Proceedings from an RLG Symposium, P. A. McClung, ed., HH22

"RLIN CJK and East Asian Library Community," H. Kaneko, CBI5

Robertson, Michael. "Foreign Concepts: Indexing and Indexes on the Continent," EAB17

Robinson, Peter. *The Digitization of Primary Textual Sources*, HH23

Robinson, William C. "Academic Library Collection Development and Management Positions: Announcements in *College and Research Libraries News* from 1980–1991," GB4

———. "The Price of Materials and Collection Development in Larger Public Libraries," GE17

Roca, Joan. *See* Richter, Linda

Roche, Marilyn M. *ARL/RLG Interlibrary Loan Cost Study: A Joint Effort by the Association of Research Libraries and the Research Libraries Group*, IE19

Rogers, Margaret N. "Are We on Equal Terms Yet? Subject Headings Concerning Women in *LCSH,* 1975–1991," DCB8

Rogers, Sally. *See* O'Neill, Edward T.

Rogers, Shelley L. "Technical Services Librarians in the 21st Century," AHA10

"The Role of the *Art & Architecture Thesaurus* in Communicating about Visual Art," P. Molholt and T. Petersen, DCE2

"The Role of the Collection Development Librarian in the 90s and Beyond," M. Otero-Boisvert, GAE5

"The Role of Training in the Reorganization of Cataloging Services," M. E. Clack, CBB2

Romero, Lisa. "Original Cataloging Errors: A Comparison of Errors Found in Entry-Level Cataloging with Errors Found in OCLC and RLIN," CBI9

Roosa, Mark, and Jane Gottlieb, comps. *Knowing the Score: Preserving Collections of Music,* HJ5

Rosenblatt, Lisa A. "Collection Development Guidelines for Selective Federal Depository Libraries," GE12

Ross, Rosemary E. "A Comparison of OCLC and WLN Hit Rates for Monographs and an Analysis of the Types of Records Retrieved," CBI10

Rossol, Monona. *The Artist's Complete Health and Safety Guide,* HC11

Rossol, Monona. *See also* Shaw, Susan D.

Rothenberg, Jeff. "Ensuring the Longevity of Digital Documents," HH24

Rouzer, Steven M. "A Firm Order Vendor Evaluation Using a Stratified Sample," BF4

Rowley, Gordon, comp. *Organization of Collection Development,* GB3

Rowley, Jennifer. *Organizing Knowledge,* EAB18

Ruschoff, Carlen. "Perspectives on the Cataloging Literature: Cataloging's Prospects: Responding to Austerity with Innovation," CBD2

Russell, Gordon, and Karin den Beyker. "Technical Services/Public Services: New Wine in Old Bottles," AHC7

Rutstein, Joel S., Anna L. DeMiller, and Elizabeth A. Fuseler. "Ownership Versus Access: Shifting Perspectives for Libraries," GE5

Rutstein, Joel S. *See also* Cochenour, Donnice

S

Sack, Nancy. *See* Fiscella, Joan B.

Safe Handling, Storage, and Destruction of Nitrate-Based Motion Picture Films, HJ13

SAFETY, IAC10

Saffady, William. "Automated Acquisitions and Serials Control," FDF9

———. "The Bibliographic Utilities in 1993: A Survey of Cataloging Support and Other Services," AIB5, CBI11

———. "Integrated Library Systems for Minicomputers and Mainframes: A Vendor Study," AID6

St. Lifer, Evan, et al. "How Safe Are Our Libraries?," IF4

Salminen, Airi, Jean Tague-Sutcliffe, and Charles McClellan. "From Text to Hypertext by Indexing," EAB19

Sam, Sherrie, and Jean A. Major. "Compact Shelving of Circulating Collections," IB5

Samuelson, Pamela. "Copyright and Digital Libraries," IAA8

Sapp, Gregg, ed. *Access Services in Libraries: New Solutions for Collection Management,* IA2

Sassé, Margo. *See* Bush, Carmel C.

Satija, Mohinder P. *See* Chan, Lois Mai

Saunders, Laverna M. "Transforming Acquisitions to Support Virtual Libraries," BG7

Saunders, Laverna M. *See also* Parang, Elizabeth

Saunders, Richard. "Collection- or Archival-Level Description for Monographic Collections," CBE8

SAVE-IT, IDE6

SAVEIT-L, IAC11

Savic, Dusko. "Designing an Expert System for Classifying Office Documents," EAB20

Saving the Past to Enrich the Future: A Plan for Preserving Information Resources in Kansas, K. L. Walter, HD18

"Saving Those Historic Videotapes: It May Already Be Too Late," J. W. Palmer, HJ7

Saye, Jerry D. *See* Haynes, Kathleen J. M.

Sayles, Jeremy. "The Textbooks-in-College-Libraries Mystery," GD53

Schaffner, Ann C., ed. "Perspectives on the Future of Union Listing," FF2

Schellinger, Merry. *See* Kiegel, Joseph

Schenck, William Z. *See* Johns, Cecily

Schimizzi, Anthony J. "The Library of Congress and the 'Bibliography' Note," CCB7

Schmidt, Diane E., Elisabeth B. Davis, and Rudy Jahr. "Biology Journal Use at an Academic Library: A Comparison of Use Studies," GD44

Schmidt, Karen A. *See* Chrzastowski, Tina E.

"Scholar or Librarian? How Academic Libraries' Dualistic Concept of the Bibliographer Affects Recruitment," J. Haar, GB2

"School Library Media Centers and Networks," P. J. Van Order and A. W. Wilkes, GE22

"School Media Center and Public Library Collections and the High School Curriculum," C. A. Doll, GE20

Schottlaender, Brian. *Retrospective Conversion: History, Approaches, Considerations*, AID7

Schrock, Nancy Carlson, and Mary Campbell Cooper. *Records in Architectural Offices: Suggestion for the Organization, Storage and Conservation of Architectural Office Archives*, HJ14

Schultz, Lois Massengale. *A Beginner's Guide to Copy Cataloging on OCLC/Prism*, CBE9

Schuneman, Anita, and Deborah A. Mohr. "Team Cataloging in Academic Libraries: An Exploratory Survey," CBD3

Schuyler, Michael. *OPAC Directory: An Annual Guide to OnLine Public Access Catalogs and Databases*, IDC5

Schwartz, Werner. *The European Register of Microform Masters—Supporting International Cooperation*, HK2

"Sci/Tech Book Approval Plans Can Be Effective," H. L. Franklin, BB5

Scott, Mona L., and Christine E. Alvey. *Conversion Tables: LCC-Dewey, Dewey-LCC*, DBA6

Scott, Patricia A. *See* Paries, Cindy

Scott, Sally J. "Preservation Needs of Oversized Illustrations in Geology Master's Theses," HC12

Seaman, Scott. "Impact of Basic Books v. Kinko's Graphics on Reserve Services at the University of Colorado, Boulder," IC10

"Searching for the Human Good: Some Suggestions for a Code of Ethics for Technical Services," E. G. Bierbaum, AA2

"Searching the World-Wide Web: Lycos, WebCrawler and More," G. R. Notess, EF11

Sebera, Donald K. *Isoperms: An Environmental Management Tool*, HG9

Security and Crime Prevention in Libraries, M. Chaney and I. MacDougall, IF2

"Seeking the 99% Chemistry Library: Extending the Serials Collection through the Use of Decentralized Document Delivery," T. E. Chrzastowski and M. A. Anthes, GD37

"Selecting a Preservation Photocopy Machine," D. W. Wright, HL3

"Selecting Electronic Journals," H. Grochmal, GD41

"Selecting Microfilm for Digital Preservation: A Case Study from Project Open Book," P. Conway, HH4

Selection and Evaluation of Electronic Resources, G. K. Dickinson, GD27

"Selection Criteria for Internet Resources," R. Cassel, GD24

"Selection for Preservation: A Digital Solution for Illustrated Texts," J. Gertz, HH11

"Selection of General Collection Materials for Transfer to Special Collection," ACRL Rare Books and Manuscripts Section, GE28

Selection of Library Materials for Area Studies: Part II. Australia, Canada, and New Zealand, C. Johns and W. Z. Schenck, eds., GE9

Sellers, Minna, and Joan Beam. "Subsidizing Unmediated Document Delivery: Current Models and a Case Study," IE20

Senkevitch, Judith J., and James H. Sweetland. "Evaluating Adult Fiction in the Smaller Public Library," GG6

Sercan, Cecilia S. "Where Has All the Copy Gone: Latin American Imprints in the RLIN Database," CBI12

Serebnick, Judith, and Frank Quinn. "Measuring Diversity of Opinion in Public Library Collections," GD21

"Serial Cancellations and Interlibrary Loan: The Link and What It Reveals," M. J. Crump and L. Freund, GE34

SERIALIST, FAE1

Serials Acquisitions Glossary, ALCTS, Serials Section, FAA2

Serials Canada: Aspects of Serials Work in Canadian Libraries, W. Jones, ed., FBA5

"Serials Citations and Holdings Correlation," J. C. Calhoun, GD36

Serials Control Systems for Libraries, T. Davis and J. Huesmann, FDF4

Serials in Cyberspace: Collections, Resources, and Services, B. MacLennan, FDA2

"Serials in Strategic Planning and Reorganization," E. D. Ten Have, FBA6

Serials Management: A Practical Guide, C. D. Chen, FAA1

Serials Management in Australia and New Zealand: Profile of Excellence, T. Burrows and P. G. Kent, FBA2

Serials Publishing and Acquisitions in Australia, A. Gans, FCA1

Serials to the Tenth Power: Tradition, Technology, and Transition, M. A. Sheble and B. Holley, FAF5

"Serials Use Study Raises Questions about Cooperative Ventures," A. L. Price and K. R. Carey, GH6

"Serials *vs.* the Dollar Dilemma: Currency Swings and Rising Costs Play Havoc with Prices," L. Ketcham and K. Born, FE11

"Service Quality: An Unobtrusive Investigation of Interlibrary Loan in Large Public Libraries," F. Herbert, IDA7

Serving the Difficult Customer: A How-to-Do-It Manual for Library Staff, K. Smith, IA3

Sessions, Judith, et al. "OhioLINK Inter-Institutional Lending Online: The Miami University Experience," IDA15

The Setup Phase of Project Open Book: A Report to the Commission on Preservation and Access on the Status of an Effort to Convert Microfilm to Digital Imagery, P. Conway and S. Weaver, HH5

Sha, Vianne Tang. *Internet Resources for Cataloging*, CAA4, DAB3

Shapiro, Beth. *See* Bosseau, Don L.

"Shared Authority Control: Governance and Training," B. G. Preece and B. Henigman, CBF12

Sharp, S. Celine. "A Library Shelver's Performance Evaluation as It Relates to Reshelving Accuracy," IB6

Shaw, Susan D., and Monona Rossol. *Overexposure: Health Hazards in Photography*, HC13

Sheble, Mary Ann, and Beth Holley. *Serials to the Tenth Power: Tradition, Technology, and Transition*, FAF5

Sheble, Mary Ann. *See also* Holley, Beth

Sheehy, Carolyn A., ed. *Managing Performing Arts Collections in Academic and Public Libraries*, GE13

Sherayko, Carolyn C., ed. *Cataloging Government Publications Online*, CCD27

Shirk, Gary M. "Contract Acquisitions: Change, Technology, and the New Library/Vendor Partnership," BG8

"The Shrinking National Collection: A Study of the Effects of the Diversion of Funds from Monographs to Serials on the Monograph Collections of Research Libraries," A. H. Perrault, GD51

"The Shrink-Wrap Project at Rutgers University Special Collections and Archives," J. Stagnitto, HI15

Silberman, Richard M. "A Mandate for Change in the Library Environment," AHA11

Silverman, Randy, and Maria Gradinette, eds. *The Changing Role of Book Repair in ARL Libraries*, HC2

Silverman, Randy. *See also* Gradinette, Maria

Silvester, Elizabeth V. *See* McGill University Libraries

Simmon, Scott. *See* Melville, Annette

"SISAC Item Contribution Identifier: New SISAC Code," FDF10

Sistema de Clasificacion Decimal Dewey, M. Dewey, DBB4

Sistrunk, Wendy. *See* Fons, Theodore A.

Sitkin, Ann. *See* Mumm, James A.

Slide Management in Libraries and Information Units, G. Sutcliffe, GE24

"Slow Revolution: The Electronic AACR2," J. K. Duke, CCA2

Smiraglia, Richard P., and Gregory H. Leazer. "Reflecting the Maturation of a Profession: Thirty-Five Years of *Library Resources & Technical Services*," AD4

Smith, Eldred, and Peggy Johnson. "How to Survive the Present While Preparing for the Future: A Research Library Strategy," GH7

Smith, Jane. "Electronic ILL at Colorado State University," IDA16

———. "Points of View: ILL PRISM Transfer," IDA17

Smith, Kitty. *Serving the Difficult Customer: A How-to-Do-It Manual for Library Staff*, IA3

Smith, Merrily A. *Managing the Preservation of Serial Literature: An International Symposium*, HI11

Smith, Patricia. *See* Bush, Carmel C.
Smith, Stephen J. "Cataloging with Copy: Methods for Increasing Productivity," CBE10
Smith, Terry. "Training Technical Services OCLC Users," AHA12
Smolian, Steven. *Preservation Appraisal of Sound Recordings Collections,* HJ15
Snowbird Leadership Institute, AGB6
Soergel, Dagobert. "Indexing and Retrieval Performance: The Logical Evidence," EAB21
"Some Implications of the Canon Debate for Collection Development," B. Quinn, GD3
"Some Indexing Decisions in the Cambridge Encyclopedia Family," D. Crystal, EF5
"Some Issues in the Indexing of Images," S. S. Layne, ED1
Sorury, Kathryn. *See* Byrd, Jacqueline
South Carolina. State Historical Records Advisory Board. *Palmetto Reflections: A Plan for South Carolina's Documentary Heritage,* HD17
"Spanish and Portuguese Online Cataloging: Where Do You Start from Scratch?" C. Erbolato-Ramsey and M. L. Grover, CBE3
"Speaking in Tongues: Communications between Technical Services and Public Services in an Online Environment," S. Beehler and P. G. Court, AA1
Special Libraries Association Annual Conferences, AGB7
"Special Section: Fair Use and Copyright," S. Bennett et al., IAA2
"Special Section on Outsourcing," J. Obgurn, AHB7
Spohrer, James H. *See* Lehmann, Stephen
Spornick, Charles D. *See* Greene, Robert J.

"Staff Involvement: The Key to the Successful Merger of Monograph and Serial Acquisitions Functions at Clemson University Library," D. L. Astle, BG1
"Staffing: The Art of Managing Change," N. M. Cline, GB1
Stagnitto, Janice. "The Shrink-Wrap Project at Rutgers University Special Collections and Archives," HI15
Standard for Information and Image Management: Microfiche, AIIM, HN14
Standard for Information and Image Management; Microfilm Package Labeling, AIIM, HN15
Standard for Information and Image Management: Micrographics—Splices for Image Microfilm—Dimensions and Operational Constraints, AIIM, HN16
Standard Recommended Practice—File Format for Storage and Exchange of Images—Bi-level Image File Format, Part 1, AIIM, HN17
Standard Recommended Practice: Identification of Microforms, AIIM, HN18
Standard Recommended Practice: Monitoring Image Quality of Roll Microfilm and Microfiche Scanners, AIIM, HN27
Standards for Archival Description, A Handbook: Information Systems, Data Exchange, Cataloging, Finding Aids, Authority Control, Editing and Publishing, Statistics, V. I. Walch, comp., CCD3
"Standards for College Libraries, 1995 Edition," ACRL, GAC2
"Standards for Community, Junior, and Technical College Learning Resources Programs," AECT and ACRL, GAC3
Standards for Electronic Imaging Technologies, Devices, and Systems, HN23

"Standards for Indexing: Revising the American National Standard Guidelines Z39.4," J. D. Anderson, EB2

Stankus, Tony. "Looking Trojan Gift Horses in the Mouth: Are Special Issues Special Enough to Pay Extra Money for and Bind?," FCA2

Stanley, Nancy Marke, and Lynne Branche Brown. "Reorganizing Acquisitions at the Pennsylvania State University Libraries: From Work Units to Teams," BG9

Stark, Marcella. *See* Courtois, Martin P.

Stelk, Roger Edward, Paul Metz, and Lane Rasmussen. "Departmental Profiles: A Collection Development Aid," GD54

Stierman, Jeanne Koekkoek. *See* Joswick, Kathleen E.

Stierman, John P. *See* Joswick, Kathleen E.

Stone, Alva T. "That Elusive Concept of 'Aboutness': The Year's Work in Subject Analysis, 1992," DAD2

Stop Thief! Strategies for Keeping Your Collections from Disappearing, IF5

"Strategies for Reducing Billable OCLC Searches Used in Cataloging," S. L. Tsui, CBE11

Strategies to Preserve Michigan's Historical Records, Michigan. State Historical Records Advisory Board, HD13

Strauss, Robert J., and Barkley W. Ogden. *Polyvinyl Acetate Adhesives for Double-Fan Adhesive Binding: Report on a Review and Specification Study*, HB1

Studwell, William E. "A Tale of Two Decades, or, The Decline of the Fortunes of LC Subject Headings," EE15

———. "Ten Years after the Question: Has There Been an Answer?," DCA5

———. "Who Killed the Subject Code?," EC9

"A Study of Collection Development Personnel Training and Evaluation in Academic Libraries," M. F. Casserly and J. L. Hegg, GB7

"A Study of OP and OSI Cancellations in an Academic Library," M. H. Lu, BH4

Study on the Long-term Retention of Selected Scientific and Technical Records of the Federal Government: Working Papers, HH25

Subcommittee on Contract Negotiations for Commercial Reproduction of Library & Archival Materials. "Contract Negotiations for the Commercial Microforms Publishing of Library and Archival Materials: Guidelines for Librarians and Archivists," HK3

"Subject Access Redefined: How New Technology Changes the Conception of Subject Representation," I. Wormell, EAB26

"Subject Access to African American Studies Resources," D. H. Clack, DCB4

"Subject Access to Individual Works of Fiction: Participating in the OCLC/Fiction Project," N. Down, DA5

Subject Analysis in Online Catalogs, R. Aluri et al., DA1

"Subject Analysis: The Critical First Stage in Indexing," C. M. Chu and A. O'Brien, EAB3

"Subject Authority Control on the World of the Internet, Part 1," M. Micco, DD1

"Subject Authority Control on the World of the Internet, Part 2," M. Micco, DD2

"Subject Enrichment Using Contents or Index Terms: The Australian Defense Force Academy Experience," S. Beatty, DA2

Subject Headings for Children: A List of Subject Headings Used by the Library of Congress with Dewey Numbers Added, L. Winkel, ed., DCC6

"Subject Searching in Two Online Catalogs: Authority Control vs. Non-Authority Control," A. Wilkes and A. Nelson, DCD1

"Subjects of Concern: Selected Examples Illustrating Problems Affecting Information Retrieval on Iran and Related Subjects Using *LCSH*," D. Gitistan, DCB6

"Subsidizing Unmediated Document Delivery: Current Models and a Case Study," M. Sellers and J. Beam, IE20

Sugnet, Chris. *See* Bosch, Stephen

Sukiasyan, Eduard. "Information-Retrieval Systems: Systems Analysis of Problems of Quality Management," EAB22

"Support Strategies for Interactive Thesaurus Navigation," S. Jones and M. Hancock-Beaulieu, EE9

A Survey of STM Online Journals 1990–1995: The Calm before the Storm, S. Hitchcock et al., FAC3

"Surveying the Damage: Academic Library Serial Cancellations 1987–88 through 1989–90," T. E. Chrzastowski and K. A. Schmidt, GE33

Sutcliffe, Glyn. *Slide Management in Libraries and Information Units,* GE24

Svenonius, Elaine. "Access to Nonbook Materials: The Limits of Subject Indexing for Visual and Aural Languages," ED5

Swartzburg, Susan G. *Preserving Library Materials: A Manual,* HA10, IB7

Sweeney, Russell. "The International Use of the *Dewey Decimal Classification,*" DBB1

Sweetland, James H., "Adult Fiction in Medium-Sized U.S. Public Libraries: A Survey," GE18

———, and Peter G. Christensen. "Gay, Lesbian, and Bisexual Titles: Their Treatment in the Review Media and Their Selection by Libraries," GD9

Sweetland, James H. *See also* Senkevitch, Judith J.

Swindler, Luke. *See* Dominguez, Patricia Buck

Sylvia, Margaret, and Marcella Lesher. "Making Hard Choices: Canceling Print Indexes," GE19

———. "What Journals Do Psychology Graduate Students Need? A Citation Analysis of Thesis References," GD45

"Systematic Reference Weeding: A Workable Model," K. E. Joswick and J. P. Stierman, GE29

"Systems Thinking about Acquisitions and Serials Issues and Trends: A Report on the 1993 Charleston Conference," J. L. Flowers, FAF1

T

Tagler, John. *See* Fisher, Janet H.

Tague-Sutcliffe, Jean. *See* Salminen, Airi

"Taking Another Look at Minimal Level Cataloging," S. S. Intner, CBE6

"Taking the Lead: Catalogers Can't Be Wallflowers," K. L. Horny, CBA5

"A Tale of Two Decades, or, The Decline of the Fortunes of LC Subject Headings," W. E. Studwell, EE15

Task Force on Archiving of Digital Information. *Preserving Digital Information,* HH26

Taylor, Arlene G. "The Information Universe: Will We Have Chaos or Control?," AA14

———. "On the Subject of Subjects," DA9

Taylor, Arlene G. *See also* Giral, Angela

"Team Cataloging in Academic Libraries: An Exploratory Survey," A. Schuneman and D. A. Mohr, CBD3

Technical Processing Online Tools (TPOT). University of California at San Diego Libraries, CAE9, FDA3

"Technical Services," R. S. Karp, AD3

"Technical Services and the Electronic Library: Defining Our Roles and Divining the Partnership," J. McCue with D. Zhang, AID5

"Technical Services and the Faculty Client in the Digital Age," C. C. Hunt, AHA8

"Technical Services and the Internet," P. S. McCoy, AA10

"Technical Services Functionality in Integrated Library Systems," R. W. Boss, AID2

"Technical Services Librarians in the 21st Century," S. L. Rogers, AHA10

"A Technical Services Perspective of Implementing an Organizational Review while Simultaneously Installing an Integrated Library System," E. L. Cook and P. Farthing, AHC3, FBA3

"Technical Services Productivity Alternatives," M. L. Johnson, AHB6

"Technical Services/Public Services: New Wine in Old Bottles," G. Russell and K. den Beyker, AHC7

"Technical Services Report," B. B. Baker, ed., AG1

Technical Services Workstations, J. M. Brugger et al., comps., AIC4

"Technical Services Workstations Improve Productivity," M. Kaplan, AIC2

Technicalities, AF4

"Technology and Library Organizational Structure," J. J. Williams, AHC8

Tehrani, Farideh, ed. "Library Security: Special Issue," IF6

"Telecommuting for Original Cataloging at the Michigan State University Libraries," L. Black and C. Hyslop, CBD1

Ten Have, Elizabeth Davis. "Serials in Strategic Planning and Reorganization," FBA6

"Ten Years after the Question: Has There Been an Answer?" W. E. Studwell, DCA5

"Ten Years of Preservation in New York State: The Comprehensive Research Libraries," J. Gertz, HD1

Tennant, Tracy. *See* Williamson, Nancy J.

"Test Reports on 15 Photocopiers," Buyers Laboratory, Inc., HL2

A Testbed for Advancing the Role of Digital Technologies for Library Preservation and Access: Final Report by Cornell University to the Commission on Preservation and Access, A. R. Kenney and L. K. Personius, HH17

"The Textbooks-in-College-Libraries Mystery," J. Sayles, GD53

"That Elusive Concept of 'Aboutness': The Year's Work in Subject Analysis, 1992," A. T. Stone, DAD2

Thawley, John, and Philip G. Kent, eds. *Amalgamations & the Centralisation of Technical Services: Profit or Loss*, AHA13

Theodore-Shusta, Eileen. *See* Pierce, Darlene M.

Le Thésaurus: Conception, elaboration, gestion, M. Hudon, DCE7

Thesaurus for Graphic Materials, Library of Congress, Prints and Photographs Division, EE12

Thomas, Alan. *Classification: Options and Opportunities*, DBA3

Thornton, Ann. *See* Bonario, Steve

Thurlow, H. "LCSH Online at the State Library of Queensland," DCB1

Tibbits, Edie. "Binding Conventions for Music Materials," HB2

Tibbo, Helen R. "The Epic Struggle: Subject Retrieval from Large Bibliographic Databases," EC10
_____. "Indexing for Information," EAB23
Tillotson, Joy. "Is Keyword Searching the Answer?," EC11
Tjoumas, Renee. "Native American Literature for Young People: A Survey of Collection Development Methods in Public Libraries," GD10
Tjoumas, Renee. *See also* Blake, Virgil L. P.
To Outwit Time: Preserving Materials in Ohio's Libraries and Archives, Ohio Historical Records Advisory Board, HD14
To Save Our Past: A Strategic Plan for Preserving Oklahoma's Documentary Heritage, Oklahoma Historical Records Advisory Board, HD15
Tonkery, Dan. "Reshaping the Serials Vendor Industry," FBB5
"A Tool to Guide the Logical Process of Conceptual Structuring," G. Massimiliano and G. Negrini, EE13
Tools for Serials Catalogers: A Collection of Useful Sites and Sources, CCD21
Topics in Photographic Preservation, HAA7
"Toward a Code of Ethics for Acquisitions Librarians," B. Dean, BI3
"Toward a New World Order: A Survey of Outsourcing Capabilities of Vendors for Acquisitions, Cataloging and Collection Development Services," C. C. Bush et al., AHB1
"Toward a Theory of Collection Development," D. P. Carrigan, GAE2
Towards a New Beginning in Cooperative Cataloging: The History, Progress, and Future of the Cooperative Cataloging Council, D. W. Reser, comp., CBI13

"Towards Finding More Catalog Copy: The Possibility of Using OCLC and the Internet to Supplement RLIN Searching," J. D. LeBlanc, CBE7
Toyofuku, Anthony, and Colby Riggs. "The Ant*ill: Using Perl to Automate the ILL Lending Process," IDA18
Trader, Margaret P. "Preservation Technologies: Photocopies, Microforms, and Digital Imaging—Pros and Cons," HH27
Training Catalogers in the Electronic Era: Essential Elements of a Training Program for Entry-Level Professional Catalogers, CBB4
"Training Opportunities for Interlibrary Loan and Document Supply Staff," G. Cornish, IDA5
"Training Staff for Preservation," S. S. Intner, HM6
"Training Technical Services OCLC Users," T. Smith, AHA12
Transborder Interlibrary Loan: Shipping Interlibrary Loan Materials from the U.S. to Canada, ARL, IDD4
"Transforming Acquisitions to Support Virtual Libraries," L. M. Saunders, BG7
Trinkaus-Randall, Gregor. *Protecting Your Collections: A Manual of Archival Security*, HF4
Trinkley, Michael. *Preservation Concerns in Construction and Remodeling of Libraries: Planning for Preservation*, HG10
Tsao, Jai-hsya. "The Quality and Timeliness of Chinese and Japanese Monographic Records in the RLIN Database," CBI14
Tsui, Susan L., "Periodical Records Conversion: From Union List to Statewide Network," FF3
_____. "Strategies for Reducing Billable OCLC Searches Used in Cataloging," CBE11
Tucker, Betty E. "The Journal Deselection Project: The LSUMC-S Experience," GE36

333

Turko, Karen. *Preservation Activities in Canada: A Unifying Theme in a Decentralized Country*, HD7
Turner, Anne. *It Comes with the Territory: Handling Problem Situations in Libraries*, IF7
Turner, Fay. "Document Ordering Standards: The ILL Protocol and Z39.50 Item Order," IDD6
Turner, J. "Indexing Film and Video Images for Storage and Retrieval," ED6
Tuten, Jane H., and Beverly Jones, comps. *Allocation Formulas in Academic Libraries*, GF4
Tutorial: Digital Resolution Requirements for Replacing Text-Based Material: Methods for Benchmarking Image Quality, A. R. Kenney and S. Chapman, HH16
Tuttle, Craig A. *An Ounce of Preservation: A Guide to the Care of Papers and Photographs*, HJ16

U

"U.K. Books and Their U.S. Imprints: A Cost and Duplication Study," B. Kruger, BB6
U.S. Copyright Office, Library of Congress, IAA9
"U.S. Periodical Price Index for 1993," A. W. Alexander and K. H. Carpenter, FE4
"U.S. Periodical Price Index for 1994," A. W. Alexander and K. H. Carpenter, FE5
"U.S. Periodical Price Index for 1995," A. W. Alexander and K. H. Carpenter, FE6
"Undifferentiated Names: A Cataloging Rule Overlooked by Catalogers, Reference Librarians, and Library Users," J. C. Lin, CBF8
"United Kingdom Approval Plans and United States Academic Libraries: Are They Necessary and Cost Effective?," M. Eldredge, BB4
University of California at San Diego Libraries. Technical Processing Online Tools (TPOT), CAE9, FDA3
University of Miami Libraries, Acquisitions Department, BAD6
University of North Carolina–Chapel Hill, Acquisitions Department, BAD7
University of Virginia Cataloging Services Department, CAE10
University of Washington Libraries, Acquisitions Division, BAD8
Urbanski, Verna, with B. C. Chang and B. L. Karon. *Cataloging Unpublished Nonprint Materials: A Manual of Suggestions, Comments and Examples*, CCD25
"Use-Based Selection for Preservation Microfilming," P. DeStefano, HI5
"Use of Faculty Publication Lists and ISI Citation Data to Identify a Core List of Journals with Local Importance," J. Hughes, GD42
"User-Centered Indexing," R. Fidel, EAB4
"A User-Centered View of Document Delivery and Interlibrary Loan," H. S. Martin and C. L. Kendrick, IE15
"User Satisfaction and Interlibrary Loan Service: A Study at Louisiana State University," A. H. Perrault and M. Arseneau, IDA14
"User Views of Compact Shelving in an Open Access Library," D. East, IB3
Uses of Document Delivery Services, M. E. Jackson and K. Croneis, comps., IE21
"Using Article Photocopy Data in Bibliographic Models for Journal Collection Management," M. D. Cooper and G. F. McGregor, GD39
"Using SPSS/PC+ and NOTIS Downloaded Files of Current Subscription Records at the University of Pittsburgh," R. C. Carter and P. B. Kohnberger, FDF2

Using Subject Headings for Online Retrieval: Theory and Practice and Potential, K. M. Drabenstott and D. Vizine-Goetz, DCA2
"Using the Internet in Serials Management," M. Kascus and F. Merriman, FDE2
"Using the OCLC/AMIGO Collection Analysis Compact Disk to Evaluate Art and Art History Collections," M. Findley, GG10
USMARC Format for Authority Data, Update No. 1, DCA6
USMARC Format for Bibliographic Data, Update No. 1, CCE3
USMARC Format for Classification Data, Update No. 1, DBA9

V

"The Value-Added Acquisitions Librarian: Defining Our Role in a Time of Change," A. Bloss, BA2
Van Bogart, John W. C. *Magnetic Tape Storage and Handling: A Guide for Libraries and Archives*, HJ17
———. *Media Stability Studies: Final Report*, HJ18
Van Houwelling, Douglas E., and Michael J. McGill. *The Evolving National Information Network: Background and Challenges*, HH28
Van Orden, Phyllis J., and Adeline W. Wilkes. "School Library Media Centers and Networks," GE22
"Variation in Place of Publication: A Model for Cataloging Simplification," R. Marker and M. A. Reagor, CCB4
Varner, Carroll H. "Outsourcing Library Production: The Leader's Role," AHB9
Varughese, Lola, and Gayle Poirier. "A Brief Survey of ARL Libraries' Cataloging of Instructional Materials," CCD24
"Vendor Evaluation," L. C. B. Brown, BF1

"Vendors and Librarians Speak on Outsourcing, Cataloging, and Acquisitions," E. Duranceau, FBB2
"The Vertical File: An Overview and Guide," D. S. Jacobs, GD11
"The Vertical File: Retain or Discard?" E. Payson, GD12
"Veterinary Subject Headings and Classification: A Critical Analysis," N. L. Pelzer, DCC4
Video Collection Development in Multi-Type Libraries: A Handbook, G. Handman, ed., GD29
Video Collections and Multimedia in ARL Libraries, K. Brancolini and R. E. Provine, comps., GD35
Video Preservation: Securing the Future of the Past, D. Boyle, HJ1
Violence in the Library: Prevention, Preparedness and Response, IF8
"Virginia Tech Sets Policy on Controversial Materials," P. Metz and B. Obenhaus, GD18
"Virtual Dreams Give Way to Digital Reality," C. LaGuardia, GD33
"Virtual Support: Evolving Technical Services," J. A. Younger, AA16
Vitzansky, Winnie. *See* Massis, Bruce
Vizine-Goetz, Diane. *See* Drabenstott, Karen M.
"Vocabulary and Health Care Information Technology: State of the Art," J. J. Cimino, DCC1
"A Voucher Scheme to Simplify Payment for International Interlibrary Transactions," S. Gould, IDA6

W

Waite, Ellen J. "Reinvent Catalogers!," CBG7
Walch, Victoria Irons, comp. *Standards for Archival Description, A Handbook: Information Systems, Data Exchange, Cataloging, Finding Aids, Authority Control, Editing and Publishing, Statistics*, CCD3

Walden, Barbara, et al. "Western European Political Science: An Acquisition Study," GE10
Walling, Linda Lucas, and Marilyn M. Irwin. *Information Services for People with Developmental Disabilities,* IA4
Walsh, Jim, Barbara Hukyk, and George Barnum, comps. *Rare and Valuable Government Documents: A Resource Packet on Identification, Preservation and Security Issues for Government Documents Collections,* HA11
Walter, Katherine L. *Saving the Past to Enrich the Future: A Plan for Preserving Information Resources in Kansas,* HD18
Walters, Sheila. "The Direct Doc Pilot Project at Arizona State: User Behavior in a Non-Mediated Document Delivery Environment," IE22
Walters, Sheila. *See also* Mitchell, Eleanor
Walters, Suzanne. *Customer Service: A How-to-Do-It Manual for Librarians,* IA5
Walters, Tyler O. "Breaking New Ground in Fostering Preservation: The Society of American Archivists' Preservation Management Training Program," HM7
Warzala, Martin. "The Evolution of Approval Services," BB9
Waters, Donald J., and Anne Kenney. *The Digital Preservation Consortium: Mission and Goals,* HH29
Watt, Marcia, and Lisa Biblo. "CD-ROM Longevity: A Select Bibliography," HJ19
Wayne State Gopher Menu, CAE11
Weaver, Shari. *See* Conway, Paul
"Weeding: A Quantitative and Qualitative Approach," L. L. Reed and R. Erikson, GE31
Wehmeyer, Susan. *See* O'Connor, Phyllis

Weibel, Stuart, Jean Godby, Eric Miller, and Rodney Daniel. "OCLC/NCSA Metadata Workshop Report," CAC9
Weihs, Jean. "Interfaces" [column], AA15
———, and Lynne C. Howarth. "Nonbook Materials: Their Occurrence and Bibliographic Description in Canadian Libraries," CCD23
Weihs, Jean. *See also* Howarth, Lynne C.
Weinberg, Bella Hass. "Why Postcoordination Fails the Searcher," EAB24
Weintraub, Tamara S., and Wayne Shimoguchi. "Catalog Record Contents Enhancement," CCB8
Weiss, Paul. "The Expert Cataloging Assistant Project at the National Library of Medicine," CBF14
Weldon, Susan. *See* O'Connor, Phyllis
Weller, Carolyn R., and J. E. Houston, eds. *ERIC Identifier Authority List (IAL) 1992,* EE16
Wellisch, Hans W. "Book and Periodical Indexing," EAB25
Wessling, Julie. "Document Delivery: A Primary Service for the Nineties," IE23
Western Association of Art Conservation Newsletter, HAA8
"Western European Political Science: An Acquisition Study," B. Walden, et al., GE10
"What Is a Work?" M. M. Yee, CCB10
"What Journals Do Psychology Graduate Students Need? A Citation Analysis of Thesis References," M. Sylvia and M. Lesher, GD45
"What Will Collection Developers Do?" M. Buckland, GAE1
"What's MIME Is Yours," N. Lombardo and P. Jobe, IDA12
"Where Has All the Copy Gone: Latin American Imprints in the RLIN Database," C. S. Sercan, CBI12
Whiffin, Jean I., ed. *International Directory of Serials Specialists,* FAB2

"Who Killed the Subject Code?" W. E. Studwell, EC9

"Who Needs to Know What? Essential Communication for Automation Implementation and Effective Reorganization," C. Coulter and L. Halpin, AHC4

"Who, What, and Where in Book Repair: Institutional Profiles of the LCCDG," M. Gradinette and R. Silverman, HC6

"Why Postcoordination Fails the Searcher," B. H. Weinberg, EAB24

Wiant, Sarah K. *See* Gasaway, Laura N.

Wicks, Scott. *See* Reich, Vicky

"A Widening Circle: Preservation Literature Review, 1992," J. M. Drewes, HAC2

Wiebe, Victor G. *See* Dworaczek, Marian

Wiedensohler, Pat. *See* Coffman, Steve

Wiemers, Eugene L. "Financial Issues for Collection Managers in the 1990s," GF8

Wilhelm, Henry, with Carol Brower. *The Permanence and Care of Color Photographs: Traditional and Digital Color Prints, Color Negatives, Slides, and Motion Pictures*, HJ20

Wilkas, Lenore Rae. *International Subscription Agents*, FAB3

Wilkes, Adeline W., and Antoinette Nelson. "Subject Searching in Two Online Catalogs: Authority Control *vs.* Non-Authority Control," DCD1

Wilkes, Adeline W. *See also* Van Orden, Phyllis J.

Williams, James F. II. *See* Allen, Nancy H.

Williams, Johnette J. "Technology and Library Organizational Structure," AHC8

Williams, Karen A. *See* Budd, John M.

Williamson, Nancy J., and Michèle Hudon, eds. *Classification Research for Knowledge Representation and Organization: Proceedings of the Fifth International Study Conference on Classification Research*, DAA3

Williamson, Nancy J., Suliang Feng, and Tracy Tennant. *The Library of Congress Classification: A Content Analysis of the Schedules in Preparation for Their Conversion into Machine-Readable Form*, DBC1

Williamson, Susan G. *See* Gerhard, Kristin H.

Willman, Carrie. *See* Kneisner, Dan

Wilson, Karen A. "Outsourcing Copy Cataloging and Physical Processing: A Review of Blackwell's Outsourcing Services for the J. Hugh Jackson Library at Stanford University," CBG8

———. "Redesigning Technical Services Work Areas for the 21st Century," AHA14

———. "Reorganization of Technical Services Staff in the 90s," AHC9

Winke, R. Conrad. "Discarding the Main Entry in an Online Cataloging Environment," CCB9

Winkel, Lois, ed. *Subject Headings for Children: A List of Subject Headings Used by the Library of Congress with Dewey Numbers Added*, DCC6

Winters, Barbara. *See* Hirshon, Arnold

Wise, Suzanne. "Making Lemonade: The Challenges and Opportunities of Forced Reference Serials Cancellations: One Academic Library's Experiences," GE37

"With Characters: Retrospective Conversion of East Asian Cataloging Records," A. H. Wu, CBI15

Wittekind, Jürgen. "The Battelle Mass Deacidification Process: A New Method for Deacidifying Books and Archival Materials," HC14

Wittenbach, Stefanie A. "Building a Better Mousetrap: Enhanced Cataloging and Access for the Online Catalog," DA10

Wittenborg, Karen. *See* Pankake, Marcia
WLN, CAE21
"Women's Studies Periodical Indexes: An In-Depth Comparison," L. Krikos, GE27
Woodruff, Allison Gyle, and Christian Plaunt. "GIPSY: Automated Geographic Indexing of Text Documents," EF12
Woodward, Hazel, and Stella Pilling, eds. *The International Serials Industry,* FCA3
"The Work of Subscription Agents," J. Merriman, FDC2
Wormell, Irene. "Subject Access Redefined: How New Technology Changes the Conception of Subject Representation," EAB26
Wright, Dorothy W. "Selecting a Preservation Photocopy Machine," HL3
———, Samuel Demas, and Walter Cybulski. "Cooperative Preservation of State-Level Publications: Preserving the Literature of New York State Agriculture and Rural Life," HD8
"Writing Collection Development Policy Statements: Format, Content, Style," P. Johnson, GC9
"Writing Collection Development Policy Statements: Getting Started," P. Johnson, GC10
Wu, Ai-Hwa. "With Characters: Retrospective Conversion of East Asian Cataloging Records," CBI15
Wu, Connie, et al. "Effective Liaison Relationships in an Academic Library," GD55

X

Xerographic Toner Adhesion Method, HL4
"An X12 Implementation in Serials: MSUS/PALS and Faxon," L. Richter and J. Roca, FDF8

Y

"The Year's Work in Collection Development, 1992," S. Lehmann and J. H. Spohrer, GAB4
"The Year's Work in Descriptive Cataloging, 1992," G. Knutson, CAB1
Yee, Martha M. "The Concept of Work for Moving Image Materials," CCD5
———, ed. *Headings for Tomorrow: Public Access Display of Subject Headings,* DCA3, EC12
———. "Manifestations and Near-Equivalents of Moving-Image Works: A Research Project," CCD7
———. "Manifestations and Near-Equivalents: Theory, with Special Attention to Moving-Image Materials," CCD6
———. "What Is a Work?," CCB10
"You Call It Corn, We Call It Syntax-Independent Metadata for Document-Like Objects," P. Caplan, CAC3
Young, Arthur P. *See* Kelland, John Laurence
Younger, Jennifer A. "After Cutter: Authority Control in the Twenty-first Century," CBF15
———. "Virtual Support: Evolving Technical Services," AA16

Z

Zhang, Dongming. *See* McCue, Janet
Zhou, Yuan. "From Smart Guesser to Smart Navigator: Changes in Collection Development for Research Libraries in a Network Environment," GE6
Zimmerman, Carole. *Material Published by Members of the Library of Congress Preservation Directorate: A Bibliography,* HAC3
Zinn, Edward. *See* Reilly, James M.

SUBJECT INDEX

The numbers following entries in this index are page numbers, not citation numbers.

A
access services, 237–67
acquisitions, 28–45
   serials, 140
African American studies
   collection development, 174
   subject headings, 109
approval plans, 31, 36–38
area studies, 185–86
art
   classification, 105
   indexing, 132, 133
   subject headings, 111, 135
Art & Architecture Thesaurus, 111, 113, 115, 127, 136–37
audiovisual materials
   cataloging, 83. *See also* nonbook materials
authority control, 65–68
   subject headings, 112–13
automated indexing, 136–38

B
backlogs, 70–71
bibliographic classification, 87
bibliographic networks, 23
bibliographies
   access services, 245
   acquisitions, 34
   collection management, 165
   descriptive cataloging, 49
   preservation, 208
   serials, 143–44
   subject analysis, 95
   technical services, 11–12
book numbers, 103–4
budgeting, 163, 192–93

C
censorship, 162, 175
change, 4
Charleston Conference, 30
children's subject headings, 111, 112
circulation, 247–51
circulation services, 238
claiming serials, 155–56
classification research, 87
classification systems, 96–105
collection evaluation, 163, 194–95
collection maintenance, 184–85, 238, 246–47
collection management, 161–97
conferences
   access services, 244
   acquisitions, 35
   serials, 146–47
   technical services, 14–15

Subject Index

conservation, 199, 209–11
conspectus, 195
controlled vocabularies, 120, 133–36. *See also* thesauri
cooperative cataloging, 71–74
cooperative collection development, 196–97
cooperative preservation, 199, 211–14
copyright, 140, 199–200, 214–15, 240–41
costs, 141
Cutter numbers, 103–4

D

database management, 23–27
descriptive cataloging, 46–85
Dewey Decimal Classification, 99–101
digital imaging, 218–23
directories
    indexers, 126
    interlibrary loan, 257–58
    serials, 142–43
    technical services, 11
disaster control and recovery, 200, 215–16
diversity, 162
document delivery, 261–65
downsizing, 4, 5

E

electronic access, 5
electronic data interchange, 30, 140
electronic discussion groups
    access services, 244–45
    indexing, 127–28
    preservation, 204–5
    serials, 145, 155
    subject analysis, 95–96
electronic publishing, 139
electronic resources, 162. *See also* electronic discussion groups; World Wide Web sites
    cataloging, 80–82
electronic serials, 140, 150–51
environmental control, 216–18
equipment standards, 236
ERIC thesaurus, 136

ethics, 31, 43–45
expert systems, 24–25

F

faceted classification systems, 104–5
Feather River conference, 30
fiction
    classification, 90, 98
    collection development, 187, 195
filing, 121
French subject headings, 112
fund-raising, 193–94

G

gay and lesbian studies, 174
gifts and exchanges, 40
government information
    cataloging, 84
    collection development, 186
graphic materials. *See* art
grey literature, 175

H

hardware, 25–26
health care subject headings, 110–11

I

imaging, 200, 218–23
indexing, 121–38
intellectual freedom, 162, 175
interlibrary loan, 238, 251–60
International Society for Knowledge Organization, 87
International Study Conference on Classification Research, 87
Internet, 177–79
    indexing, 132
    subject access, 117–18
Iranian subject headings, 110

K

knowledge organization, 87

L

legal materials, 84
library automation
    acquisitions, 28, 38–39
    collection management, 168
    interlibrary loan, 259–60
    serials, 156–58
library binding, 199, 208
    standards, 234
Library of Congress Classification, 101–3
Library of Congress Subject Headings, 108–10, 133–34, 136
licensing agreements, 29
local systems, 26–27

M

magnetic tape preservation, 229–30
management
    acquisitions, 41–42
    cataloging, 56–74
    preservation programs, 224–26
    serials, 147–50, 152–53
    technical services, 15–23
management decisions, 29
manuscripts
    cataloging, 79
    preservation, 229
microforms, 200, 218–23, 230–31
    standards, 234–35
motion pictures and video recordings
    cataloging, 79–80
    preservation, 228
    subject headings, 110
multiculturalism, 162
multimedia
    collections, 177–79
    standards, 234–36
music and sound recordings
    cataloging, 79
    preservation, 227, 228, 229
music thesauri, 135

N

Native American studies, 175
networking technologies, 5
newspapers
    cataloging, 82
    indexing, 132
nonbook materials. *See also* audiovisual materials; multimedia
    cataloging, 82–83
    preservation, 227–30

O

online periodicals. *See* electronic serials; periodicals
OPACs, 130–31
    classification, 98–99
organization, 29
    collection management, 169
out-of-print material, 39–40
outsourcing, 4, 19, 68–70

P

performing arts, 186
periodicals. *See also* electronic serials
    access services, 242
    acquisitions, 34–35
    descriptive cataloging, 51–52
    indexing, 128
    preservation, 203–4
    serials, 144–45
    subject analysis, 94–95
    technical services, 12–13
pest management, 216–18
photocopying, 231–32
policies
    collection management, 171–73
preservation, 190–91, 198–236
pricing, serials, 158–59
professional associations
    access services, 242–43
    acquisitions, 35
    serials, 145
    technical services, 14
public libraries, 187–88
publishing, 43–44

R

rare books, 78–79
realia, 83

reference collections, 188
reorganization, 4, 21–23
reproductions, 84
reserve services, 248–50
resource sharing, 159–60, 196–97
restructuring, 5, 141

S
school libraries, 188
security, 215–16, 266–67
selection, 173
serials, 139–60
   cancellation, 163, 191–92
   cataloging, 82
   collection development, 180–81
software, 25–26
sound recordings. *See* music and sound recordings
special collections, 189
standards, 120
   cataloging, 74–85
   circulation, 250–51
   classification systems, 99
   collection management, 165
   indexing, 128–29
   interlibrary loan, 258–59
   preservation, 233–36
   subject headings, 107–8
subdivisions, 109
subject analysis, 86–118
subject headings, 86, 105–16
subscription agents, 153–55

T
Telecommunications Act of 1996, 163
thesauri, 113–16, 133–36. *See also* controlled vocabularies

training, 5
   cataloging, 59–60
   collection management, 169–70
   preservation, 232–33

U
Universal Decimal Classification, 104

V
vendors, 28–29, 40–41, 47, 139, 140
vertical files, 175
veterinary science, 112
videotapes. *See* Motion pictures and video recordings

W
weeding, 190
women's studies, 162
   collection development, 174, 189–90
   subject headings, 110
workstations, 60–61
World Wide Web sites
   acquisitions, 35–36
   collection management, 166–67
   descriptive cataloging, 52–56
   indexing, 128
   preservation, 207
   serials, 152
   subject analysis, 95–96

X
X12 standard, 140